Cancer Prevention with Molecular Target Therapies 4.0

Cancer Prevention with Molecular Target Therapies 4.0

Editor

Laura Paleari

Basel • Beijing • Wuhan • Barcelona • Belgrade • Novi Sad • Cluj • Manchester

Editor
Laura Paleari
Research, Innovation and HTA
Liguria Health Authority, A.Li.Sa.
Genova
Italy

Editorial Office
MDPI AG
Grosspeteranlage 5
4052 Basel, Switzerland

This is a reprint of articles from the Special Issue published online in the open access journal *International Journal of Molecular Sciences* (ISSN 1422-0067) (available at: www.mdpi.com/journal/ijms/special_issues/4V0LCM6P7E).

For citation purposes, cite each article independently as indicated on the article page online and as indicated below:

Lastname, A.A.; Lastname, B.B. Article Title. *Journal Name* **Year**, *Volume Number*, Page Range.

ISBN 978-3-7258-1952-2 (Hbk)
ISBN 978-3-7258-1951-5 (PDF)
doi.org/10.3390/books978-3-7258-1951-5

© 2024 by the authors. Articles in this book are Open Access and distributed under the Creative Commons Attribution (CC BY) license. The book as a whole is distributed by MDPI under the terms and conditions of the Creative Commons Attribution-NonCommercial-NoDerivs (CC BY-NC-ND) license.

Contents

Laura Paleari
Personalized Assessment for Cancer Prevention, Detection, and Treatment
Reprinted from: *Int. J. Mol. Sci.* 2024, 25, 8140, doi:10.3390/ijms25158140 1

Francesca Marino-Merlo, Sandro Grelli, Antonio Mastino, Michele Lai, Paola Ferrari, Andrea Nicolini, et al.
Human T-Cell Leukemia Virus Type 1 Oncogenesis between Active Expression and Latency: A Possible Source for the Development of Therapeutic Targets
Reprinted from: *Int. J. Mol. Sci.* 2023, 24, 14807, doi:10.3390/ijms241914807 4

Norwin Kubick, Justyna Paszkiewicz, Irmina Bieńkowska, Michał Ławiński, Jarosław Olav Horbańczuk, Mariusz Sacharczuk and Michel Edwar Mickael
Investigation of Mutated in Colorectal Cancer (MCC) Gene Family Evolution History Indicates a Putative Role in Th17/Treg Differentiation
Reprinted from: *Int. J. Mol. Sci.* 2023, 24, 11940, doi:10.3390/ijms241511940 20

Connor H. O'Meara, Zuhayr Jafri and Levon M. Khachigian
Immune Checkpoint Inhibitors, Small-Molecule Immunotherapies and the Emerging Role of Neutrophil Extracellular Traps in Therapeutic Strategies for Head and Neck Cancer
Reprinted from: *Int. J. Mol. Sci.* 2023, 24, 11695, doi:10.3390/ijms241411695 34

Miriam Dellino, Marco Cerbone, Antonio Simone Laganà, Amerigo Vitagliano, Antonella Vimercati, Marco Marinaccio, et al.
Upgrading Treatment and Molecular Diagnosis in Endometrial Cancer—Driving New Tools for Endometrial Preservation?
Reprinted from: *Int. J. Mol. Sci.* 2023, 24, 9780, doi:10.3390/ijms24119780 56

Enrico Gurreri, Giannicola Genovese, Luigi Perelli, Antonio Agostini, Geny Piro, Carmine Carbone and Giampaolo Tortora
KRAS-Dependency in Pancreatic Ductal Adenocarcinoma: Mechanisms of Escaping in Resistance to KRAS Inhibitors and Perspectives of Therapy
Reprinted from: *Int. J. Mol. Sci.* 2023, 24, 9313, doi:10.3390/ijms24119313 80

Elżbieta Pawluczuk, Marta Łukaszewicz-Zajac and Barbara Mroczko
The Comprehensive Analysis of Specific Proteins as Novel Biomarkers Involved in the Diagnosis and Progression of Gastric Cancer
Reprinted from: *Int. J. Mol. Sci.* 2023, 24, 8833, doi:10.3390/ijms24108833 96

Bai Zhang, Xueyi Li, Kai Tang, Ying Xin, Guanshuo Hu, Yufan Zheng, et al.
Adhesion to the Brain Endothelium Selects Breast Cancer Cells with Brain Metastasis Potential
Reprinted from: *Int. J. Mol. Sci.* 2023, 24, 7087, doi:10.3390/ijms24087087 110

Raed Sulaiman, Pradip De, Jennifer C. Aske, Xiaoqian Lin, Adam Dale, Kris Gaster, et al.
Characterization and Clinical Relevance of Endometrial CAFs: Correlation between Post-Surgery Event and Resistance to Drugs
Reprinted from: *Int. J. Mol. Sci.* 2023, 24, 6449, doi:10.3390/ijms24076449 126

Alessandro Parente, Hwui-Dong Cho, Ki-Hun Kim and Andrea Schlegel
Association between Hepatocellular Carcinoma Recurrence and Graft Size in Living Donor Liver Transplantation: A Systematic Review
Reprinted from: *Int. J. Mol. Sci.* 2023, 24, 6224, doi:10.3390/ijms24076224 155

Dalila Boi, Elisabetta Rubini, Sara Breccia, Giulia Guarguaglini and Alessandro Paiardini
When Just One Phosphate Is One Too Many: The Multifaceted Interplay between Myc and Kinases
Reprinted from: *Int. J. Mol. Sci.* **2023**, *24*, 4746, doi:10.3390/ijms24054746 . . . . **174**

Tomasz Górnicki, Jakub Lambrinow, Monika Mrozowska, Hanna Romanowicz, Beata Smolarz, Aleksandra Piotrowska, et al.
Expression of RBMS3 in Breast Cancer Progression
Reprinted from: *Int. J. Mol. Sci.* **2023**, *24*, 2866, doi:10.3390/ijms24032866 . . . . **209**

Editorial

Personalized Assessment for Cancer Prevention, Detection, and Treatment

Laura Paleari

Research, Innovation and HTA Unit, Liguria Health Authority, A.Li.Sa., 16121 Genova, Italy; laura.paleari@alisa.liguria.it

Citation: Paleari, L. Personalized Assessment for Cancer Prevention, Detection, and Treatment. *Int. J. Mol. Sci.* **2024**, *25*, 8140. https://doi.org/10.3390/ijms25158140

Received: 14 July 2024
Accepted: 23 July 2024
Published: 26 July 2024

Copyright: © 2024 by the author. Licensee MDPI, Basel, Switzerland. This article is an open access article distributed under the terms and conditions of the Creative Commons Attribution (CC BY) license (https://creativecommons.org/licenses/by/4.0/).

The intention of this Special Issue is to highlight research that aims to recognize cancer's complexity to better prevent or treat its occurrence.

Personalized (or precision) medicine in oncology is grounded in the idea that cancers are not all the same and people may respond differently to treatments due to their genetic, environmental, and lifestyle characteristics [1]. Before the advent of precision medicine, cancer treatments were mainly based on standardized approaches, such as chemotherapy and radiotherapy, with significant side effects and variable results. Precision medicine has essentially changed this landscape, enabling more personalized, effective, and targeted care [1]. The improvement in the technologies used for the analysis of cancer cell DNA (the so-called NGS, next generation sequencing technologies) has led to a change in the cancer treatment paradigm, thanks to continuous studies. They are designed to address the following: identify specific genetic mutations that drive tumor growth; quantify the mutational burden, which can lead to the discovery of resistance to treatment mechanisms; and develop molecularly targeted drugs designed to inhibit specific alterations in neoplastic cells [2]. An example of personalized research is represented by Suliman and colleagues' work, which aimed to correlate chemo-resistance with biological variables in endometrial cancer (EC) patients [3]. To reach this goal, they explored the functional association between aggressive cancer-associated fibroblast cells (CAFs), derived from EC patients, and resistance to chemo/targeted drugs [3]. Their results showed a positive correlation between grade 3 ($p = 0.025$) and stage 3/4 diseases ($p = 0.0106$) with aggressive CAFs and the post-surgery event (PSE). Furthermore, aggressive CAFs, derived from patients with PSE, displayed resistance to paclitaxel and lenvatinib [3].

Furthermore, the clinical utilization of measurable biomarkers has been possible by personalized medicine. This approach aims to avoid ineffective treatments by the evaluation of specific genetic characteristics of each patient. In the clinical setting, the Ki-67 proliferation marker is well known; it often correlates with an unfavorable prognosis in tumors [4,5], even if its use is to date has mainly been restricted to breast cancers (BCs) [6–9]. Recently, a retrospective study on a cohort of women affected by EC has been published, with the intention of exploring Ki-67's predictive and prognostic role [9]. The results of this research suggest a positive role for the Ki-67 index as a prognostic and predictive marker of response to hormone treatments in early-stage ECs with a statistically significant benefit in terms of disease-free survival (DFS) [HR = 0.25 (95% CI; 0.09–0.69), $p = 0.008$] and overall survival (OS) [HR = 0.30 (95% CI; 0.10–0.87), $p = 0.03$] in the high-expressing Ki 67 group treated with hormone therapy [10]. Moreover, a comprehensive analysis to detect novel biomarkers has been published, with a focus on the diagnosis and progression of gastric cancer (GC) [11]. It has been demonstrated that specific chemokines and their receptors, the vascular endothelial growth factor (VEGF) and epidermal growth factor receptor (EGFR), interleukin 6 (IL-6) and C-reactive protein (CRP), matrix metalloproteinases (MMPs) and their tissue inhibitors (TIMPs), a disintegrin and metalloproteinase with thrombospondin motifs (ADAMTS), as well as DNA- and RNA-based biomarkers, and c-MET (tyrosine-protein kinase Met) play a role in the pathogenesis of GC [11].

Another important aspect of precision medicine is represented by predictive medicine, which uses the analysis of genetic characteristics to evaluate the individual risk of developing cancer or the individual response to a treatment. This can allow for early diagnosis and targeted preventive interventions, bringing patients' life expectancy and quality of life to a higher level. For instance, taste and smell disorders (TSDs), which are common side effects in patients undergoing cancer treatments, have been studied to recognize which treatments specifically cause them with the aim to improve patients' quality of life [12]. Buttiron Webber and colleagues identify, among cancer treatments, those that principally lead to taste and smell changes and provide evidence for wider studies, including those focusing on prevention [12]. The results of this research definitely highlight how patients' quality of life is a crucial issue in oncology and how its advancement can be reached only with the active participation of all the professional health figures involved in cancer management.

To date, the advent of new technologies, the omics sciences, and the use of artificial intelligence are significantly contributing to the progress of precision oncology. For instance, the clinical utility of comprehensive genome profiling (CGP) tests, used in precision oncology to guide therapeutic choices, remains controversial [13,14]. Very recently, the results of a study on a learning program for treatment recommendations by molecular tumor boards (MTBs) and artificial intelligence (AI) were published [15]. The aim of this study was to examine the effectiveness of a learning program aimed at improving treatment recommendations, focusing on genomic alterations with low levels of evidence. For this purpose, simulated cases of advanced cancer were used and the efficiency of an AI-based annotation system to improve clinical decisions was examined [15]. The findings of this quality improvement study suggest that use of a learning program improved the concordance of treatment recommendations provided by MTBs compared to central ones. Treatment recommendations made by an AI system showed higher concordance than that for MTBs, indicating the potential clinical utility of the AI system.

Despite the substantial progress achieved thanks to precision medicine, it is important to underline that this field is still constantly evolving. Research studies are ongoing; nonetheless, precision medicine has already demonstrated its enormous potential in improving the effectiveness of cancer treatments and providing new hope for patients suffering from cancer.

Conflicts of Interest: The author declares no conflicts of interest.

References

1. Tsimberidou, A.M.; Fountzilas, E.; Nikanjam, M.; Kurzrock, R. Review of precision cancer medicine: Evolution of the treatment paradigm. *Cancer Treat Rev.* **2020**, *86*, 102019. [CrossRef] [PubMed]
2. Morganti, S.; Tarantino, P.; Ferraro, E.; D'Amico, P.; Duso, B.A.; Curigliano, G. Next Generation Sequencing (NGS): A Revolutionary Technology in Pharmacogenomics and Personalized Medicine in Cancer. *Adv. Exp. Med. Biol.* **2019**, *1168*, 9–30. [PubMed]
3. Sulaiman, R.; De, P.; Aske, J.C.; Lin, X.; Dale, A.; Gaster, K.; Espaillat, L.R.; Starks, D.; Dey, N. Characterization and Clinical Relevance of Endometrial CAFs: Correlation between Post-Surgery Event and Resistance to Drugs. *Int. J. Mol. Sci.* **2023**, *24*, 6449. [CrossRef] [PubMed]
4. Li, L.T.; Jiang, G.; Chen, Q.; Zheng, J.N. Ki67 is a promising molecular target in the diagnosis of cancer (review). *Mol. Med. Rep.* **2015**, *11*, 1566–1572. [CrossRef] [PubMed]
5. Menon, S.S.; Guruvayoorappan, C.; Sakthivel, K.M.; Rasmi, R.R. Ki-67 protein as a tumor proliferation marker. *Clin. Chim. Acta* **2019**, *491*, 39–45. [CrossRef] [PubMed]
6. Gerdes, J.; Lelle, R.J.; Pickartz, H.; Heidenreich, W.; Schwarting, R.; Kurtsiefer, L.; Stauch, G.; Stein, H. Growth fractions in breast cancers determined in situ with monoclonal antibody Ki-67. *J. Clin. Pathol.* **1986**, *39*, 977–980. [CrossRef] [PubMed]
7. Nielsen, T.O.; Leung, S.C.Y.; Rimm, D.L.; Dodson, A.; Acs, B.; Badve, S.; Denkert, C.; Ellis, M.J.; Fineberg, S.; Flowers, M.; et al. Assessment of Ki67 in Breast Cancer: Updated Recommendations from the International Ki67 in Breast Cancer Working Group. *J. Natl. Cancer Inst.* **2021**, *113*, 808–819. [CrossRef] [PubMed]
8. Hashmi, A.A.; Hashmi, K.A.; Irfan, M.; Khan, S.M.; Edhi, M.M.; Ali, J.P.; Hashmi, S.K.; Asif, H.; Faridi, N.; Khan, A. Ki67 index in intrinsic breast cancer subtypes and its association with prognostic parameters. *BMC Res. Notes* **2019**, *12*, 605. [CrossRef] [PubMed]

9. Whelan, T.J.; Smith, S.; Nielsen, T.O.; Parpia, S.; Fyles, A.W.; Bane, A.; Liu, F.-F.; Grimard, L.; Stevens, C.; Bowen, J.; et al. LUMINA: A prospective trial omitting radiotherapy (RT) following breast conserving surgery (BCS) in T1N0 luminal A breast cancer (BC). *J. Clin. Oncol.* **2022**, *40*, LBA501. [CrossRef]
10. Paleari, L.; Rutigliani, M.; D'Ecclesiis, O.; Gandini, S.; Briata, I.M.; Webber, T.B.; Provinciali, N.; DeCensi, A. Exploring the Prognostic and Predictive Roles of Ki-67 in Endometrial Cancer. *Int. J. Transl. Med.* **2023**, *3*, 479–486. [CrossRef]
11. Pawluczuk, E.; Łukaszewicz-Zając, M.; Mroczko, B. The Comprehensive Analysis of Specific Proteins as Novel Biomarkers Involved in the Diagnosis and Progression of Gastric Cancer. *Int. J. Mol. Sci.* **2023**, *24*, 8833. [CrossRef] [PubMed]
12. Buttiron Webber, T.; Briata, I.M.; DeCensi, A.; Cevasco, I.; Paleari, L. Taste and Smell Disorders in Cancer Treatment: Results from an Integrative Rapid Systematic Review. *Int. J. Mol. Sci.* **2023**, *24*, 2538. [CrossRef] [PubMed]
13. Ida, H.; Koyama, T.; Mizuno, T.; Sunami, K.; Kubo, T.; Sudo, K.; Tao, K.; Hirata, M.; Yonemori, K.; Kato, K.; et al. Clinical utility of comprehensive genomic profiling tests for advanced or metastatic solid tumor in clinical practice. *Cancer Sci.* **2022**, *113*, 4300–4310. [CrossRef] [PubMed]
14. Tojota, M.Y.; Segal, J.P.; Wang, P. Clinical Utility and Benefits of Comprehensive Genomic Profiling in Cancer. *J. Appl. Lab. Med.* **2024**, *9*, 76–91. [CrossRef] [PubMed]
15. Sunami, K.; Naito, Y.; Saigusa, Y.; Amano, T.; Ennishi, D.; Imai, M.; Kage, H.; Kanai, M.; Kenmotsu, H.; Komine, K.; et al. A Learning Program for Treatment Recommendations by Molecular Tumor Boards and Artificial Intelligence. *JAMA Oncol.* **2023**, *10*, 76–91. [CrossRef] [PubMed]

Disclaimer/Publisher's Note: The statements, opinions and data contained in all publications are solely those of the individual author(s) and contributor(s) and not of MDPI and/or the editor(s). MDPI and/or the editor(s) disclaim responsibility for any injury to people or property resulting from any ideas, methods, instructions or products referred to in the content.

Review

Human T-Cell Leukemia Virus Type 1 Oncogenesis between Active Expression and Latency: A Possible Source for the Development of Therapeutic Targets

Francesca Marino-Merlo [1,†], Sandro Grelli [2,†], Antonio Mastino [3], Michele Lai [4], Paola Ferrari [5], Andrea Nicolini [6,*], Mauro Pistello [4,‡] and Beatrice Macchi [7,‡]

1. Department of Chemical, Biological, Pharmaceutical and Environmental Sciences, University of Messina, 98166 Messina, Italy; fmarino@unime.it
2. Department of Experimental Medicine, University of Rome "Tor Vergata", 00133 Rome, Italy; grelli@med.uniroma2.it
3. The Institute of Translational Pharmacology, CNR, 00133 Rome, Italy; antonio.mastino@ift.cnr.it
4. Retrovirus Center and Virology Section, Department of Translational Research, University of Pisa, 56100 Pisa, Italy; michele.lai@unipi.it (M.L.); mauro.pistello@unipi.it (M.P.)
5. Unit of Oncology, Department of Medical and Oncological Area, Azienda Ospedaliera—Universitaria Pisana, 56125 Pisa, Italy; p.ferrari@ao-pisa.toscana.it
6. Department of Oncology, Transplantations and New Technologies in Medicine, University of Pisa, 56126 Pisa, Italy
7. Department of Chemical Science and Technology, University of Rome "Tor Vergata", 00133 Rome, Italy; macchi@med.uniroma2.it
* Correspondence: andrea.nicolini@med.unipi.it
† These authors contributed equally to this work.
‡ These authors contributed equally to this work.

Abstract: The human T-cell leukemia virus type 1 (HTLV-1) is the only known human oncogenic retrovirus. HTLV-1 can cause a type of cancer called adult T-cell leukemia/lymphoma (ATL). The virus is transmitted through the body fluids of infected individuals, primarily breast milk, blood, and semen. At least 5–10 million people in the world are infected with HTLV-1. In addition to ATL, HTLV-1 infection can also cause HTLV-I-associated myelopathy (HAM/TSP). ATL is characterized by a low viral expression and poor prognosis. The oncogenic mechanism triggered by HTLV-1 is extremely complex and the molecular pathways are not fully understood. However, viral regulatory proteins Tax and HTLV-1 bZIP factor (HBZ) have been shown to play key roles in the transformation of HTLV-1-infected T cells. Moreover, several studies have shown that the final fate of HTLV-1-infected transformed Tcell clones is the result of a complex interplay of HTLV-1 oncogenic protein expression with cellular transcription factors that subvert the cell cycle and disrupt regulated cell death, thereby exerting their transforming effects. This review provides updated information on the mechanisms underlying the transforming action of HTLV-1 and highlights potential therapeutic targets to combat ATL.

Keywords: HTLV-1; adult T-cell leukemia; viral oncogenesis; Tax; HBZ; apoptosis

1. Introduction

The human T-cell leukemia virus type 1 (HTLV-1), also known as human T-lymphotropic virus type 1, was the first human retrovirus discovered. It belongs to subtype Delta of the subfamily *Orthoretrovirinae* and is endemic to the southwestern part of Japan, South America, the Caribbean, the Middle East, Australo-Melanesia [1], and Western and sub-Saharan Africa [2]. HTLV-1 has been recognized as the causative agent of adult T-cell leukemia/lymphoma (ATL), HTLV-1-associated myelopathy/tropical spastic paraparesis (HAM/TSP), and a number of inflammatory diseases. ATL is an aggressive malignancy that usually occurs in approximately 5% of infected adult individuals, 30–50 years after

HTLV-1 infection. The prognosis of the aggressive subtype of ATL is very poor, with median survival ranging from 6 to 24 months. It is estimated that 5–10 million people are infected with HTLV-1 worldwide, although this number is most likely underestimated due to the lack of broader epidemiological studies. In vivo human-to-human transmission occurs through breastfeeding, blood transfusions, needle sharing, and sexual intercourse [3,4]. In addition, the rapid spread of HTLV-1 has been demonstrated in transplant recipients [5]. To understand the HTLV-1 infection/transformation that occurs in vivo, several noteworthy questions need to be addressed. The clarification of these questions may provide a representation of the complex interplay between HTLV-1 and the host gene expression that characterizes leukemogenesis and viral persistence in ATL [6]. This review provides detailed information on the current knowledge of the interplay between oncogenic viral proteins and cellular factors underlying the transforming action of HTLV-1 and provides a brief overview of the potential therapeutic targets arising from this interplay and possible future therapeutic approaches.

2. HTLV-1 Genome and Virus Transmission and Spread

The HTLV-1 genome consists of a small positive ss(+) RNA with a size of approximately 9 kB. After infection, HTLV-1 is integrated into the host DNA as a provirus, similarly to HIV-1. Both the sense strand and the anti-sense strand of the integrated provirus can be transcribed [7]. The viral genome is flanked by long terminal repeats (LTRs) at both the $5'$ and $3'$ ends. These direct repeats consist of three regions: the unique 3 (U3), the repeated (R), and the unique 5 (U5) region. The LTRs contain important elements necessary for viral transcription, polyadenylation, and integration [8]. The HTLV-1 genome includes the standard structural and enzyme genes *gag*, *pro*, *pol*, and *env*, which encode proteins essential for viral replication [9], as well as the coding potential for the accessory/regulatory proteins Tax, Rex, p12, p13, and p30 [10–13]. All these proteins are encoded by the sense strand, whereas the regulatory protein HTLV-1 bZIP factor (HBZ) is encoded by the anti-sense strand. HTLV-1 is suitable to be studied by means of in vitro models of infection. Because HTLV-1 is strongly cell-associated, unlike HIV-1, cell-free transmission is very limited in favor of transmission by cell-to-cell contact. The cell-to-cell transmission of HTLV-1 is facilitated by several transport strategies that allow the virus to spread, bypassing host immune responses. These strategies include transcytosis through epithelial barriers [14] and induction of membrane structures, such as virological synapses [15], cellular conduits [16], biofilm-like extracellular viral assemblies [17], and extracellular vesicles [18,19]. Viral infection in vitro usually requires cocultivation between the recipient lymphocytes and irradiated infected cells in the presence of IL-2 as a growth factor. After several passages in vitro, the number of which varies depending on the different cocultures the newly infected cells can be immortalized, as indicated by the fact that they no longer require growth factors. During the period between infection and immortalization in vitro, different cells in the infected cultures undergo different and opposite fates simultaneously: some of them undergo strong proliferation and others undergo apoptosis. This might depend on the balance between the differential expression of viral and cellular proteins at the cellular level and the influence on the survival/death pathways induced by viral gene expression in the infected cells [4], as detailed in the next paragraph. Although this process has been described in detail in vitro, it is reasonable to assume that a similar pattern could occur in vivo. The spread of the virus in vivo, similar for all retroviruses, occurs through two distinct pathways: the infectious pathway, from cell to cell, and the mitotic pathway, once the HTLV-1 genome is integrated into the host cell genome [20]. It has been clarified that infectious spread persists in parallel with mitotic spread even during the chronic phase of HTLV-1 infection [21]. HTLV-1 primarily infects CD4+ T cells, the population that is preferentially selected after in vitro infection. The clonal expansion of HTLV-1-infected cells is influenced by the host immune response [22], and fluctuating levels of the virus LTR in vivo seem to indicate that the virus actively replicates during chronic infections [23]. Studies aimed at comparing the clone abundance and distribution of the provirus in CD4+

and CD8+ T-cell subpopulations in HTLV-1-infected individuals have shown that CD8+ T cells carried only 5% of the provirus load, whereas the provirus was present in a greater number of CD4+ clones [24]. Nevertheless, CD8+ T cells expand strongly in vivo, but their role has not been fully elucidated [24]. The mechanisms of cell-to-cell transmission of HTLV-1 are summarized in Figure 1. Essentially, ATL is the end result of selecting an infected clone, in most cases CD4+, from the many infected clones. The selected clone gradually acquires the characteristics of the transformed cells over many years. The mechanisms underlying this long and complex process have been unraveled in recent years, largely thanks to advances in genomics and single-cell technology. See the subsequent sections and reference [25] for further information.

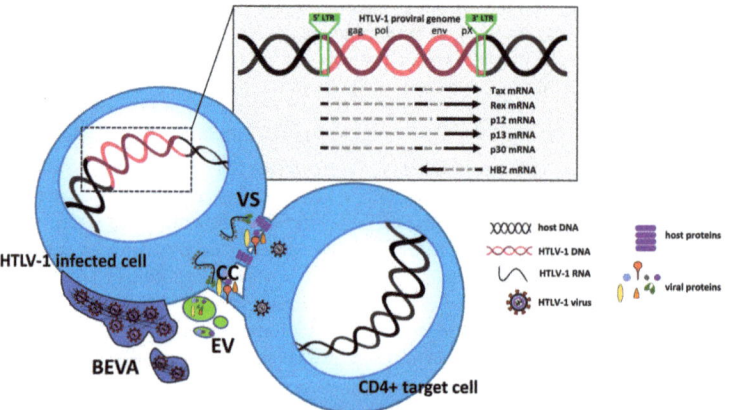

Figure 1. Schematic representation of HTLV-1 cell-to-cell transmission. VS = virological synapses. CC = cellular conduits. EV = extracellular vesicles. BEVA = biofilm-like extracellular viral assemblies. The mechanism of transcytosis is not reported due to difficulties in graphical representation.

3. ATL- and HTLV-1-Driven Transformation: Generalities

ATL develops in 5% of individuals infected with HTLV-1 and is characterized by lymphadenopathy, skin lesions, hypercalcemia, and severe involvement of organs, such as the central nervous system, lungs, liver, spleen, and bone marrow. Survival prognosis is poor and may be as short as 6 months. ATL patients respond poorly to traditional anti-leukemia chemotherapy. Rather, more favorable responses have been achieved by combined treatment with zidovudine (AZT)plus interferon (IFN) [26] or allogeneic hematopoietic stem-cell transplantation [27]. In addition, biologic therapy has relied on mogamulizumab, which targets chemokine receptor 4 [28], and the immunomodulator and antitumorallenalinomide [29], a human CD30-directed chimeric antibody bound to the microtubule-disrupting agent [30]; the EZH1 and EZH2 dual inhibitor valemetostat, which disrupts the hypermethylation of histone H3 lysine 27 (H3K27) and allows the re-expression of repressed genes [31]; and histone deacetylase (HDAC) inhibitors [32,33]. The results of the above studies, although not very informative, encourage finding new potential biological targets in the mechanism of viral transformation (see Sections 6 and 7). Several studies have shown that the final fate of transformed T-cell clones infected with HTLV-1 is the result of a finely tuned regulation of viral gene expression. In particular, the balance between the two major oncogenic viral proteins Tax and HBZ determines the fate of HTLV-1 infection [34]. Therefore, a regulated interplay between cell survival and cell death plays a key role in the selection of malignant clones.

One of the most important factors in the transformation and pathogenesis of HTLV-1 is the multifunctional viral protein Tax. This protein has the ability to interact directly with a variety of cellular proteins, including transcription factors, cell signaling proteins, cell cycle regulators, apoptotic proteins, and DNA damage response factors. Relevant to the aim of this review is the assumption that Tax positively modulates the expression of Bcl3 through activating the phosphatidylinositol 3-kinase (PI3K)/Akt signaling pathway [35]. Even the sustained activation of the cellular transcription factor Nuclear Factor-κB(NF-κB) by Tax [36] adds to its multifunctionality towards cellular targets (see Section 4). In addition, Tax is known to play a role as a transactivator and regulator of transcription of HTLV-1 structural and enzymatic proteins and of its own transcription through interaction with the promoter within the 5′ LTR. Tax-mediated transactivation requires binding to a repeated sequence of 21 nucleotides rich in G/C, representing the Tax-responsive element (TRE) located in the U3 region of the 5′ LTR [37]. Tax recruits the cAMP response element-binding protein (CREB)/activating transcription factor (ATF) to the cyclic AMP response elements (CREs) [38]. This leads to the formation of a nucleoprotein complex that recruits other CREB-binding proteins (CBP) and p300 [39]. This multiprotein complex is a potent activator of the 5′ LTR promoter and of viral mRNA transcription. Even if initial studies were unsuccessful in detecting a direct interaction between Tax and host DNA, it was cleared that, when in complex with CREB, Tax actually binds to DNA on the 21 bp repeats [40,41]. On the other hand, Tax can also upregulate viral transcription by directly binding to an epigenetic repressor, histone deacetylase 1 (HDAC1). Indeed, Tax inhibits or dissociates the binding of HDAC1 to the HTLV-1 promoter, thereby regulating viral protein transcription [42,43].

4. Tax and Cell Signaling: Role of the Transcription Nuclear factor NF-κB

In addition to regulating viral transcription, Tax regulates the expression of several signaling molecules involved in the processes of oncogenesis, proliferation, immune response, and apoptosis. Among these, the nuclear transcription factor NF-κB plays a central role in coordinating the various cellular signals. The prototypical NF-κB complex corresponds to a heterodimer of the p50 (NFKB1) and RelA (p65) members of the NF-κB/Rel family of transcription factors [44]. Tax induces the phosphorylation and degradation of both IκBα and IκBβ, suggesting that this HTLV-1 regulatory protein can induce the nuclear translocation of NF-κB by acting upstream or at the level of IκB phosphorylation [45,46]. The activation of NF-κB is involved in a number of events involving cytokine production in inflammatory contexts. Evidence that Tax regulates the expression of multiple cytokines is that its inhibition is dependent on the inhibition of NF-κB in HTLV-1-infected cells [47]. The sustained activation of NF-κB has a remarkable impact on the immortalization and transformation of HTLV-I-infected cells. For additional mechanisms of Tax-dependent or Tax-independent activation of the NF-kB pathways during HTLV-1 infection, please refer to a recent review [48]. The reduction in Tax in primary ATL cells rapidly abrogates NF-κB activation, leading to the induction of apoptosis [49]. One distinctive feature of immortalized HTLV-1 cells is the robust expression of anti-apoptotic genes. It has been reported that the Tax-mediated upregulation of Bfl-1, a member of the Bcl-2 family, is strongly expressed in HTLV-1-infected T-cell lines [50]. Direct evidence for the role of Bcl-2 family genes comes from studies showing that the silencing of Bfl-1 and Bcl-xL decreased the survival of HTLV-1 cells [50]. In vitro studies have shown that Tax transactivates the anti-apoptotic survive in promoter via the activation of NF-κB [51]. Nevertheless, Tax has not been exclusively associated with the inhibition of apoptosis, but also with its activation (see Section 5).

5. Latency and Leukemogenesis: Role of Tax, HBZ, and Apoptosis

HTLV-1 infection can remain latent for years before full disease onset, suggesting that there are sophisticated mechanisms regulating the on/off switching of viral protein expression, but also the activation of genes related to the T-cell receptor/NF-κB and signaling related to immune surveillance, such as HLA and FAS [52]. In addition to the progression of infection, the expression of Tax is also critical for stimulating the cytotoxic T-lymphocyte immune response, which is thought to play a role in the viability of ATL long-term survivors [53]. The loss or reduction in Tax expression in immortalized cells protects them from immune response attacks and renders them more prone to survival, expansion, and proliferation through a process of continuous transient expression [54].

Tax is not the only factor responsible for the various changes observed in viral expression and cell fate during infection. As already indicated in Section 3, HBZ also plays an important role by counteracting Tax to some extent and, on the other hand, integrating the process of virus-induced leukemogenesis [55]. Initially, HBZ was not thought to directly impact HTLV-1-mediated immortalization; instead, it was considered to regulate the establishment and maintenance of chronic infection [56]. The expression of hbz gene has been shown to upregulate JunD abundance; HBZ heterodimerizes with JunD, in turn recruiting the transcription factor Sp1 to the 3′ LTR of the provirus to enhance its activity [57]. Successively, HBZ was shown to be indirectly involved in leukemogenesis by increasing the expression of two oncogenic miRNA, miR-17 and miR-21. These miRNAs, in turn, downregulate the expression of the single-stranded DNA-binding protein hSSB2, thus promoting genomic instability [58]. Other mechanisms relevant to how HBZ can promote genomic instability include its capacity to induce the accumulation of double-stranded DNA breaks (DSBs) by attenuating the nonhomologous end-joining (NHEJ) repair pathway [59] and its ability to suppress the transcription of MutS homologue 2 (MSH2), an essential mismatch repair factor, by inhibiting nuclear respiratory factor 1 (NRF-1) [60]. Another indication of the possible involvement of HBZ in neoplastic transformation is the detection of its translocation to the nucleus in cells from ATL patients, but not in those from asymptomatic carriers or from patients affected by non-neoplastic pathologies, where HBZ is exclusively localized in the cytoplasm [61]. In fact, the nuclear localization of HBZ in ATL cells could favor its functions on the cellular gene promoters of infected cells or its interaction with the host transcription factors involved in the leukemogenesis process. Moreover, HBZ was shown to delay or prevent Tax-induced senescence by modulating the hyper-activation of NF-kB by Tax, thus facilitating HTLV-1-induced leukemogenesis [34]. An additional mechanism by which HBZ can promote leukemogenesis is the inhibition of regulated cell death (RCD). It was shown that HBZ inhibits both the intrinsic and extrinsic pathways of apoptotic RCD by suppressing the transcription of Bim and FasL via targeting FoxO3a [62] and that level of hbz transcription were associated with the expression of the apoptosis inhibitor surviving [63].

Other accessory proteins play a selective role in HTLV-1 leukemogenesis. The p30 counteracts viral transformation by inhibiting the export of tax/rex mRNA from the nucleus [64], while p13 is involved in transformation by increasing the intracellular content of reactive oxygen species [65]. On the other hand, p12 enhances STAT5 activation in transduced peripheral blood mononuclear cells, allowing them to proliferate even in the presence of a low IL-2 concentration [66]. Thus, a number of different mechanisms underlie HTLV-1 latency/leukemogenesis involving both viral and cellular signals, although the main players are Tax and HBZ proteins. In brief, Tax is involved in the initiation of immortalization/transformation, whereas HBZ is required for the maintenance of the transformed stage. Nevertheless, the modulation of Tax and hbz expression at different stages of infection appears to favor the establishment of a dynamic, rather than static, state of latency and persistence. Indeed, HBZ has been shown to inhibit viral sense transcription and favor the entry of HTLV-1 provirus into the latency phase [67–69]. Ex vivo experiments have shown that, after HTLV-1 provirus integration into the cell, its replication essentially consists of a series of successive phases of burst and reactivation alternating with viral latency controlled by Tax, with the expression of pro- and anti-apoptotic genes playing a central role over cell cycle gene expression [70]. Experimental in vitro models have investigated the role of Tax in the switch between life and death, ultimately leading to the selection of immortalized clones in the final phase of HTLV-1 transformation (see below). An important point in this context is that seemingly conflicting results suggest that the role of Tax in controlling apoptosis is not clear: several studies have supported an effect preventing apoptosis via HTLV-1 gene expression in infected cells and have shown reduced susceptibility to apoptosis induced by anti-Fas [71] and TNFα [72]. This resistance to apoptosis has been associated with the activation of NF-κB via the Tax protein in HTLV-1-infected cells [51] or with the Tax-induced repression of pro-apoptotic genes [73] or the expression of anti-apoptotic genes [74]. However, despite the evidence for the prevention of cell death during HTLV-1 infection, other studies reported that the viral protein Tax not only exerts a somewhat anti-apoptotic effect but is also responsible for promoting apoptosis in HTLV-1-infected cells under certain experimental conditions, as mentioned in the previous paragraph. Clear evidence for this phenomenon was first provided by a study reporting that the expression of Tax in an inducible system turned out to induce, rather than inhibit, cell death [75]. The same and other authors even provided mechanistic details for apoptosis induction by Tax expression. Tax-triggered death was shown to be: i. dependent on ICE protease [76], ii. related to the upregulation of the Fas ligand (FasL) gene [77], and iii. promoted by the nuclear expression of the CREB-binding protein (CBP)/p300-binding domain of Tax [78]. In another study, Tax was shown to induce cell death through the NF-κB-mediated activation of the TNF-related apoptosis-inducing ligand (TRAIL) [79].

Taken together, these data fit with the apparent paradox of a dual role of Tax in apoptotic cell death in the early phases preceding the completion of the transformation process triggered by HTLV-1. To shed light into this subject, some of us pioneered the experimental model of HTLV-1 infection in vitro several years ago. Our model consisted of a long-term culture of PBMCs from healthy donors after exposure to irradiated HTLV-1-infected cells as viral donors [80]. We felt that such a model was the best strategy to recapitulate in vitro what occurs naturally in vivo in HTLV-1-infected patients during the long period of immortalization and transformation that could lead to ATL. Indeed, the results of this study showed that the in vitro HTLV-1 infection of mononuclear cells induces a high rate of cell proliferation during the first weeks accompanied by a heightened susceptibility to RCD through apoptosis. Moreover, there is a progressive decrease in cell death rates observed during the long-term culture phase, ultimately leading to cell immortalization [80]. Additionally, in situ hybridization showed that the cells undergoing apoptosis were indeed the infected ones, with HTLV-1 proviral DNA, rather than residual uninfected cells [80]. Although this was observed in an entire population, we hypothesized that some infected cells died by apoptosis because pro-apoptotic signals prevailed in response to infection triggered by viral gene expression, whereas the infected cells that survived were protected by the prevalent expression of anti-apoptotic genes, which were also triggered by viral proteins. Based on our results, we suggested, for the first time, that the induction of massive apoptosis in response to infection might act as selective pressure for the emergence of anti-apoptotic, well-endowed infected clones prone to immortalization [80]. A more recent study, performed at the single-cell level in the MT-1 ATL cell line, attempted to further elucidate the opposing effects of Tax on cell death by highlighting a situation in which Tax expression affects susceptibility to apoptosis. The results showed that a balance between antiapoptotic and proapoptotic genes depended on the on/off switching of Tax expression in the cells used in this study. A high expression of Tax was preferentially associated with an antiapoptotic gene expression scenario, whereas a low or absent expression of Tax was associated with a greater susceptibility to apoptosis. This switch was continuously active in culture and due to the coexistence of different expanding clones [81]. To explain the dual role of Tax in ATL, when cells progress in the cell cycle despite a low Tax expression, the results of another study hypothesized that ATL cells acquire genetic/epigenetic alterations during the transformation process. These allow bypassing the Tax/NF-kB-dependent induction of senescence [82]. However, it should be considered that the above studies were performed on ATL cells where the HTLV-1 transformation process had already been completed. Thus, the role of Tax in these cells could explain what happens in the cells of ATL patients in an advanced phase of the disease to maintain the leukemic state. It does not describe the role of Tax in the long premalignant phase of HTLV-1 infection, characterized by the oligo-/polyclonal expansion of non-malignant HTLV-1-infected cells. This precedes overt ATL in patients and actually involves HTLV-1-driven oncogenesis.

A recent study investigated the association between the early transcription of the positive viral strand, viral burst, and expression of pro- and anti-apoptotic genes in two naturally infected T-cell clones from patients transduced with a Tax-responsive timer protein [70]. The results showed that anti-apoptotic genes were expressed during the early positive-strand virus burst, followed by a phase in which proapoptotic genes outlasted the virus burst [70]. Another study of naturally HTLV-1-infected Tax-expressing T-cell clones from patients showed that a high Tax expression occurred during the burst phase, immediately followed by a phase of Tax expression heterogeneity associated with a poor proliferation, slow cell cycle, and high susceptibility to apoptosis [83]. Conversely, Tax-expressing clones showed long-term increases in proliferation and decreases in apoptosis [83].

Interestingly, apoptosis may play a role in the awakening of the virus from latency. It has been provocatively hypothesized that, in cells that are prone to apoptotic RCD in response to infection, apoptosis itself may awaken viral latency and promote viral replication through the direct upregulation of caspase 9. This protein, in addition to having a proapoptotic effect, may form a complex with Sp1-p53 and activate viral LTR [84]. On the other hand, viral reactivation from latency could also be triggered by metabolic changes, such as hypoxia, which has been shown to increase the transcription of the HTLV-1 proviral plus strand; conversely, the inhibition of glycolysis and mitochondrial electron transport chain hinders the transcription of the proviral plus strand [85]. Interestingly, the results of these recent detailed studies appear to be essentially consistent with the observations reported above. Overall, the results of these studies suggest that both the apoptotic RCD signaling pathway and the metabolic pathway in HTLV-1 infection may represent potential targets for the development of molecules to reverse latency.

Simplified schematics of the complex and elusive mechanisms leading to immortalization and transformation processes induced in HTLV-1-infected cells are summarized in Figure 2.

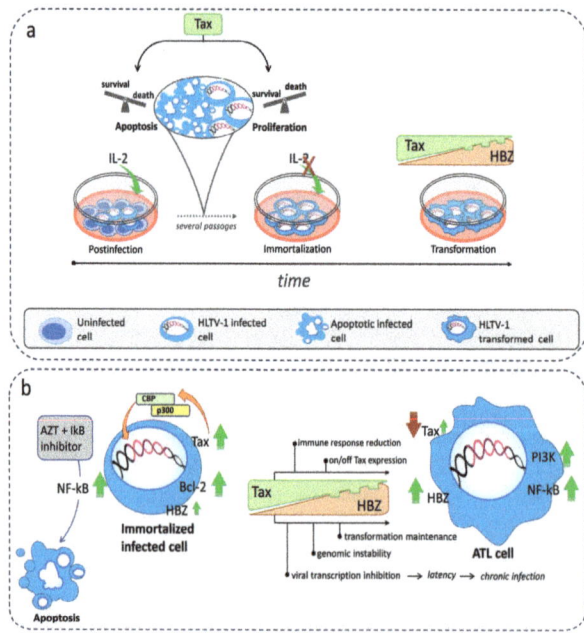

Figure 2. Schematic representation of (**a**) the outcome of HTLV-1 infection in vitro leading to immortalization/transformation and (**b**) the essential processes of leukemogenesis leading to ATL in HTLV-1-infected cells. The fragmented lower edge of Tax expression at low levels and the small green arrow for Tax in ATL cells in (**b**) show that Tax can be sporadically turned on and off during leukemogenesis.

6. Proposals for HTLV-1/ATL-Targeted Therapy

Although the mechanisms of HTLV-1 immortalization/transformation have been extensively studied, it has not been possible to find a unique, suitable candidate treatment. This may be due to the large number of targets that could potentially be hit. The disease starts after the specific HTLV-1 infection of CD4+ cells. Therefore, hypothetically, a wide range of potential targets could be found within the viral genome/proteins, as well as among the host cell genes transcribed during infection that are highly subjected to virus-induced dysregulation. In particular, the specific cell signaling activated by HTLV-1 in infected cells includes multifunctional factors/complexes that could serve as targets for pharmacological intervention. Accordingly, several inhibitors of host cell signaling have recently been proposed in in vitro/ex vivo studies as potential anti-ATL therapeutics. These include apigenin, as an inhibitor of the aryl hydrocarbon receptor and transcription factor that enhances the cytotoxicity of antiretroviral drugs, and dimethyl fumarate as an inhibitor of the so-called CBM complex [86,87]. Another potential cellular target that has been studied for several years is PI3K. PI3K is involved in several processes of oncogenesis and is preferentially expressed in hematopoietic cells [88,89]. It has been demonstrated that the process of multiple nucleation and cell proliferation in ATL strictly depends on changes in the PI3K cascade [90]. Recently, a PI3K-δ/AKT inhibitor, idelalisib, was shown to specifically decrease the proliferation of ATL cells in vitro [91]. In addition, a dual PI3K and HDAC inhibitor, CUDC-907, previously used in patients with hematologic malignancies, has been considered as a potential candidate for ATL treatment. It was found to exert a cytotoxic effect in HTLV-1-infected cells by inhibiting the phosphorylation of downstream PI3K targets, such as AKT, REL A, and p70 S6K, and by decreasing HSP90 chaperone activity in ATL cells in vitro. Moreover, in HTLV-1-infected cells, CUDC-907 induced the upregulation of caspases, concomitant with the down-regulation of anti-apoptotic gene expression and the suppression of NF-κB activation by inhibiting IKKα/ß phosphorylation [92]. The suppression of NF-κB signaling proved to be one of the most important means to counteract the growth of HTLV-1-infected cells. This was demonstrated in HTLV-1-infected ATL cells subjected to mono-treatment with butein, a polyphenol that possesses pro-apoptotic and anti-proliferative properties by down-regulating AKT, AP1, and NF-κB activation [93]. Given the signaling mediated by NF-κB, a suitable approach may be to combine an inhibitor of NF-κB signaling with a chemotherapeutic or antiviral drug to promote susceptibility to cell death. The results of a recent study conducted by some of us showed how the pharmacological inhibition of IκBα could enhance the pro-apoptotic effect of AZT in chronically infected HTLV-1 cell lines [94]. Using a similar approach, a combination treatment was reported to be the most fruitful approach for ATL treatment, even in the case of co-treatment with biological drugs. The combination of ruxolitinib, an inhibitor of the JAK/STAT pathway constitutively activated in HTLV-1-transformed T cells [95], and the Bcl-2/Bcl-xL inhibitor navitoclax showed antitumor efficacy in an additive/synergistic manner on IL-2-dependent ATL cell lines and ex vivo on lymphocytes from ATL patients [96]. Recently, a triple combination of NF-κB and PI3K inhibitors with an inhibitor of the oncogenic driver Bromine and Extra-Terminal domain (BET) motif family, involved in the down-regulation of MYC transcription, was used synergistically to achieve an antiproliferative effect in ATL cells in vitro and ex vivo [97]. To evaluate the response of ATL patients to treatment in vivo, it may be critical to determine the efficacy of therapy on viral replication in addition to its effects on cell survival. A recent study in ATL patients subjected to long-term therapy with AZT and IFNα demonstrated that the combination treatment resulted in a complete inhibition of RT activity, a reduction in two other virological parameters, and a dramatic change in clonality pattern, as observed in the short-term cultures of PBMCs from patients who responded to the therapy [98].

Studies conducted in ATL patients have shown that an important goal is to induce cells to undergo apoptosis or generally move toward RCD, in response to therapy. Among the various ways to undergo RCD, interaction with the autophagy pathway should be considered. Autophagy may actually play a dual role in RCD, either initiating cell death or maintaining it [99]. The autophagy protein Beclin-1 appears to be involved in maintaining the activation of the factors NF-κB and STAT3 in HTLV-1-transformed cell lines [100]. Therefore, in the case of HTLV-1 infection, it seems that the suppression of autophagy may be an appropriate approach to treat ATL. Indeed, autophagy appears to be a "self-feeding" mechanism in tumor progression that supports cancer cell growth. Recently, chloroquine/hydroxychloroquine was shown to inhibit autophagic flux in ATL cells ex vivo. The mechanism involved the accumulation of p47 together with the autophagic protein LC3IIand led to increases in IkBα, resulting in the inhibition of NF-κB activation and susceptibility to apoptosis [101]. On the other hand, small-molecule inhibitors of sirtuin 2, the nicotinamide adenine dinucleotide-dependent deacylase involved in the control of cell cycle modulation, inhibited the growth of patient ATL cell lines not only by inducing caspase-dependent and -independent cell death, but also by increasing autophagosome accumulation and inhibiting autophagosome degradation [102]. Therefore, in this case, the upregulation of autophagic flux in ATL cells appears to be associated with mechanisms that induce cell death. These seemingly contradictory results encourage further studies on the role of autophagy in HTLV-1 infection and on a possible pharmacological modulation of autophagy as a novel strategy to target ATL.

Potential targets could be identified not only in genes directly involved in cell signaling but also among regulators of gene expression. Recently, 12 miRNA associated with the regulation of key cell signaling genes in ATL cases have been identified, thanks to a bioinformatics approach [103]. Based on the seemingly conflicting results obtained in defining a satisfactory treatment for ATL, it seems plausible that combination treatments may be the most appropriate approaches to inhibit a complex network of cell signaling as induced by HTLV-1 infection. Indeed, such a therapeutic strategy is likely to be the best weapon to avoid the possible feedback control aimed at restoring leukemic cell survival.

As mentioned in the previous section, even latency reversion may represent a potential strategy for a novel HTLV-1/ATL-targeted therapy. Latency could possibly be reversed by activating the expression of viral antigens that are likely to be recognized by the immune system. This would allow the recruitment of effector cells capable of knocking out the virus. Recently, a treatment to reverse latency has been proposed by using the histone deacetylase inhibitors (HDACi) panobinostat and romidepsin [104]. These inhibitors were shown to repress the transcription of Tax and of Tax-targeted genes, although only slightly.

A summary of the main suggestions for HTLV-1-targeted therapy is provided in Table 1.

Table 1. Proposals for HTLV-1/ATL-targeted therapy.

Proposed Therapeutic Treatment	Target		Available Results	
	Viral	Cellular	In Vitro/Ex Vivo	In Vivo
AZT + IFNα (98)	RT	IFN-receptor other? [1]	Samples from patients: (a) Complete inhibition of RT activity and (b) reduction in virus parameters in responding patients; (c) Dramatic change in the clonality pattern.	Prolonged survival with respect to untreated patients
Idelalisib (91)	?	PI3K-δ/AKT	Inhibition of proliferation in ATL cells.	No
CUDC-907 (92)	?	PI3K/HDAC	(a) Induction of cytotoxicity in HTLV-1-infected cells; (b) Inhibition of HSP90 activity; (c) Increased caspase activity in ATL cells.	No

Table 1. Cont.

Proposed Therapeutic Treatment	Target		Available Results	
	Viral	Cellular	In Vitro/Ex Vivo	In Vivo
Butein (93)	?	AKT/AP1 NF-kB	(a) Induction of apoptosis; (b) Inhibition of proliferation of HTLV-1-infected and ATL cells.	No
AZT+ Bay 11-7085 (94)	RT-	IκBα phosphorylation	(a) Increased apoptosis; (b) Up-reg. pro-apoptotic and down-reg. anti-apoptotic genes in HTLV-1-infected/transformed cells.	No
Ruxolitinib+ Navitoclax (96)	?	JAK/STAT Bcl-2/Bcl-xL	Cytotoxicity in IL-2-dependent ATL cell lines and ex vivo in lymphocytes from ATL patients.	No
I-BET762+ Copanlisib+ bardoxolone methyl (97)	?	BET NF-κB PI3K	Inhibition of proliferation in ATL cells in vitro and ex vivo samples from patients.	Prolonged survival of ATL-bearing xenograft mice
Chloroquine/ Hydroxy chloroquine (101)	?	Autophagic flux	Ex vivo from ATL patients: (a) Inhibition of autophagy; (b) Accumulation of p47 with LC3IIand inhibition of NF-κB activation; (c) Proneness to apoptosis.	No
NCO-90/141 (102)	?	Sirtuin 2	(a) Increased apoptosis; (b) Autophagy in ATL cells.	No
? (103)	?	12 miRNA	In silico analysis identified 12 miRNA deregulated in HTLV-1 samples predicted to interact with 90 genes.	No

^{1} ? = unknown/not investigated.

7. Potential of Gene Editing Technology in the Eradication of Persistent HTLV-1 Infection and ATL Therapy

CRISPR/Cas9 genome editing is a novel technology that uses a guide RNA (gRNA) to precisely cleave double-stranded DNA at a specific site. After cleavage, double-stranded DNA breaks in human cells are normally repaired by the error-prone NHEJ pathway [105]. This can lead to insertions and deletions that alter gene-reading frames, disrupt DNA regulatory motifs, or damage RNA structures [106]. CRISPR technology has the potential to be a therapeutic strategy for HTLV-1 disease, as demonstrated by the use of zinc finger nucleases (ZFNs) to disrupt LTR promoter function [107] and inhibit the proliferation of HTLV-1-positive cell lines [108]. However, CRISPR/Cas9 has advantages over ZFNs and other gene editing approaches, such as simplicity, cost-effectiveness, and efficiency, and has been shown to reduce ATL cell proliferation in vitro by targeting HBZ [109]. Since the advent of gene editing, several studies have focused on targeting the HIV-1 provirus. This work may shed light on the efficacy, safety, and limitations of these approaches for targeting HTLV-1 [110]. The goal of gene editing approaches against HIV-1 is primarily to remove proviral DNA from the host genome. This can be achieved by targeting both LTRs and causing their disruption, followed by excision of the proviral DNA from the host genome. However, effective delivery and low off-target effects are critical for successful application in clinical trials [111,112]. In this sense, CRISPR technology is a promising gene editing tool, but it has some drawbacks that need to be addressed before it can be effectively used for antiretroviral therapy. In vitro, CRISPR/Cas9 can remove the integrated HIV-1 genome from cellular DNA, but in vivo, off-target activity, gene rearrangements, target selection limitations, and a limited number of effective transporters complicate the process. To overcome these problems, multiple RNA guide structures should be introduced into a single cell to ensure the cleavage of the provirus. However, the use of multiple guide structures may increase off-target activity and unpredictable DNA rearrangements. One of our recent studies examined the fate of HIV-1 provirus and cellular repair mechanisms triggered by CRISPR/Cas9. The study was conducted in two parts: the first part examined the fate of HIV-1 provirus in 293T cells and the second part confirmed the results in a human T-cell leukemia line latently transduced with HIV-1-GFP and in T-cell leukemia cells infected with a clinical lymphotropic isolate of HIV-1. The study found that, after CRISPR-mediated LTR ablation, the excised HIV-1 provirus remains in the cells for an extended period of time and can circulate as single molecules or concatemers that remain as episomes in the infected cells [113]. Non-integrated HIV-1 is abundant in resting, non-proliferating CD4+ T cells and leads to de novo virus production after the exposure to cytokine of the resting cells. The results of this study raise concerns about the persistence of CRISPR/Cas9-excised proviral DNA in the absence of antiretroviral therapy [113].

Although retroviral proviruses are largely restricted to HIV-1, HTLV-1 offers more sites for gRNA targeting than HIV-1 due to its highly conserved viral genome with remarkable sequence homogeneity. Moreover, as shown in Figure 3, CRISPR/Cas9 can disable both latent and actively replicating HTLV-1 and abrogate the function or expression of viral Tax and HBZ, which are the main drivers of HTLV-1-mediated transformation and proliferation. Targeting the viral LTRs involved in viral genome integration and gene expression may allow the effective treatment of HTLV-1-infected individuals, asymptomatic carriers, and ATL and HAM/TSP patients. The careful design of gRNAs that disrupt two viral elements simultaneously can disrupt overlapping reading frames between HBZ and the 3'LTR and Tax and the 3' LTR [110].

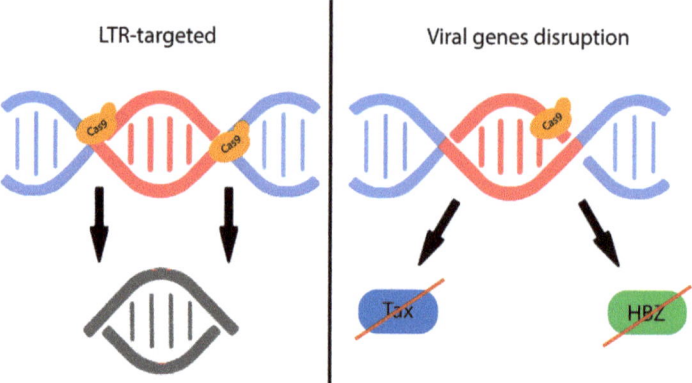

Figure 3. Schematic representation of how CRISPR-Cas9 gene editing could affect HTLV-1 integration and cell transformation.

8. Conclusions

Unfortunately, although some progress has been made in slowing its progression, ATL is still an incurable disease. However, in recent years, considerable progress has been made in defining the mechanisms at the molecular and genetic levels associated with the events underlying the complex viral/cellular system that leads to the selection of the transformed clone determining the onset of ATL in HTLV-1-infected patients. The definition of these mechanisms has already led to the identification of new therapeutic targets and corresponding agents that can act on them. Therefore, this seems to be an indication of how new therapeutic strategies may be found to counteract ATL in the near future. Nevertheless, a precise prediction of if and when it will actually be possible to use therapeutic treatments that can prevent or control ATL and potentially transform HTLV-1 infection into a chronic disease is currently not possible.

Funding: This r study was partly funded by The Italian Ministry of University and Research, P.R.I.N. Project code: 2017M8R7N9.

Conflicts of Interest: The authors declare no conflict of interest.

References

1. Gessain, A.; Cassar, O. Epidemiological aspects and world distribution of HTLV-1 infection. *Front. Microbiol.* **2012**, *3*, 388. [CrossRef] [PubMed]
2. Gessain, A.; Ramassamy, J.L.; Afonso, P.V.; Cassar, O. Geographic distribution, clinical epidemiology and genetic diversity of the human oncogenic retrovirus HTLV-1 in Africa, the world's largest endemic area. *Front. Immunol.* **2023**, *14*, 1043600. [CrossRef] [PubMed]
3. Percher, F.; Jeannin, P.; Martin-Latil, S.; Gessain, A.; Afonso, P.V.; Vidy-Roche, A.; Ceccaldi, P.E. Mother-to-child transmission of HTLV-1 epidemiological aspects, mechanisms and determinants of mother-to-child transmission. *Viruses* **2016**, *8*, 40. [CrossRef] [PubMed]
4. Lairmore, M.D.; Haines, R.; Anupam, R. Mechanisms of human t-lymphotropic virus type 1 transmission and disease. *Curr. Opin. Virol.* **2012**, *2*, 474–481. [CrossRef] [PubMed]
5. Cook, L.B.; Melamed, A.; Demontis, M.A.; Laydon, D.J.; Fox, J.M.; Tosswill, J.H.; de Freitas, D.; Price, A.D.; Medcalf, J.F.; Martin, F.; et al. Rapid dissemination of human t-lymphotropic virus type 1 during primary infection in transplant recipients. *Retrovirology* **2016**, *13*, 3. [CrossRef]
6. Nasr, R.; El Hajj, H.; Kfoury, Y.; de Thé, H.; Hermine, O.; Bazarbachi, A. Controversies in targeted therapy of adult t cell leukemia/lymphoma: On target or off target effects? *Viruses* **2011**, *3*, 750–769. [CrossRef]
7. Laverdure, S.; Polakowski, N.; Hoang, K.; Lemasson, I. Permissive sense and antisense transcription from the 5' and 3' long terminal repeats of human t-cell leukemia virus type 1. *J. Virol.* **2016**, *90*, 3600–3610. [CrossRef]
8. Fujisawa, J.; Seiki, M.; Kiyokawa, T.; Yoshida, M. Functional activation of the long terminal repeat of human t-cell leukemia virus type i by a trans-acting factor. *Proc. Natl. Acad. Sci. USA* **1985**, *82*, 2277–2281. [CrossRef]
9. Boxus, M.; Willems, L. Mechanisms of HTLV-1 persistence and transformation. *Br. J. Cancer* **2009**, *101*, 1497–1501. [CrossRef]
10. Baydoun, H.; Duc-Dodon, M.; Lebrun, S.; Gazzolo, L.; Bex, F. Regulation of the human t-cell leukemia virus gene expression depends on the localization of regulatory proteins tax, rex and p30ii in specific nuclear subdomains. *Gene* **2007**, *386*, 191–201. [CrossRef]
11. Sarkis, S.; Galli, V.; Moles, R.; Yurick, D.; Khoury, G.; Purcell, D.F.J.; Franchini, G.; Pise-Masison, C.A. Role of HTLV-1 orf-i encoded proteins in viral transmission and persistence. *Retrovirology* **2019**, *16*, 43. [CrossRef] [PubMed]
12. Silic-Benussi, M.; Marin, O.; Biasiotto, R.; D'Agostino, D.M.; Ciminale, V. Effects of human t-cell leukemia virus type 1 (HTLV-1) p13 on mitochondrial k+ permeability: A new member of the viroporin family? *Febs Lett.* **2010**, *584*, 2070–2075. [CrossRef] [PubMed]
13. Younis, I.; Khair, L.; Dundr, M.; Lairmore, M.D.; Franchini, G.; Green, P.L. Repression of human t-cell leukemia virus type 1 and type 2 replication by a viral mrna-encoded posttranscriptional regulator. *J. Virol.* **2004**, *78*, 11077–11083. [CrossRef] [PubMed]
14. Martin-Latil, S.; Gnadig, N.F.; Mallet, A.; Desdouits, M.; Guivel-Benhassine, F.; Jeannin, P.; Prevost, M.C.; Schwartz, O.; Gessain, A.; Ozden, S.; et al. Transcytosis of HTLV-1 across a tight human epithelial barrier and infection of subepithelial dendritic cells. *Blood* **2012**, *120*, 572–580. [CrossRef]
15. Mulherkar, T.H.; Gomez, D.J.; Sandel, G.; Jain, P. Co-infection and cancer: Host-pathogen interaction between dendritic cells and hiv-1, HTLV-1, and other oncogenic viruses. *Viruses* **2022**, *14*, 2037. [CrossRef]
16. Van Prooyen, N.; Gold, H.; Andresen, V.; Schwartz, O.; Jones, K.; Ruscetti, F.; Lockett, S.; Gudla, P.; Venzon, D.; Franchini, G. Human t-cell leukemia virus type 1 p8 protein increases cellular conduits and virus transmission. *Proc. Natl. Acad. Sci. USA* **2010**, *107*, 20738–20743. [CrossRef]
17. Pais-Correia, A.M.; Sachse, M.; Guadagnini, S.; Robbiati, V.; Lasserre, R.; Gessain, A.; Gout, O.; Alcover, A.; Thoulouze, M.I. Biofilm-like extracellular viral assemblies mediate HTLV-1 cell-to-cell transmission at virological synapses. *Nat. Med.* **2010**, *16*, 83–89. [CrossRef]
18. Pinto, D.O.; Al Sharif, S.; Mensah, G.; Cowen, M.; Khatkar, P.; Erickson, J.; Branscome, H.; Lattanze, T.; DeMarino, C.; Alem, F.; et al. Extracellular vesicles from HTLV-1 infected cells modulate target cells and viral spread. *Retrovirology* **2021**, *18*, 6. [CrossRef]
19. Kim, Y.; Mensah, G.A.; Al Sharif, S.; Pinto, D.O.; Branscome, H.; Yelamanchili, S.V.; Cowen, M.; Erickson, J.; Khatkar, P.; Mahieux, R.; et al. Extracellular vesicles from infected cells are released prior to virion release. *Cells* **2021**, *10*, 781. [CrossRef]
20. Overbaugh, J.; Bangham, C.R. Selection forces and constraints on retroviral sequence variation. *Science* **2001**, *292*, 1106–1109. [CrossRef]
21. Laydon, D.J.; Sunkara, V.; Boelen, L.; Bangham, C.R.M.; Asquith, B. The relative contributions of infectious and mitotic spread to HTLV-1 persistence. *PLoS Comput. Biol.* **2020**, *16*, e1007470. [CrossRef] [PubMed]
22. Izaki, M.; Yasunaga, J.I.; Nosaka, K.; Sugata, K.; Utsunomiya, H.; Suehiro, Y.; Shichijo, T.; Yamada, A.; Sugawara, Y.; Hibi, T.; et al. In vivo dynamics and adaptation of HTLV-1-infected clones under different clinical conditions. *PLoS Pathog.* **2021**, *17*, e1009271. [CrossRef] [PubMed]

23. Fox, J.M.; Hilburn, S.; Demontis, M.A.; Brighty, D.W.; Rios Grassi, M.F.; Galvao-Castro, B.; Taylor, G.P.; Martin, F. Long terminal repeat circular DNA as markers of active viral replication of human t lymphotropic virus-1 in vivo. *Viruses* **2016**, *8*, 80. [CrossRef] [PubMed]
24. Melamed, A.; Laydon, D.J.; Al Khatib, H.; Rowan, A.G.; Taylor, G.P.; Bangham, C.R. HTLV-1 drives vigorous clonal expansion of infected cd8$^+$ t cells in natural infection. *Retrovirology* **2015**, *12*, 91. [CrossRef] [PubMed]
25. Yasunaga, J. Viral, genetic, and immune factors in the oncogenesis of adult t-cell leukemia/lymphoma. *Int. J. Hematol.* **2023**, *117*, 504–511. [CrossRef]
26. El Hajj, H.; Tsukasaki, K.; Cheminant, M.; Bazarbachi, A.; Watanabe, T.; Hermine, O. Novel treatments of adult t cell leukemia lymphoma. *Front. Microbiol.* **2020**, *11*, 1062. [CrossRef] [PubMed]
27. Cook, L.B.; Fuji, S.; Hermine, O.; Bazarbachi, A.; Ramos, J.C.; Ratner, L.; Horwitz, S.; Fields, P.; Tanase, A.; Bumbea, H.; et al. Revised adult t-cell leukemia-lymphoma international consensus meeting report. *J. Clin. Oncol.* **2019**, *37*, 677–687. [CrossRef] [PubMed]
28. Ishida, T.; Joh, T.; Uike, N.; Yamamoto, K.; Utsunomiya, A.; Yoshida, S.; Saburi, Y.; Miyamoto, T.; Takemoto, S.; Suzushima, H.; et al. Defucosylated anti-ccr4 monoclonal antibody (kw-0761) for relapsed adult t-cell leukemia-lymphoma: A multicenter phase ii study. *J. Clin. Oncol.* **2012**, *30*, 837–842. [CrossRef]
29. Tanaka, T.; Inamoto, Y.; Ito, A.; Watanabe, M.; Takeda, W.; Aoki, J.; Kim, S.W.; Fukuda, T. Lenalidomide treatment for recurrent adult t-cell leukemia/lymphoma after allogeneic hematopoietic cell transplantation. *Hematol. Oncol.* **2022**, *41*, 389–395. [CrossRef] [PubMed]
30. Baba, Y.; Sakai, H.; Kabasawa, N.; Harada, H. Successful treatment of an aggressive adult t-cell leukemia/lymphoma with strong cd30 expression using brentuximabvedotin as combination and maintenance therapy. *Intern. Med.* **2023**, *62*, 613–616. [CrossRef] [PubMed]
31. Izutsu, K.; Makita, S.; Nosaka, K.; Yoshimitsu, M.; Utsunomiya, A.; Kusumoto, S.; Morishima, S.; Tsukasaki, K.; Kawamata, T.; Ono, T.; et al. An open-label, single-arm phase 2 trial of valemetostat for relapsed or refractory adult t-cell leukemia/lymphoma. *Blood* **2023**, *141*, 1159–1168. [CrossRef]
32. Utsunomiya, A.; Izutsu, K.; Jo, T.; Yoshida, S.; Tsukasaki, K.; Ando, K.; Choi, I.; Imaizumi, Y.; Kato, K.; Kurosawa, M.; et al. Oral histone deacetylase inhibitor tucidinostat (hbi-8000) in patients with relapsed or refractory adult t-cell leukemia/lymphoma: Phase IIb results. *Cancer Sci.* **2022**, *113*, 2778–2787. [CrossRef] [PubMed]
33. Katsuya, H. Current and emerging therapeutic strategies in adult t-cell leukemia-lymphoma. *Int. J. Hematol.* **2023**, *117*, 512–522. [CrossRef]
34. Zhi, H.J.; Yang, L.P.; Kuo, Y.L.; Ho, Y.K.; Shih, H.M.; Giam, C.Z. Nf-κb hyper-activation by HTLV-1 tax induces cellular senescence, but can be alleviated by the viral anti-sense protein HBZ. *PLoS Pathog.* **2011**, *7*, e1002025. [CrossRef] [PubMed]
35. Saito, K.; Saito, M.; Taniura, N.; Okuwa, T.; Ohara, Y. Activation of the pi3k-akt pathway by human t cell leukemia virus type 1 (HTLV-1) oncoprotein tax increases bcl3 expression, which is associated with enhanced growth of HTLV-1-infected t cells. *Virology* **2010**, *403*, 173–180. [CrossRef]
36. Sun, S.C.; Ballard, D.W. Persistent activation of nf-κb by the tax transforming protein of HTLV-1: Hijacking cellular iκb kinases. *Oncogene* **1999**, *18*, 6948–6958. [CrossRef]
37. Shimotohno, K.; Takano, M.; Teruuchi, T.; Miwa, M. Requirement of multiple copies of a 21-nucleotide sequence in the u3 regions of human t-cell leukemia virus type I and type II long terminal repeats for trans-acting activation of transcription. *Proc. Natl. Acad. Sci. USA* **1986**, *83*, 8112–8116. [CrossRef] [PubMed]
38. Giam, C.Z.; Xu, Y.L. HTLV-i tax gene product activates transcription via pre-existing cellular factors and camp responsive element. *J. Biol. Chem.* **1989**, *264*, 15236–15241. [CrossRef] [PubMed]
39. Adya, N.; Giam, C.Z. Distinct regions in human t-cell lymphotropic virus type I tax mediate interactions with activator protein creb and basal transcription factors. *J. Virol.* **1995**, *69*, 1834–1841. [CrossRef] [PubMed]
40. Lenzmeier, B.A.; Giebler, H.A.; Nyborg, J.K. Human t-cell leukemia virus type 1 tax requires direct access to DNA for recruitment of creb binding protein to the viral promoter. *Mol. Cell Biol.* **1998**, *18*, 721–731. [CrossRef] [PubMed]
41. Kimzey, A.L.; Dynan, W.S. Specific regions of contact between human t-cell leukemia virus type i tax protein and DNA identified by photocross-linking. *J. Biol. Chem.* **1998**, *273*, 13768–13775. [CrossRef]
42. Lu, H.; Pise-Masison, C.A.; Linton, R.; Park, H.U.; Schiltz, R.L.; Sartorelli, V.; Brady, J.N. Tax relieves transcriptional repression by promoting histone deacetylase 1 release from the human t-cell leukemia virus type 1 long terminal repeat. *J. Virol.* **2004**, *78*, 6735–6743. [CrossRef] [PubMed]
43. Ego, T.; Ariumi, Y.; Shimotohno, K. The interaction of HTLV-1 tax with hdac1 negatively regulates the viral gene expression. *Oncogene* **2002**, *21*, 7241–7246. [CrossRef] [PubMed]
44. Mitchell, S.; Vargas, J.; Hoffmann, A. Signaling via the nfκb system. *Wiley Interdiscip. Rev. Syst. Biol. Med.* **2016**, *8*, 227–241. [CrossRef]
45. Kanno, T.; Brown, K.; Franzoso, G.; Siebenlist, U. Kinetic analysis of human t-cell leukemia virus type i tax-mediated activation of nf-κb. *Mol. Cell Biol.* **1994**, *14*, 6443–6451. [CrossRef] [PubMed]
46. Geleziunas, R.; Ferrell, S.; Lin, X.; Mu, Y.; Cunningham, E.T., Jr.; Grant, M.; Connelly, M.A.; Hambor, J.E.; Marcu, K.B.; Greene, W.C. Human t-cell leukemia virus type 1 tax induction of nf-κb involves activation of the iκb kinase α (ikkα) and ikkβ cellular kinases. *Mol. Cell Biol.* **1998**, *18*, 5157–5165. [CrossRef]

47. Fuggetta, M.P.; Bordignon, V.; Cottarelli, A.; Macchi, B.; Frezza, C.; Cordiali-Fei, P.; Ensoli, F.; Ciafre, S.; Marino-Merlo, F.; Mastino, A.; et al. Downregulation of proinflammatory cytokines in HTLV-1-infected t cells by resveratrol. *J. Exp. Clin. Cancer Res.* **2016**, *35*, 118. [CrossRef]
48. Mohanty, S.; Harhaj, E.W. Mechanisms of oncogenesis by HTLV-1 tax. *Pathogens* **2020**, *9*, 543. [CrossRef]
49. Hleihel, R.; Skayneh, H.; de The, H.; Hermine, O.; Bazarbachi, A. Primary cells from patients with adult t cell leukemia/lymphoma depend on HTLV-1 tax expression for nf-κb activation and survival. *Blood Cancer J.* **2023**, *13*, 67. [CrossRef] [PubMed]
50. Macaire, H.; Riquet, A.; Moncollin, V.; Biemont-Trescol, M.C.; DucDodon, M.; Hermine, O.; Debaud, A.L.; Mahieux, R.; Mesnard, J.M.; Pierre, M.; et al. Tax protein-induced expression of antiapoptotic bfl-1 protein contributes to survival of human t-cell leukemia virus type 1 (HTLV-1)-infected t-cells. *J. Biol. Chem.* **2012**, *287*, 21357–21370. [CrossRef] [PubMed]
51. Kawakami, H.; Tomita, M.; Matsuda, T.; Ohta, T.; Tanaka, Y.; Fujii, M.; Hatano, M.; Tokuhisa, T.; Mori, N. Transcriptional activation of survivin through the nf-κb pathway by human t-cell leukemia virus type I Tax. *Int. J. Cancer* **2005**, *115*, 967–974. [CrossRef] [PubMed]
52. Kogure, Y.; Kataoka, K. Genetic alterations in adult t-cell leukemia/lymphoma. *Cancer Sci.* **2017**, *108*, 1719–1725. [CrossRef] [PubMed]
53. Jo, T.; Noguchi, K.; Sakai, T.; Kubota-Koketsu, R.; Irie, S.; Matsuo, M.; Taguchi, J.; Abe, K.; Shigematsu, K. HTLV-1 tax-specific memory cytotoxic t lymphocytes in long-term survivors of aggressive-type adult t-cell leukemia/lymphoma. *Cancer Med.* **2022**, *11*, 3238–3250. [CrossRef] [PubMed]
54. Yasunaga, J.I. Strategies of human t-cell leukemia virus type 1 for persistent infection: Implications for leukemogenesis of adult t-cell leukemia-lymphoma. *Front. Microbiol.* **2020**, *11*, 979. [CrossRef] [PubMed]
55. Akkouche, A.; Moodad, S.; Hleihel, R.; Skayneh, H.; Chambeyron, S.; El Hajj, H.; Bazarbachi, A. In vivo antagonistic role of the human t-cell leukemia virus type 1 regulatory proteins tax and hbz. *PLoS Pathog.* **2021**, *17*, e1009219. [CrossRef]
56. Arnold, J.; Yamamoto, B.; Li, M.; Phipps, A.J.; Younis, I.; Lairmore, M.D.; Green, P.L. Enhancement of infectivity and persistence in vivo by hbz, a natural antisense coded protein of HTLV-1. *Blood* **2006**, *107*, 3976–3982. [CrossRef] [PubMed]
57. Gazon, H.; Lemasson, I.; Polakowski, N.; Cesaire, R.; Matsuoka, M.; Barbeau, B.; Mesnard, J.M.; Peloponese, J.M., Jr. Human t-cell leukemia virus type 1 (HTLV-1) bzip factor requires cellular transcription factor jund to upregulate HTLV-1 antisense transcription from the 3' long terminal repeat. *J. Virol.* **2012**, *86*, 9070–9078. [CrossRef]
58. Vernin, C.; Thenoz, M.; Pinatel, C.; Gessain, A.; Gout, O.; Delfau-Larue, M.H.; Nazaret, N.; Legras-Lachuer, C.; Wattel, E.; Mortreux, F. HTLV-1 bzip factor hbz promotes cell proliferation and genetic instability by activating oncomirs. *Cancer Res.* **2014**, *74*, 6082–6093. [CrossRef]
59. Rushing, A.W.; Hoang, K.; Polakowski, N.; Lemasson, I. The human t-cell leukemia virus type 1 basic leucine zipper factor attenuates repair of double-stranded DNA breaks via nonhomologous end joining. *J. Virol.* **2018**, *92*, e00672. [CrossRef]
60. Sakurada-Aono, M.; Sakamoto, T.; Kobayashi, M.; Takiuchi, Y.; Iwai, F.; Tada, K.; Sasanuma, H.; Hirabayashi, S.; Murakawa, Y.; Shirakawa, K.; et al. HTLV-1 bzip factor impairs DNA mismatch repair system. *Biochem. Biophys. Res. Commun.* **2023**, *657*, 43–49. [CrossRef] [PubMed]
61. Forlani, G.; Shallak, M.; Tedeschi, A.; Cavallari, I.; Marcais, A.; Hermine, O.; Accolla, R.S. Dual cytoplasmic and nuclear localization of HTLV-1-encoded hbz protein is a unique feature of adult t-cell leukemia. *Haematologica* **2021**, *106*, 2076–2085. [CrossRef] [PubMed]
62. Tanaka-Nakanishi, A.; Yasunaga, J.; Takai, K.; Matsuoka, M. HTLV-1 bzip factor suppresses apoptosis by attenuating the function of foxo3a and altering its localization. *Cancer Res.* **2014**, *74*, 188–200. [CrossRef] [PubMed]
63. Mitobe, Y.; Yasunaga, J.; Furuta, R.; Matsuoka, M. HTLV-1 bzip factor rna and protein impart distinct functions on t-cell proliferation and survival. *Cancer Res.* **2015**, *75*, 4143–4152. [CrossRef]
64. Nicot, C.; Dundr, M.; Johnson, J.M.; Fullen, J.R.; Alonzo, N.; Fukumoto, R.; Princler, G.L.; Derse, D.; Misteli, T.; Franchini, G. HTLV-1-encoded p30ii is a post transcriptional negative regulator of viral replication. *Nat. Med.* **2004**, *10*, 197–201. [CrossRef] [PubMed]
65. Silic-Benussi, M.; Cavallari, I.; Vajente, N.; Vidali, S.; Chieco-Bianchi, L.; Di Lisa, F.; Saggioro, D.; D'Agostino, D.M.; Ciminale, V. Redox regulation of t-cell turnover by the p13 protein of human t-cell leukemia virus type 1: Distinct effects in primary versus transformed cells. *Blood* **2010**, *116*, 54–62. [CrossRef] [PubMed]
66. Nicot, C.; Mulloy, J.C.; Ferrari, M.G.; Johnson, J.M.; Fu, K.; Fukumoto, R.; Trovato, R.; Fullen, J.; Leonard, W.J.; Franchini, G. HTLV-1 p12(i) protein enhances stat5 activation and decreases the interleukin-2 requirement for proliferation of primary human peripheral blood mononuclear cells. *Blood* **2001**, *98*, 823–829. [CrossRef] [PubMed]
67. Gazon, H.; Chauhan, P.S.; Porquet, F.; Hoffmann, G.B.; Accolla, R.; Willems, L. Epigenetic silencing of HTLV-1 expression by the hbzrna through interference with the basal transcription machinery. *Blood Adv.* **2020**, *4*, 5574–5579. [CrossRef] [PubMed]
68. Gaudray, G.; Gachon, F.; Basbous, J.; Biard-Piechaczyk, M.; Devaux, C.; Mesnard, J.M. The complementary strand of the human t-cell leukemia virus type 1 rna genome encodes a bzip transcription factor that down-regulates viral transcription. *J. Virol.* **2002**, *76*, 12813–12822. [CrossRef] [PubMed]
69. Clerc, I.; Polakowski, N.; Andre-Arpin, C.; Cook, P.; Barbeau, B.; Mesnard, J.M.; Lemasson, I. An interaction between the human t cell leukemia virus type 1 basic leucine zipper factor (hbz) and the kix domain of p300/cbp contributes to the down-regulation of tax-dependent viral transcription by hbz. *J. Biol. Chem.* **2008**, *283*, 23903–23913. [CrossRef] [PubMed]

70. Kiik, H.; Ramanayake, S.; Miura, M.; Tanaka, Y.; Melamed, A.; Bangham, C.R.M. Time-course of host cell transcription during the HTLV-1 transcriptional burst. *PLoS Pathog.* **2022**, *18*, e1010387. [CrossRef] [PubMed]
71. Copeland, K.F.; Haaksma, A.G.; Goudsmit, J.; Krammer, P.H.; Heeney, J.L. Inhibition of apoptosis in t cells expressing human t cell leukemia virus type I Tax. *AIDS Res. Hum. Retrovir.* **1994**, *10*, 1259–1268. [CrossRef] [PubMed]
72. Yang, Y.C.; Hsu, T.Y.; Lin, R.H.; Su, I.J.; Chen, J.Y.; Yang, C.S. Resistance to tumor necrosis factor-α-induced apoptosis in human t-lymphotropic virus type i-infected t cell lines. *AIDS Res. Hum. Retrovir.* **2002**, *18*, 207–212. [CrossRef]
73. Brauweiler, A.; Garrus, J.E.; Reed, J.C.; Nyborg, J.K. Repression of bax gene expression by the HTLV-1 tax protein: Implications for suppression of apoptosis in virally infected cells. *Virology* **1997**, *231*, 135–140. [CrossRef]
74. Mori, N.; Fujii, M.; Cheng, G.; Ikeda, S.; Yamasaki, Y.; Yamada, Y.; Tomonaga, M.; Yamamoto, N. Human t-cell leukemia virus type i tax protein induces the expression of anti-apoptotic gene bcl-xl in human t-cells through nuclear factor-κb and c-amp responsive element binding protein pathways. *Virus Genes.* **2001**, *22*, 279–287. [CrossRef] [PubMed]
75. Chlichlia, K.; Moldenhauer, G.; Daniel, P.T.; Busslinger, M.; Gazzolo, L.; Schirrmacher, V.; Khazaie, K. Immediate effects of reversible HTLV-1 tax function: T-cell activation and apoptosis. *Oncogene* **1995**, *10*, 269–277.
76. Chlichlia, K.; Busslinger, M.; Peter, M.E.; Walczak, H.; Krammer, P.H.; Schirrmacher, V.; Khazaie, K. Ice-proteases mediate HTLV-i tax-induced apoptotic t-cell death. *Oncogene* **1997**, *14*, 2265–2272. [CrossRef] [PubMed]
77. Chen, X.; Zachar, V.; Zdravkovic, M.; Guo, M.; Ebbesen, P.; Liu, X. Role of the fas/fas ligand pathway in apoptotic cell death induced by the human t cell lymphotropic virus type i tax transactivator. *J. Gen. Virol.* **1997**, *78 Pt 12*, 3277–3285. [CrossRef]
78. Nicot, C.; Harrod, R. Distinct p300-responsive mechanisms promote caspase-dependent apoptosis by human t-cell lymphotropic virus type 1 tax protein. *Mol. Cell Biol.* **2000**, *20*, 8580–8589. [CrossRef] [PubMed]
79. Rivera-Walsh, I.; Waterfield, M.; Xiao, G.; Fong, A.; Sun, S.C. Nf-κb signaling pathway governs trail gene expression and human t-cell leukemia virus-i tax-induced t-cell death. *J. Biol. Chem.* **2001**, *276*, 40385–40388. [CrossRef]
80. Matteucci, C.; Balestrieri, E.; Macchi, B.; Mastino, A. Modulation of apoptosis during HTLV-1-mediated immortalization process in vitro. *J. Med. Virol.* **2004**, *74*, 473–483. [CrossRef]
81. Mahgoub, M.; Yasunaga, J.I.; Iwami, S.; Nakaoka, S.; Koizumi, Y.; Shimura, K.; Matsuoka, M. Sporadic on/off switching of HTLV-1 tax expression is crucial to maintain the whole population of virus-induced leukemic cells. *Proc. Natl. Acad. Sci. USA* **2018**, *115*, E1269–E1278. [CrossRef] [PubMed]
82. Shudofsky, A.M.D.; Giam, C.Z. Cells of adult t-cell leukemia evade HTLV-1 tax/nf-κb hyperactivation-induced senescence. *Blood Adv.* **2019**, *3*, 564–569. [CrossRef] [PubMed]
83. Ramanayake, S.; Moulding, D.A.; Tanaka, Y.; Singh, A.; Bangham, C.R.M. Dynamics and consequences of the HTLV-1 proviral plus-strand burst. *PLoS Pathog.* **2022**, *18*, e1010774. [CrossRef] [PubMed]
84. Abou-Kandil, A.; Chamias, R.; Huleihel, M.; Godbey, W.T.; Aboud, M. Role of caspase 9 in activation of HTLV-1 ltr expression by DNA damaging agents. *Cell Cycle* **2011**, *10*, 3337–3345. [CrossRef] [PubMed]
85. Kulkarni, A.; Mateus, M.; Thinnes, C.C.; McCullagh, J.S.; Schofield, C.J.; Taylor, G.P.; Bangham, C.R.M. Glucose metabolism and oxygen availability govern reactivation of the latent human retrovirus HTLV-1. *Cell Chem. Biol.* **2017**, *24*, 1377–1387.e3. [CrossRef] [PubMed]
86. Sales, D.; Lin, E.; Stoffel, V.; Dickson, S.; Khan, Z.K.; Beld, J.; Jain, P. Apigenin improves cytotoxicity of antiretroviral drugs against HTLV-1 infected cells through the modulation of ahr signaling. *NeuroImmune Pharm. Ther.* **2023**, *2*, 49–62. [CrossRef]
87. Sato, T.; Maeta, T.; Ito, S. Dimethyl fumarate suppresses the proliferation of HTLV-1-infected t cells by inhibiting cbm complex-triggered nf-κb signaling. *Anticancer. Res.* **2023**, *43*, 1901–1908. [CrossRef] [PubMed]
88. Thorpe, L.M.; Yuzugullu, H.; Zhao, J.J. Pi3k in cancer: Divergent roles of isoforms, modes of activation and therapeutic targeting. *Nat. Rev. Cancer* **2015**, *15*, 7–24. [CrossRef]
89. Hemmati, S.; Sinclair, T.; Tong, M.; Bartholdy, B.; Okabe, R.O.; Ames, K.; Ostrodka, L.; Haque, T.; Kaur, I.; Mills, T.S.; et al. Pi3k alpha and delta promote hematopoietic stem cell activation. *JCI Insight* **2019**, *4*, e125832. [CrossRef] [PubMed]
90. Fukuda, R.; Hayashi, A.; Utsunomiya, A.; Nukada, Y.; Fukui, R.; Itoh, K.; Tezuka, K.; Ohashi, K.; Mizuno, K.; Sakamoto, M.; et al. Alteration of phosphatidylinositol 3-kinase cascade in the multilobulated nuclear formation of adult t cell leukemia/lymphoma (atll). *Proc. Natl. Acad. Sci. USA* **2005**, *102*, 15213–15218. [CrossRef] [PubMed]
91. Katsuya, H.; Cook, L.B.M.; Rowan, A.G.; Satou, Y.; Taylor, G.P.; Bangham, C.R.M. Phosphatidylinositol 3-kinase-delta (pi3k-delta) is a potential therapeutic target in adult t-cell leukemia-lymphoma. *Biomark. Res.* **2018**, *6*, 24. [CrossRef] [PubMed]
92. Ishikawa, C.; Mori, N. The role of cudc-907, a dual phosphoinositide-3 kinase and histone deacetylase inhibitor, in inhibiting proliferation of adult t-cell leukemia. *Eur. J. Haematol.* **2020**, *105*, 763–772. [CrossRef]
93. Ishikawa, C.; Senba, M.; Mori, N. Butein inhibits nf-κb, ap-1 and akt activation in adult t-cell leukemia/lymphoma. *Int. J. Oncol.* **2017**, *51*, 633–643. [CrossRef]
94. Matteucci, C.; Marino-Merlo, F.; Minutolo, A.; Balestrieri, E.; Valletta, E.; Macchi, B.; Mastino, A.; Grelli, S. Inhibition of iκbα phosphorylation potentiates regulated cell death induced by azidothymidine in HTLV-1 infected cells. *Cell Death Discov.* **2020**, *6*, 9. [CrossRef] [PubMed]
95. Migone, T.S.; Lin, J.X.; Cereseto, A.; Mulloy, J.C.; O'Shea, J.J.; Franchini, G.; Leonard, W.J. Constitutively activated jak-stat pathway in t cells transformed with HTLV-1. *Science* **1995**, *269*, 79–81. [CrossRef] [PubMed]

96. Zhang, M.; Mathews Griner, L.A.; Ju, W.; Duveau, D.Y.; Guha, R.; Petrus, M.N.; Wen, B.; Maeda, M.; Shinn, P.; Ferrer, M.; et al. Selective targeting of jak/stat signaling is potentiated by bcl-xl blockade in il-2-dependent adult t-cell leukemia. *Proc. Natl. Acad. Sci. USA* **2015**, *112*, 12480–12485. [CrossRef] [PubMed]
97. Daenthanasanmak, A.; Bamford, R.N.; Yoshioka, M.; Yang, S.M.; Homan, P.; Karim, B.; Bryant, B.R.; Petrus, M.N.; Thomas, C.J.; Green, P.L.; et al. Triple combination of bet plus pi3k and nf-κb inhibitors exhibit synergistic activity in adult t-cell leukemia/lymphoma. *Blood Adv.* **2022**, *6*, 2346–2360. [CrossRef] [PubMed]
98. Macchi, B.; Balestrieri, E.; Frezza, C.; Grelli, S.; Valletta, E.; Marcais, A.; Marino-Merlo, F.; Turpin, J.; Bangham, C.R.; Hermine, O.; et al. Quantification of HTLV-1 reverse transcriptase activity in atl patients treated with zidovudine and interferon-α. *Blood Adv.* **2017**, *1*, 748–752. [CrossRef] [PubMed]
99. Zhang, T.; Yu, J.; Cheng, S.; Zhang, Y.; Zhou, C.H.; Qin, J.; Luo, H. Research progress on the anticancer molecular mechanism of targets regulating cell autophagy. *Pharmacology* **2023**, *108*, 224–237. [CrossRef] [PubMed]
100. Chen, L.; Liu, D.; Zhang, Y.; Zhang, H.; Cheng, H. The autophagy molecule beclin 1 maintains persistent activity of nf-κb and stat3 in HTLV-1-transformed t lymphocytes. *Biochem. Biophys. Res. Commun.* **2015**, *465*, 739–745. [CrossRef]
101. Fauzi, Y.R.; Nakahata, S.; Chilmi, S.; Ichikawa, T.; Nueangphuet, P.; Yamaguchi, R.; Nakamura, T.; Shimoda, K.; Morishita, K. Antitumor effects of chloroquine/hydroxychloroquine mediated by inhibition of the nf-κb signaling pathway through abrogation of autophagic p47 degradation in adult t-cell leukemia/lymphoma cells. *PLoS ONE* **2021**, *16*, e0256320. [CrossRef] [PubMed]
102. Kozako, T.; Mellini, P.; Ohsugi, T.; Aikawa, A.; Uchida, Y.I.; Honda, S.I.; Suzuki, T. Novel small molecule sirt2 inhibitors induce cell death in leukemic cell lines. *BMC Cancer* **2018**, *18*, 791. [CrossRef] [PubMed]
103. Machado, C.B.; da Cunha, L.S.; Maues, J.H.D.; Pessoa, F.M.C.D.; de Oliveira, M.B.; Ribeiro, R.M.; Lopes, G.S.; de Moraes, M.O.; de Moraes, M.E.A.; Khayat, A.S.; et al. Role of MIRANs in human t cell leukemia virus type 1 induced t cell leukemia: A literature review and bioinformatics approach. *Int. J. Mol. Sci.* **2022**, *23*, 5486. [CrossRef] [PubMed]
104. Schnell, A.P.; Kohrt, S.; Aristodemou, A.; Taylor, G.P.; Bangham, C.R.M.; Thoma-Kress, A.K. Hdac inhibitors panobinostat and romidepsin enhance tax transcription in HTLV-1-infected cell lines and freshly isolated patients' t-cells. *Front. Immunol.* **2022**, *13*, 978800. [CrossRef] [PubMed]
105. Huang, J.; Cook, D.E. The contribution of DNA repair pathways to genome editing and evolution in filamentous pathogens. *FEMS Microbiol. Rev.* **2022**, *46*, fuac035. [CrossRef]
106. Xue, C.; Greene, E.C. DNA repair pathway choices in crispr-cas9-mediated genome editing. *Trends Genet.* **2021**, *37*, 639–656. [CrossRef] [PubMed]
107. Tanaka, A.; Takeda, S.; Kariya, R.; Matsuda, K.; Urano, E.; Okada, S.; Komano, J. A novel therapeutic molecule against HTLV-1 infection targeting provirus. *Leukemia* **2013**, *27*, 1621–1627. [CrossRef]
108. Rojo-Romanos, M.; Karpinski, J.; Millen, S.; Beschorner, N.; Simon, F.; Paszkowski-Rogacz, M.; Lansing, F.; Schneider, P.M.; Sonntag, J.; Hauber, J.; et al. Precise excision of HTLV-1 provirus with a designer-recombinase. *Mol. Ther.* **2023**, *31*, 2266–2285. [CrossRef] [PubMed]
109. Nakagawa, M.; Shaffer, A.L., 3rd; Ceribelli, M.; Zhang, M.; Wright, G.; Huang, D.W.; Xiao, W.; Powell, J.; Petrus, M.N.; Yang, Y.; et al. Targeting the HTLV-I-regulated BATF3/IRF4 transcriptional network in adult t cell leukemia/lymphoma. *Cancer Cell* **2018**, *34*, 286–297.e210. [CrossRef] [PubMed]
110. Panfil, A.R.; Green, P.L.; Yoder, K.E. Crispr genome editing applied to the pathogenic retrovirus HTLV-1. *Front. Cell Infect. Microbiol.* **2020**, *10*, 580371. [CrossRef] [PubMed]
111. Kaminski, R.; Chen, Y.; Fischer, T.; Tedaldi, E.; Napoli, A.; Zhang, Y.; Karn, J.; Hu, W.; Khalili, K. Corrigendum: Elimination of hiv-1 genomes from human t-lymphoid cells by crispr/cas9 gene editing. *Sci. Rep.* **2016**, *6*, 28213. [CrossRef] [PubMed]
112. Xiao, Q.; Guo, D.; Chen, S. Application of crispr/cas9-based gene editing in hiv-1/aids therapy. *Front. Cell Infect. Microbiol.* **2019**, *9*, 69. [CrossRef] [PubMed]
113. Lai, M.; Maori, E.; Quaranta, P.; Matteoli, G.; Maggi, F.; Sgarbanti, M.; Crucitta, S.; Pacini, S.; Turriziani, O.; Antonelli, G.; et al. Crispr/cas9 ablation of integrated hiv-1 accumulates proviral DNA circles with reformed long terminal repeats. *J. Virol.* **2021**, *95*, e0135821. [CrossRef] [PubMed]

Disclaimer/Publisher's Note: The statements, opinions and data contained in all publications are solely those of the individual author(s) and contributor(s) and not of MDPI and/or the editor(s). MDPI and/or the editor(s) disclaim responsibility for any injury to people or property resulting from any ideas, methods, instructions or products referred to in the content.

Article

Investigation of Mutated in Colorectal Cancer (MCC) Gene Family Evolution History Indicates a Putative Role in Th17/Treg Differentiation

Norwin Kubick [1], Justyna Paszkiewicz [2], Irmina Bieńkowska [3], Michał Ławiński [3,4], Jarosław Olav Horbańczuk [3], Mariusz Sacharczuk [3,5,*,†] and Michel Edwar Mickael [3,6,*,†]

[1] Department of Biology, Institute of Plant Science and Microbiology, Univeristy of Hamburg, Ohnhorststr. 18, 22609 Hamburg, Germany; n.kubick@uke.de

[2] Department of Health, John Paul II University of Applied Sciences in Biala Podlaska, Sidorska 95/97, 21-500 Biała Podlaska, Poland; j.paszkiewicz@dyd.akademiabialska.pl

[3] Institute of Animal Biotechnology and Genetics, Polish Academy of Science, Postępu 36A, 05-552 Jastrzębiec, Poland; i.bienkowska@igbzpan.pl (I.B.); michal-lawinski@wp.pl (M.Ł.); j.horbanczuk@igbzpan.pl (J.O.H.)

[4] Department of General Surgery, Gastroenterology and Oncology, Medical University of Warsaw, 02-091 Warsaw, Poland

[5] Department of Pharmacodynamics, Faculty of Pharmacy, Medical University of Warsaw, l Banacha 1, 02-697 Warsaw, Poland

[6] PM Research Center, Väpnaregatan 22, 58649 Linköping, Sweden

* Correspondence: m.mickael@igbzpan.pl (M.S.); m.sacharczuk@igbzpan.pl (M.E.M.)

† These authors contributed equally to this work.

Citation: Kubick, N.; Paszkiewicz, J.; Bieńkowska, I.; Ławiński, M.; Horbańczuk, J.O.; Sacharczuk, M.; Mickael, M.E. Investigation of Mutated in Colorectal Cancer (MCC) Gene Family Evolution History Indicates a Putative Role in Th17/Treg Differentiation. *Int. J. Mol. Sci.* **2023**, *24*, 11940. https://doi.org/10.3390/ijms241511940

Academic Editor: Laura Paleari

Received: 13 April 2023
Revised: 14 July 2023
Accepted: 21 July 2023
Published: 26 July 2023

Copyright: © 2023 by the authors. Licensee MDPI, Basel, Switzerland. This article is an open access article distributed under the terms and conditions of the Creative Commons Attribution (CC BY) license (https://creativecommons.org/licenses/by/4.0/).

Abstract: The MCC family of genes plays a role in colorectal cancer development through various immunological pathways, including the Th17/Treg axis. We have previously shown that MCC1 but not MCC2 plays a role in Treg differentiation. Our understanding of the genetic divergence patterns and evolutionary history of the MCC family in relation to its function, in general, and the Th17/Treg axis, in particular, remains incomplete. In this investigation, we explored 12 species' genomes to study the phylogenetic origin, structure, and functional specificity of this family. In vertebrates, both MCC1 and MCC2 homologs have been discovered, while invertebrates have a single MCC homolog. We found MCC homologs as early as Cnidarians and Trichoplax, suggesting that the MCC family first appeared 741 million years ago (Ma), whereas MCC divergence into the MCC1 and MCC2 families occurred at 540 Ma. In general, we did not detect significant positive selection regulating MCC evolution. Our investigation, based on MCC1 structural similarity, suggests that they may play a role in the evolutionary changes in Tregs' emergence towards complexity, including the ability to utilize calcium for differentiation through the use of the EFH calcium-binding domain. We also found that the motif NPSTGE was highly conserved in MCC1, but not in MCC2. The NPSTGE motif binds KEAP1 with high affinity, suggesting an Nrf2-mediated function for MCC1. In the case of MCC2, we found that the "modifier of rudimentary" motif is highly conserved. This motif contributes to the regulation of alternative splicing. Overall, our study sheds light on how the evolution of the MCC family is connected to its function in regulating the Th17/Treg axis.

Keywords: Th17; Treg; differentiation; colorectal cancer; evolution

1. Introduction

The Th17/Treg axis is crucial for both promoting and repressing colorectal cancer. Th17 and Treg belong to the CD4+ T cell population [1], and while they share a large portion of their transcriptome [2], they have functionally diverged. Th17 cells can be proinflammatory by producing several cytokines, such as IL17A, IL17F, IL1, and IL6 [3]. It has been shown that Th17 cells stimulate the infiltration of cytotoxic CD8+ T lymphocytes into colorectal cancer tissues, thus supporting the body in the fight against cancer [4]. In contrast to Th17, Treg cells support colorectal cancer growth [5,6]. Treg cells are also known to be able to inhibit Th17 cells through both direct and indirect pathways.

MCC/MCC1 ("mutated in colorectal cancer") plays a significant role in colorectal cancer progression. MCC1 was found to be linked to the familial adenomatous polyposis susceptibility locus on chromosome 5q. Familial adenomatous polyposis is a colorectal cancer risk factor because hereditary precancerous colorectal polyps can evolve into colorectal cancer [7]. Recently, it was shown that MCC1 could be contributing to cancer progression through the dysregulation of the WNT pathway [8]. Additionally, recent research has revealed that the nuclear factor-B (NF-κB) pathway and cell cycle regulation, two crucial cellular processes related to carcinogenesis, may be affected by MCC1 expression [9]. MCC1 belongs to the MCC family, which includes another homolog known as MCC2/USHBP1 (Usher syndrome type 1C Harmonin-binding protein 1). However, the function of MCC2 is still unclear [10]. Previously, we found, upon an inspection of microarray/RNAseq data that compare Th17 and Treg differentiation, that MCC1 and not MCC2 is upregulated in Treg, but not in Th17, along with a host of genes that are associated with cell cycle regulation [11]. However, the phylogenetic relationship between the two proteins is currently unknown [11].

In this report, we conducted an extensive phylogenetic analysis of the MCC family to investigate its role in the context of the Th17/Treg axis function in colorectal cancer. We found a single MCC homolog in various invertebrate species, including Spirlai, Arthopoda, Cnidria, and Trichoplx. During the Cambrian explosion (e.g., during Vertebrate emergence) and the two rounds of genome duplication (2R), two homologs, namely MCC1 and MCC2, emerged. The main building blocks of MCC1 in vertebrates are two domains of MCC-PDZ and a single EF-hand domain. This structure is also found in Oedothorax gibbosus (Gibbous dwarf spider) and is partially conserved in other investigated invertebrate species, hinting toward the functional conservation between MCC and MCC1. MCC2 sequences seem to have lost EF-hand domains. Our investigation indicates that the nearest homolog to the MCC family ancestral sequence homolog is a protein containing an EF-hand domain. Moreover, motif inspection suggests that the MCC1 and MCC2 families could be playing a primary role in cell cycle regulation. Additionally, we found that the motif NPSTGE, which is known to play a role in its interaction with KEAP1, is conserved in MCC1. Our findings suggest that MCC1 could be enhancing Treg differentiation by inhibiting the KEAP1 effect.

2. Results
2.1. Phylogenetic Analysis

Phylogenetic analysis indicates that two rounds of duplications resulted in the divergence of MCC1 and MCC2. We downloaded human MCC1 and MCC2 protein sequences from the GEO protein repository. We utilized the Blastp server to acquire homologous proteins for MCC1 and MCC2 among twelve species. We employed Seaview to perform the multiple sequence alignment using the MUSCLE algorithm (Figure 1A). After that, we constructed the phylogenetic tree using the PyML method utilizing the LG model (Figure 1B). We found a single homolog of the MCC in invertebrates. Our research identified a putative MCC homolog (e.g., XP_019849192.1) in Amphimedon queenslandica. However, Blastp's E-value was lower than our threshold of 1×10^{-10}. Thus, that sequence was not accepted. Two MCC homologs were found in all vertebrate species investigated. The bony fish seems to be the first vertebrate to possess a pair of MCC homologs. These results indicate that the divergence of the MCC1 and MCC2 homologs occurred at the 2R stage during the Cambrian explosion. Our findings indicate that all vertebrate MCC1 sequences and MCC sequences in Cnidaria, Spiralia, and Arthropoda possess two MCC-PDZ domains. In contrast, all MCC2 sequences, as well as Tunicate and Trichoplx MCC sequences, feature only a single MCC-PDZ domain. It is noteworthy that the results of domain prediction for MCC1 and MCC2 exhibited variations among the three servers employed. The NIH server predicted the existence of unique SMC domains in certain MCC1 and MCC2 sequences that were investigated. Neither PFAM nor HMM servers identified this domain in the species studied. We investigated this line further by comparing MCC1 sequences against the SMC family sequences identified using the NIH server. However, the Blastp results did not indicate a significant homology between the two groups (MCC sequences and the SMC families). Therefore, we excluded the SMC domain from our subsequent analysis. However, experimental validation could shed light on this controversial aspect.

2.2. Ancestral Sequence Reconstruction and Network Split Results

Our ancestral sequence reconstruction indicates that the MCC's nearest-most ancient homolog is an EH hand domain-containing protein. We reconstruct the ancestral sequence of the MCC family using MegaX. The sequence reconstruction was based on our generated phylogenetic tree (Figures 1 and 2A). We searched for homologs for the generated ancestral sequence using BLASTP and HMM search servers. The sequences that obtained the highest scores corresponded to an EF-hand domain-containing protein, a UBZ1 type domain-containing protein, an ETS domain-containing protein, and ABHD8. The results and the reconstructed sequence were fed into the SplitsTree program. The results show that the nearest homolog to the ancestral sequence is an EF-hand domain-containing protein (Figure 2A,B).

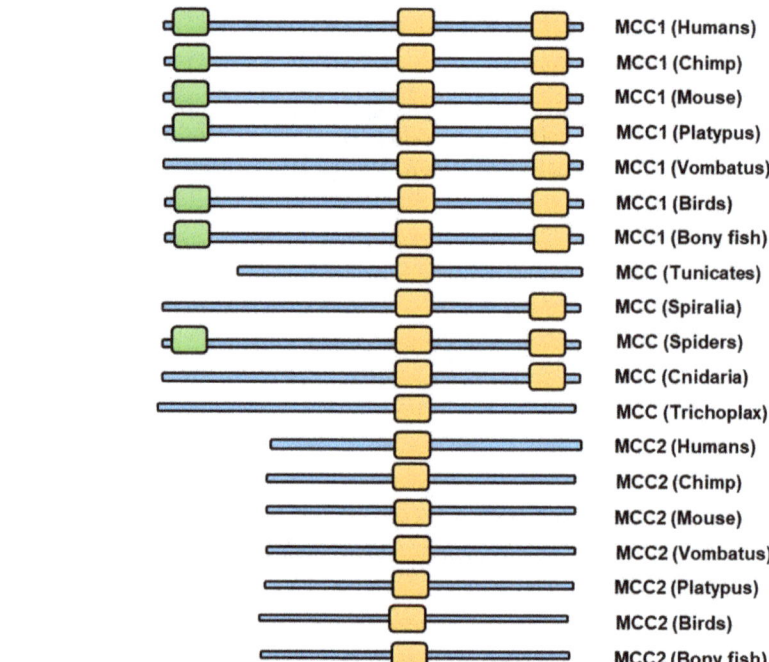

Figure 1. *Cont.*

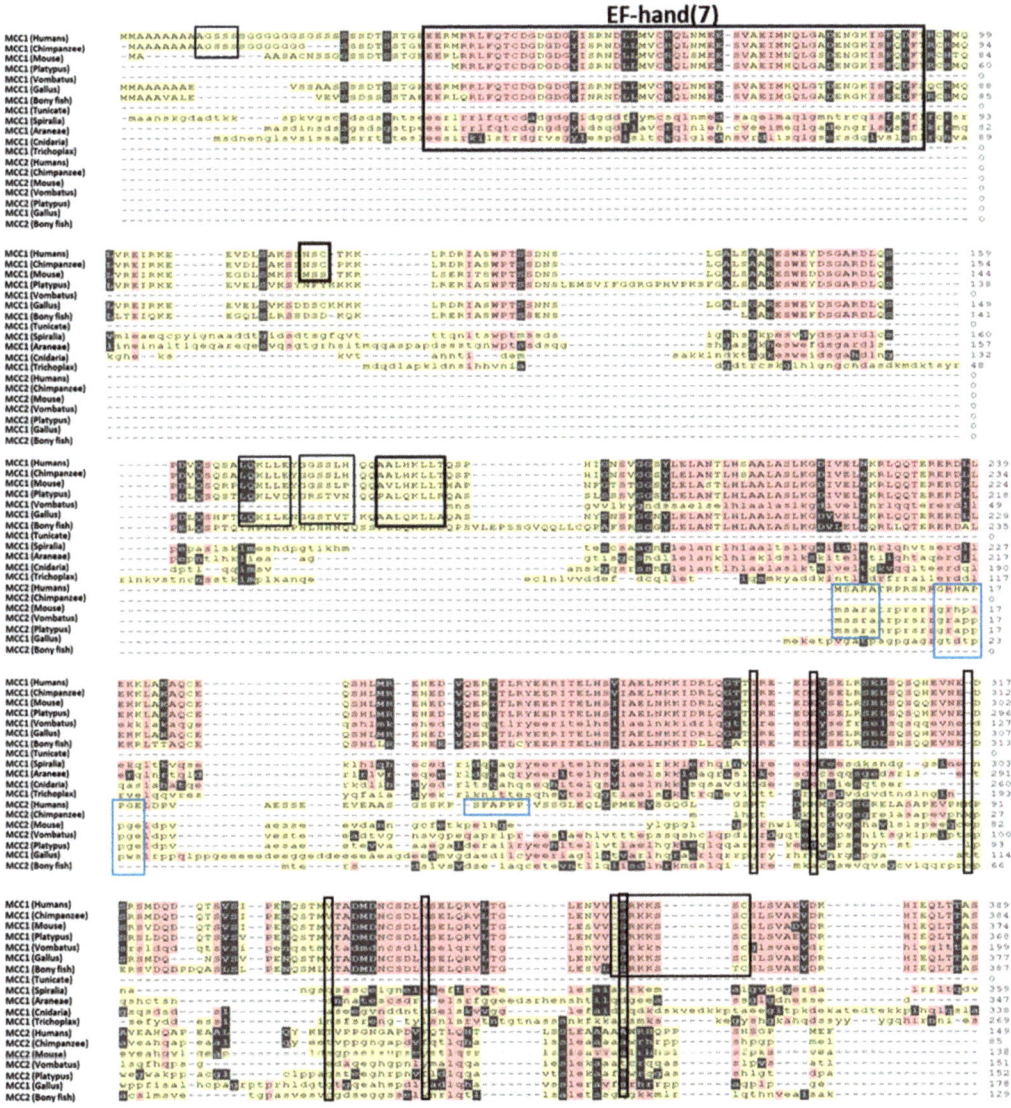

Figure 1. *Cont.*

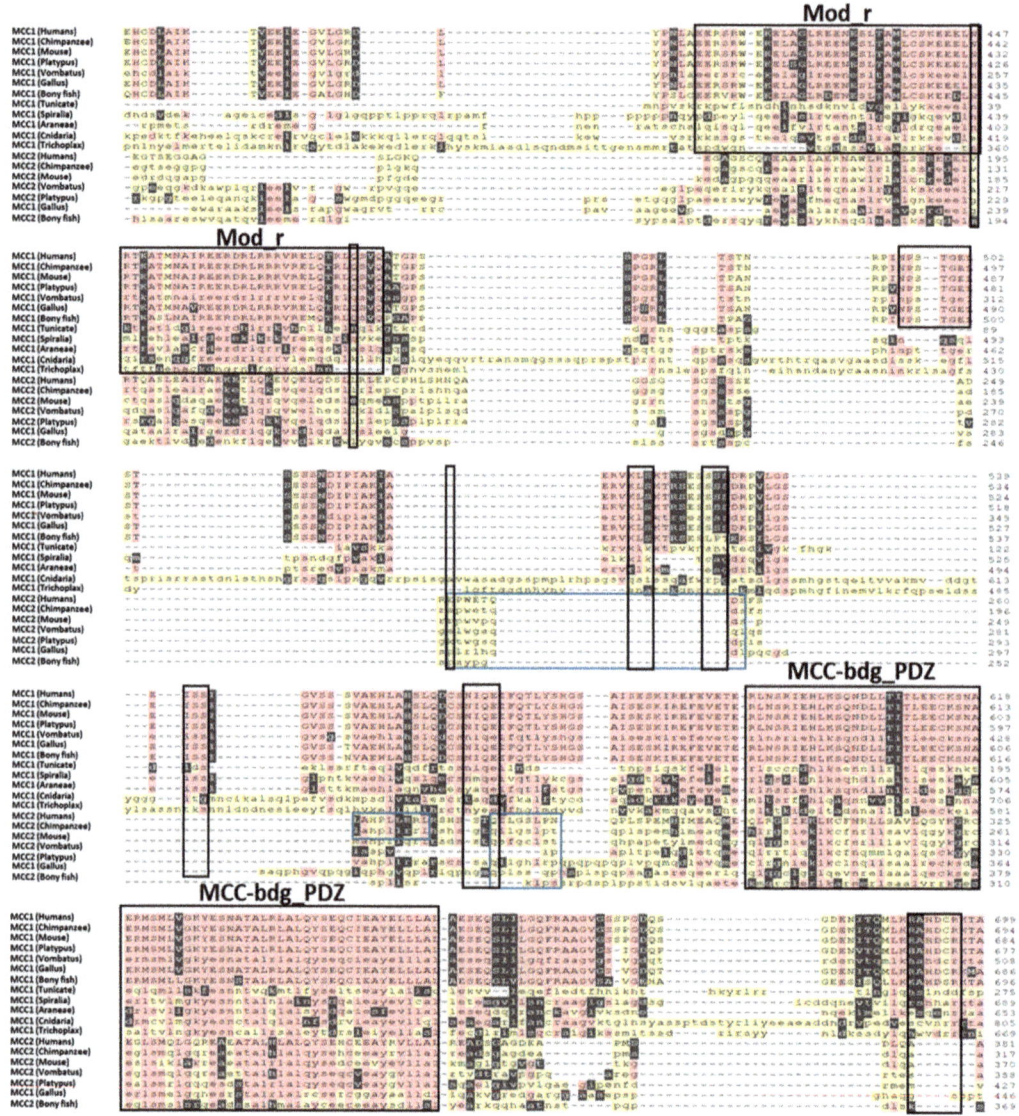

Figure 1. *Cont.*

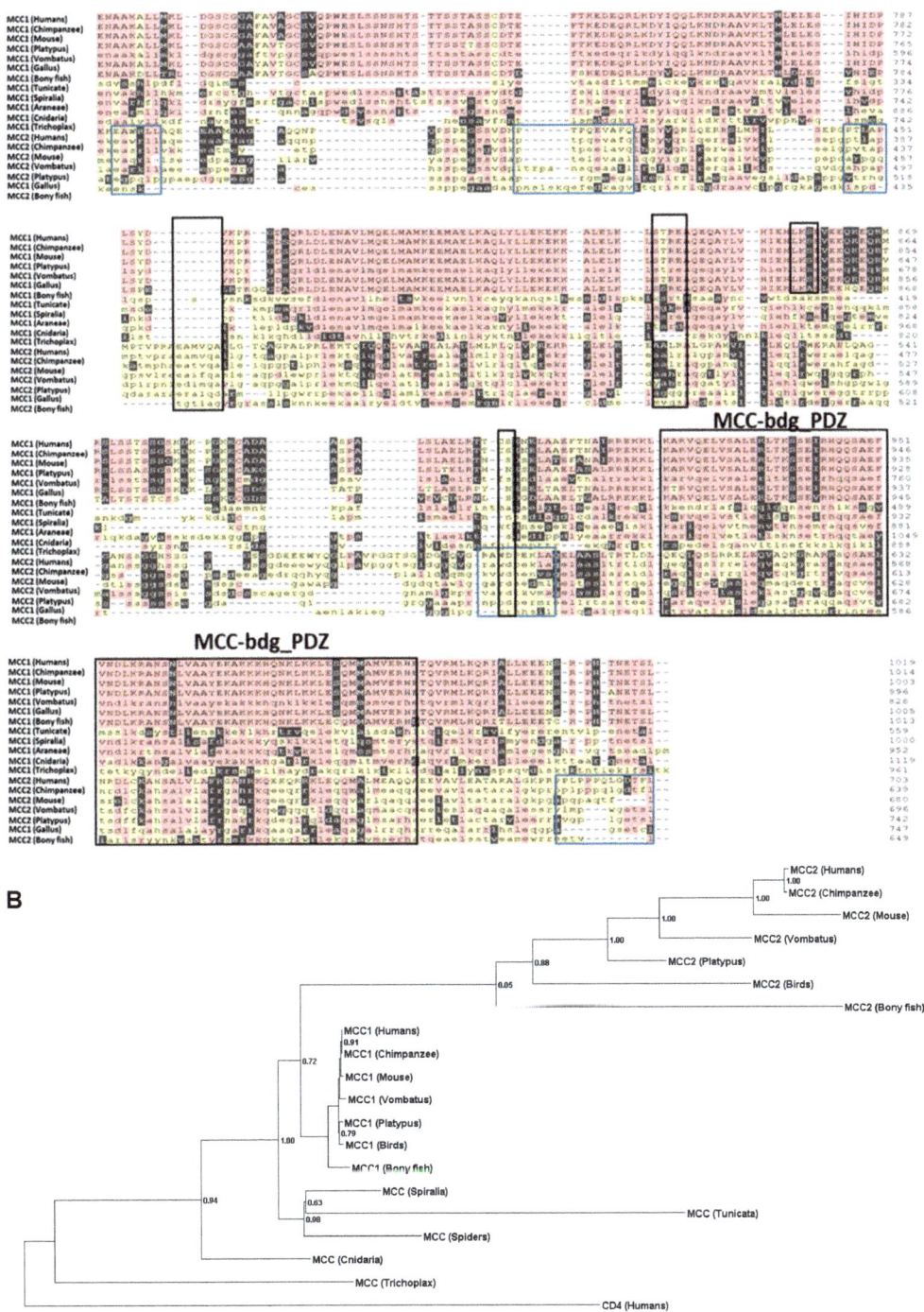

Figure 1. Phylogenetic analysis of the MCC family. (**A**) Multiple sequence alignment of the MCC family members show a high degree of homology. EFHAND(7) is shown in light red, and the first

and second MCC-BDG-PDZ are shown in light orange. (**B**) The two rounds of duplication are clear within the family species. Only one homolog for the MCC family appears in invertebrates. During the Cumbrian explosion of vertebrates, gene duplication occurred, giving rise to two distinct homologs: MCC1 and MCC2.

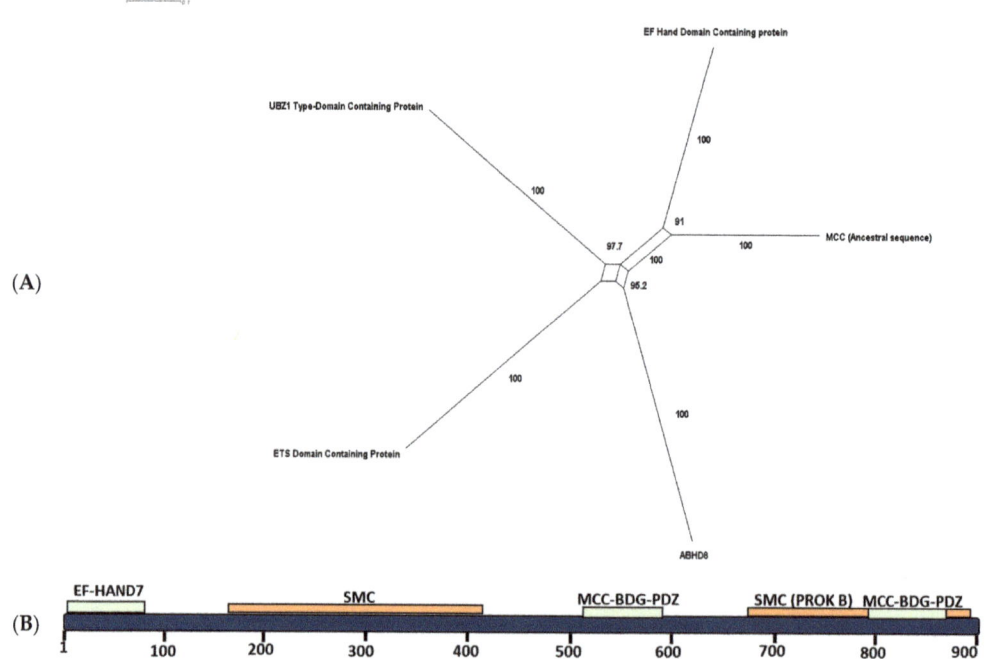

Figure 2. Ancestral sequence reconstruction of the MMC family. (**A**) The nearest homolog for the reconstructed ancestral sequence resembles that of an EF-hand domain-containing protein. (**B**) The EF-hand domain-containing protein structure contains two MCC-PDZ domains, a single EF-hand domain, and an SMC domain.

2.2.1. Sites of Functional Divergence

A study of site-specific shifts in evolutionary rates following gene duplication was performed using Sequence Harmony. Multiple sites seem to play a primary role in the specificity of the divergence between MCC/MCC1 and MCC2. One of the important residues that seems to play a critical role in determining the functional specificity between the two groups is 737-N (Asparagine) (Table 1). All the MCC1 groups have an N residue, while MCC2 has E, D, or R. This residue lies in the first MCC-bdg_PDZ domain of the MCC1 group and hints at the possible role of MCC-bdg_PDZ in the functional divergence between the two groups.

Table 1. Functional divergent sites.

Pos	Entropy				SH	Rnk	Consensus	
Ali	A	B	AB	Rel.			A	B
772	1.19	1.84	2.38	1.05	0.00	11	Qne	ATlp
770	1.21	1.84	2.39	1.05	0.00	11	Lefr	DPav
909	0.82	1.56	2.04	1.05	0.00	9	Tav	PLS
891	0.41	1.84	1.89	1.05	0.00	6	Rt	QLav
886	1.21	1.84	2.39	1.05	0.00	6	Kamp	QEgl
949	0.00	1.15	1.37	1.05	0.00	4	D	Rhk
701	0.82	0.86	1.78	1.05	0.00	4	Sct	Gr
948	0.82	1.15	1.89	1.05	0.00	4	Lfv	Pnr
684	1.42	0.59	2.06	1.05	0.00	4	Ivqt	Ph
698	0.82	1.66	2.08	1.05	0.00	4	Rms	Qchk
946	1.21	2.13	2.50	1.05	0.00	4	Qdps	Aegkt
686	1.42	1.84	2.52	1.05	0.00	4	Edkp	SAtv
1162	0.65	0.00	1.36	1.05	0.00	3	Kr	Q
900	0.41	0.99	1.57	1.05	0.00	3	Ki	RQ
1128	0.41	1.15	1.63	1.05	0.00	3	Ed	Rqt
715	0.98	0.59	1.79	1.05	0.00	3	IL	Ag
711	0.81	1.38	1.97	1.05	0.00	3	Dn	Rqe
709	0.82	1.66	2.08	1.05	0.00	3	Qde	Flrv
898	1.21	1.15	2.14	1.05	0.00	3	Qeks	Rhn
901	1.21	1.38	2.22	1.05	0.00	3	Ndks	Eaq
1190	1.42	1.15	2.27	1.05	0.00	3	Lmft	Asv
1126	1.28	1.84	2.44	1.05	0.00	3	SNl	KDat
1189	1.78	1.66	2.69	1.05	0.00	3	Aslnt	Rehq
952	0.41	0.00	1.21	1.05	0.00	2	Nt	K
954	0.00	1.15	1.37	1.05	0.00	2	V	Qae
1004	0.82	0.00	1.47	1.05	0.00	2	Vch	L
1153	1.04	0.00	1.61	1.05	0.00	2	Edk	R
1154	0.41	1.15	1.63	1.05	0.00	2	Kr	Gns
1131	0.82	0.59	1.68	1.05	0.00	2	Hay	Rq
1132	0.41	1.84	1.89	1.05	0.00	2	Qi	ASrv
410	0.82	1.38	1.97	1.05	0.00	2	Edn	Rqh
1092	0.82	1.45	2.00	1.05	0.00	2	Kqr	DEl
876	1.58	1.84	2.63	1.05	0.00	2	Tlmnv	KApr
737	0.00	1.38	1.46	1.05	0.00	1	N	Edr
1142	0.00	1.38	1.46	1.05	0.00	1	K	Cfs
1185	0.00	1.84	1.63	1.05	0.00	1	K	AEst
734	0.41	1.38	1.72	1.05	0.00	1	Yf	Rqh
730	0.00	2.13	1.73	1.05	0.00	1	L	Qekrs

2.2.2. Motifs

We investigated the existence of differential motifs between the MCC1 and MCC2 sequences (Table 2). Several distinctive motifs are found solely in one of the groups. For example, EP-WETQDSF, which is known to play a role in the coatomer construction of the vesicles coat, could be identified in primates in MCC2, but not in MCC1 (Figure 1). Another example is GRHAPPGE. This motif functions as a tankyrase-binding site by interacting with the ankyrin repeat domain region in Tankyrase-1 and Tankyrase-2 (TNKS). TNKS are modulators of the Wnt/β-catenin signaling pathway. Notably, we found the NPSTGE motif in MCC1, but not in MCC2. This motif binds to the Kelch domain of KEAP1 with high affinity. NPSTGE is required for the efficient recruitment of target proteins to the Cul3-based E3 ligase to enable KEAP1 to regulate the function of Nrf2 (NFE2L2).

Table 2. Motifs analysis and their location.

Motif	Position	Known Function	Species	Family
NPSTGE	496–501 [A]	Motif binds the Kelch domain of KEAP1 with high affinity.	H, C, M, P, V, G, D	MCC1
AGSSS	10–14 [A]	SPOP is part of the complex SPOP/Cul3 and it plays an important role in protein degradation ubiquitination.	H and C	MCC1
LQKLLE ALHKLLT	167–173 [A] 184–190 [A]	LXXLL supports binding to nuclear receptors. RORγt is a well-known nuclear receptor, but it is unknown whether MMC can bind to it.	H, C, M, P, and G	MCC1
CGRK	364–367 [A]	Peptide C-terminal amidation, possibly to protect against degradation.	H, C, M, P, V, G, D	MCC1
NSC	117–119 [A]	NSC is a motif for N-glycosylation, a post-translational modification process where glycans are added to specific Asn residues in proteins in the endoplasmic reticulum and Golgi apparatus.	H and C	MCC1
GGSSLH/P	175–180 [A]	NEK2 is a phosphorylation motif recognized and targeted by protein kinases for phosphorylation.	H, C, M	MCC1
CGRKKSSC	364–371 [A]	This motif is a site for the attachment of a fucose residue to a serine, indicating the possible involvement of fucosylation in the regulation or function of that protein.	H, C, M, P, V, G, D	MCC1
RKKSSCS KKSSCSL	366–372 [A] 367–373 [A]	The motif sequence is recognized by PKA as a specific target for phosphorylation.	H, C, M, P, V, G, D	MCC1
LKSE	857–860 [A]	SUMO-1 may attach to that motif in a Sumoylation process. Sumoylation can have transcriptional regulation, DNA repair, nuclear transport, and protein quality control.	H, C, M, P, V, G	MCC1
GRHP/APPGE	13–20 [A]	The motif may function as a docking site for Tankyrase-1 and 2, regulating protein interactions, localization, and stability.	H, P, V	MCC2
LPPP	693–696 [A]	LPPP motif acts as a docking site in calcineurin substrates to allow them to interact with the catalytic and regulatory subunits of calcineurin and facilitate participation in cellular signaling events.	H, C	MCC2
FAPP	42–45 [A]	This motif is recognized by the EVH1 domains, which are specific protein domains found in the PPP4R3 regulatory subunits of the PP4 holoenzyme.	H	MCC2
MSARA	1–5 [A]	This motif is found in pro-apoptotic proteins and it counteracts caspase inhibition by the Inhibitor of Apoptosis Proteins in apoptotic cells.	H, M, V, P	MCC2
EPWETQDSF	251–259 [A]	This motif mediates the coatomer subunit delta construction. Coatomer plays a part in forming vesicle coat.	H, C, M	MCC2
RAWDPEKLA	583–591 [A]	This short motif is found in cargo proteins and mediates kinesin-1-dependent microtubule transport by binding to the KLC TPR region.	H, C, M	MCC2
PTLAP PPLPP	447–451 [A] 691–695 [A]	PxLxP motif is recognized by a subset of MYND domain-containing proteins. MYND domain is involved in transcriptional regulation, chromatin remodeling, protein localization, and signal transduction	H, C	MCC2
QILGSLPN	276–283 [A]	The PTB RRM2 Interacting (PRI) motif is found in some splicing regulators	H, C	MCC2
PPQLGD	695–700 [A]	TRAF2-binding motif. Members of the tumor necrosis factor receptor (TNFR) superfamily recruit TNFR-associated factors (TRAFs) to initiate cell signaling.	H, C	MCC2

Table 2. Cont.

Motif	Position	Known Function	Species	Family
PPLP	691–694 [A]	PPLP is a motif recognized by WW domains of Group II that play a role in multiple cellular processes.	H, C	MCC2
M/LAHPLL	261–266 [A]	Dileucine motifs involved in the trafficking and sorting of proteins in the endosomal-basolateral-lysosomal pathway	H, C, M	MCC2
EAW/SRLL	384–389 [A]	Dileucine motifs involved in the trafficking and sorting of proteins in the endosomal-basolateral-lysosomal pathway	H, C, M	MCC2

2.3. Positive Selection

We conducted a positive selection analysis for the MCC family. Our results indicate that the MCC family does not seem to have evolved under significant positive selection. Conversely, on several levels of analysis, namely, general, branch, and site, our results indicate that the MCC evolutionary pattern followed a negative selection pattern. (i) For the global model, we used the M0 model to estimate the possibility of positive selection over the whole tree. We found the ω value calculated using PAML for the M0 model to be 0.45728, *p*-Value < 0.00001. (ii) Similarly, for the branch values, for MCC1 vertebrates, the ω value was 0.89, suggesting purifying selection, without strong evidence of statistical significance. Also, in the case of the MCC2 branch, the ω value was 0.64046. On the site level, we did not detect any sites that were subjected to significant positive selection (i.e., $\omega > 1$) (Table 3).

Table 3. Positive selection models investigated.

Model	ω
general	0.45728, *p*-value is <0.00001
branch	ω value was 0.89 *p*-value > 0.005
branch-site	ω value was 0.64046 *p*-value > 0.005
sites	No sites with $\omega > 1$

3. Discussion

Our study provides evidence supporting the ancient origin of the MCC family. In addition to MCC1 and MCC2 homologs in vertebrates, we identified a single MCC homolog in various invertebrate species (e.g., Trichoplax and Cnidarians) (Figure 1). Selection analysis did not reveal any sites subjected to significant positive selection, indicating overall structure conservation between the two families and within each family. The structure of the MCC family includes two main elements: (i) MCC-PDZ and (ii) the EFh-binding domain. The structure of MCC1 in vertebrates as well as Cnidaria, Spiralia, and Spiders includes two MCC-PDZ domains, while the MCC2 sequence exhibits a single MCC-PDZ domain (Figure 1). All vertebrates' MCC1 proteins possess an EFh-binding domain, except for Vombatus. We have located an EFh-binding domain in the elephant shark (XP_007895580.2). Similarly, the MCC homolog in Oedothorax gibbosus (Gibbous dwarf spider) contains an EFh-binding domain, while no other invertebrate species investigated contain this particular domain. Our ancestral sequence reconstruction analysis indicates that the MCC family ancestral sequence homolog had at least one MCC-PDZ domain. Our phylogenetic network indicates that the nearest homolog to the ancestral sequence of the MCC could have contained an EFh-domain that was lost in Trichoplax and Cnidaria, but reappeared in Spiders. It is also plausible that the emergence of the EP-hand domain first appeared in Spiders (Figure 2A).

Our results indicate a divergence of function between MCC1 and MCC2. We have found nine motifs conserved in MCC1, but not in the MCC2 family. Additionally, 14 motifs were specific only to MCC2. However, in the case of MCC1, these functional motifs were highly conserved in most of the species investigated. For example, Mod(r) is a motif encompassing

150 residues. It is conserved in various eukaryotic proteins investigated and is homologous to the Drosophila melanogaster Modifier of rudimentary "Mod(r)" proteins [12]. Modr's primary function is to enable lysosomal sorting [13]. This motif is highly conserved within all MCC1-expressing species. As a result, the Mod (r) proteins may play a primary role in the functional specification between MCC1 and MCC2. However, whether MCC1 could play a role in lysosomal sorting is yet to be determined. The QILGSLPN motif is highly conserved in MCC2. QILGSLPN was shown to play a role in alternative splicing (Table 2) [14,15]. However, whether MCC2 plays a role in alternative splicing is not yet known.

Our research suggests that MCC1 and not MCC2 could be contributing to Treg cells' function in colorectal cancer. In our recent publication, using microarray analysis of the Th17/Treg pathway, we revealed that several genes that regulate the cell cycle are upregulated in Treg, but not in Th17 [11]. MCC1 was upregulated in Tregs, but not in Th17. Several other genes linked to controlling the cell cycle were also differentially expressed. For example, we found that WWP2, which is known to control the cell cycle, is upregulated in Treg, but not in Th17 [16]. In this current study, we found that the CGRKKSSC (364–371) motif could be contributing to the functional specificity of MCC1. This motif is a known site for fucose residue attachment to serine and is implicated in Notch signaling [17,18]. Notch signaling plays a role in regulating the cell cycle [19]. Importantly, our results indicate that MCC1 and not MCC2 possess the motif NPSTGE. This motif binds to the Kelch domain of KEAP1 with high affinity. KEAP1 is one of the primary regulators of the cell cycle, both dependently and independently of Nrf2 [20,21]. MCC1 has been implicated in inhibiting and enhancing the cell cycle, as its function seems cell-specific. MCC1 inhibits the cell cycle in colorectal epithelial cells, but positively supports B lymphocyte malignancies [9]. Treg cells support colorectal cancer growth by inhibiting Th17 [22]. MCC1 contains an EFh-binding domain which was shown to bind calcium ions. EF-hand domain binding to calcium ions is involved in various functions, such as buffering calcium in the cytosol, signal transduction, and the contraction of fibers in vertebrates and invertebrates [23–25]. Treg uses calcium for its differentiation, and the Treg-mediated regulation of conventional T cells involves the suppression of calcium signaling. The calcium inhibition of calcineurin reduces NfkB via an IKK-mediated pathway. Based on the existence of the EFh-binding motif (known to mediate calcium signaling) in MCC1 and not in MCC2, and the upregulated expression of MCC1 in Treg cells, we can speculate on a putative role for MCC1 in Treg calcium regulation, possibly through regulating calcineurin-mediated pathways [26,27]. Thus, MCC support of Treg differentiation through enhancing the cell cycle could be supporting cancer development.

Study limitations: Our study provides valuable insights into the regulatory role of MCC1 in Treg differentiation by utilizing extensive RNA-seq data analysis. However, further experimental studies are warranted to expand upon these findings. In vitro assays could be conducted to investigate the impact of MCC1 overexpression or deletion on Treg differentiation. Additionally, in vivo studies using adoptive transfer in appropriate mouse models, such as Rag1-/-mice, could elucidate the effect of MCC on CD4+ T cell differentiation in the context of colorectal cancer genesis and development. These complementary experimental approaches will not only strengthen the evidence supporting the role of MCC in Treg function, but also provide valuable insights for potential therapeutic interventions targeting MCC in colorectal cancer.

4. Methods

4.1. Database Search

This study aimed to examine the relationship between the molecular evolution of the MCC family and its functions. We reasoned that due to the diversity of this protein family, studying protein sequences rather than DNA sequences might be more informative. Furthermore, we selected 12 species and genera that span more than 500 million years to ensure that our analysis is a fair representation of the evolutionary history of the MCC family. BLASTP searches were conducted using human MCC1 and MCC2 protein sequences against the proteomes of Chimpanzee (*Pan troglodytes*), House Mouse (*Mus musculus*), Common Wombat (*Vombatus ursinus*), Platypus (*Ornithorhynchus anatinus*), Red Junglefowl (*Gallus gallus*), Zebrafish (*Danio rerio*), Sea Squirt (*Ciona intestinalis*), Arthropoda, Spiralia, Cnidaria, Trichoplax, and Sponge (*Amphimedon queenslandica*) [28]. Whenever one protein possessed more than one transcript, only the longest transcript was used in the analysis [29]. Sequences were selected as candidate proteins if their E values were $<1 \times e^{-10}$ [28]. Conserved domains were investigated using the CDD function on the NIH website (accessed on 25 August 2022), PFAM, and HMRR. Based on the consensus domains determined by these three domain-predicting methods, sequences were filtered for the presence of conserved domains homologous to human MCC1 and MCC2 domains, respectively.

4.2. Alignment and Phylogenetic Analysis

The phylogenetic investigation was performed in three stages. First, MCC family amino acid sequences were aligned in Seaview using MUSCLE. After that, we employed PhyML to determine the best phylogenetic tree to represent the interrelationships among the MCC family homologs. We used the LG model with empirical values calculated for amino acid equilibrium frequencies [30]. Invariable sites and across-site rate variations were calculated using an optimized algorithm.

4.3. Functional Divergence Estimation and Motif Search

We used Sequence Harmony and Multi-RELIEF to identify functional differences within the MCC subfamilies that may contribute to the disparity in function within the Th17/Treg axis. Sequence Harmony compared the two groups of sequences (e.g., MCC/MCC1 and MCC2 families) to identify the variable amino acids and their distribution frequency. The positions, where the amino acid compositions differ between the two groups, were assigned low score values. A score of zero indicates distinct amino acids at that position, while a score of one signifies nearly identical amino acid compositions. Multi-RELIEF predicts the residue specificity of residues by performing two comparisons. The first comparison is between each sequence and its nearest homolog within the same group, and the second comparison involves comparing each sequence with its closest homolog in the second group. A residue is considered specific if it exhibits a high specificity score in at least one of the two comparisons. Additionally, we conducted an extensive motifs search using an ELM server http://elm.eu.org/ accessed on 30 September 2022 with a motif cut-off value of 100. We also conducted a motif search using www.genome.jp accessed on 3 October 2022 [31].

4.4. Positive Selection Analysis

We utilized the maximum likelihood approach method to investigate the selection process that governed the MCC family's evolution. We back-translated the downloaded protein sequences using the EMBOSS server (https://www.ebi.ac.uk/Tools/st/emboss_backtranseq, accessed on 5 September 2022) to estimate the cDNA of the investigated sequences [32]. After that, we used CODEML-PAML (V4.4) to estimate the substitution rate ratio (ω) given by the ratio between nonsynonymous (dN) and synonymous (dS) mutations. We utilized four models: general (basic), branch, branch-site, and sites. The main difference between these models is their level of investigation. While the global model assumes a constant ω ratio for all the trees investigated, the branch model calculates two ω values (the branch investigated and the tree, respectively). The branch-site model calculates ω values for each nucleotide on a specific branch, while the site model estimates ω values for each nucleotide in the alignment. Statistical significance is calculated based on the χ^2 test [33].

Author Contributions: N.K., M.E.M., I.B., J.P. and M.Ł. contributed to writing, creating figures, data collection, and data analysis. J.O.H. and M.S. contributed to writing and supervision. All authors have read and agreed to the published version of the manuscript.

Funding: The research was funded by the John Paul II University of Applied Sciences in Biala Podlaska—project No. PB/5/2022.

Institutional Review Board Statement: Not applicable.

Informed Consent Statement: Not applicable.

Data Availability Statement: All raw data are available upon request.

Acknowledgments: We would like to acknowledge Macarious Abraham and Meriam Yoakim for their continuous support.

Conflicts of Interest: The authors declare no conflict of interest.

References

1. Bhaumik, S.; Mickael, M.E.; Moran, M.; Spell, M.; Basu, R. RORγt Promotes Foxp3 Expression by Antagonizing the Effector Program in Colonic Regulatory T Cells. *J. Immunol.* **2021**, *207*, 2027–2038. [CrossRef]
2. Mickael, M.-E.; Basu, R.; Bhaumik, S. Retinoid-Related Orphan Receptor RORγt in CD4+ T Cell Mediated Intestinal Homeostasis and Inflammation. *Am. J. Pathol.* **2020**, *190*, 1984–1999. [CrossRef] [PubMed]
3. Mickael, M.E.; Bhaumik, S.; Chakraborti, A.; Umfress, A.A.; van Groen, T.; Macaluso, M.; Totenhagen, J.; Sorace, A.G.; Bibb, J.A.; Standaert, D.G.; et al. RORγt-Expressing Pathogenic CD4+ T Cells Cause Brain Inflammation during Chronic Colitis. *J. Immunol.* **2022**, *208*, 2054–2066. [CrossRef] [PubMed]
4. Amicarella, F.; Muraro, M.G.; Hirt, C.; Cremonesi, E.; Padovan, E.; Mele, V.; Governa, V.; Han, J.; Huber, X.; Droeser, R.A.; et al. Dual Role of Tumour-Infiltrating T Helper 17 Cells in Human Colorectal Cancer. *Gut* **2017**, *66*, 692–704. [CrossRef]
5. Bhaumik, S.; Łazarczyk, M.; Kubick, N.; Klimovich, P.; Gurba, A.; Paszkiewicz, J.; Teodorowicz, P.; Kocki, T.; Horbańczuk, J.O.; Manda, G.; et al. Investigation of the Molecular Evolution of Treg Suppression Mechanisms Indicates a Convergent Origin. *Curr. Issues Mol. Biol.* **2023**, *45*, 628–648. [CrossRef]
6. Olguín, J.E.; Medina-Andrade, I.; Rodríguez, T.; Rodríguez-Sosa, M.; Terrazas, L.I. Relevance of Regulatory T Cells during Colorectal Cancer Development. *Cancers* **2020**, *12*, 1888. [CrossRef] [PubMed]
7. Mukherjee, N.; Dasgupta, H.; Bhattacharya, R.; Pal, D.; Roy, R.; Islam, S.; Alam, N.; Biswas, J.; Roy, A.; Roychoudhury, S.; et al. Frequent Inactivation of MCC/CTNNBIP1 and Overexpression of Phospho-Beta-Catenin Y654 Are Associated with Breast Carcinoma: Clinical and Prognostic Significance. *Biochim. Biophys. Acta Mol. Basis Dis.* **2016**, *1862*, 1472–1484. [CrossRef] [PubMed]
8. Mukherjee, N.; Panda, C.K. Wnt/β-Catenin Signaling Pathway as Chemotherapeutic Target in Breast Cancer: An Update on Pros and Cons. *Clin. Breast Cancer* **2020**, *20*, 361–370. [CrossRef]
9. Edwards, S.K.E.; Baron, J.; Moore, C.R.; Liu, Y.; Perlman, D.H.; Hart, R.P.; Xie, P. Mutated in Colorectal Cancer (MCC) Is a Novel Oncogene in B Lymphocytes. *J. Hematol. Oncol.* **2014**, *7*, 56. [CrossRef]
10. Young, T.; Poobalan, Y.; Ali, Y.; Siew Tein, W.; Sadasivam, A.; Ee Kim, T.; Erica Tay, P.; Dunn, N.R. Mutated in Colorectal Cancer (Mcc), a Candidate Tumor Suppressor, Is Dynamically Expressed during Mouse Embryogenesis. *Dev. Dyn.* **2011**, *240*, 2166–2174. [CrossRef] [PubMed]
11. Mickael, M.E.; Kubick, N.; Łazarczyk, M.; Sacharczuk, M.; Marchewka, J.; Urbański, P.; Horbańczuk, J.O. Transcriptome Analysis of the Th17/Treg Axis Reveals Multiple Pathways That Ensure Distinct Differentiation Patterns. *Anim. Sci. Pap. Rep.* **2023**, *41*, 79–93.
12. Begley, D.; Murphy, A.M.; Hiu, C.; Tsubota, S.I. Modifier of Rudimentary^{P1}, Mod(r)^{P1}, a Trans-Acting Regulatory Mutation of Rudimentary. *Mol. Gen. Genet. MGG* **1995**, *248*, 69–78. [CrossRef]
13. Bache, K.G.; Slagsvold, T.; Cabezas, A.; Rosendal, K.R.; Raiborg, C.; Stenmark, H. The Growth-Regulatory Protein HCRP1/h Vps37A Is a Subunit of Mammalian ESCRT-I and Mediates Receptor down-Regulation. *Mol. Biol. Cell* **2004**, *15*, 4337–4346. [CrossRef] [PubMed]
14. Joshi, A.; Coelho, M.B.; Kotik-Kogan, O.; Simpson, P.J.; Matthews, S.J.; Smith, C.W.J.; Curry, S. Crystallographic Analysis of Polypyrimidine Tract-Binding Protein-Raver1 Interactions Involved in Regulation of Alternative Splicing. *Structure* **2011**, *19*, 1816–1825. [CrossRef] [PubMed]
15. Peng, X.; Nelson, E.S.; Maiers, J.L.; DeMali, K.A. New Insights into Vinculin Function and Regulation. In *International Review of Cell and Molecular Biology*; Elsevier: Amsterdam, The Netherlands, 2011.
16. Yang, Y.; Liao, B.; Wang, S.; Yan, B.; Jin, Y.; Shu, H.B.; Wang, Y.Y. E3 Ligase WWP2 Negatively Regulates TLR3-Mediated Innate Immune Response by Targeting TRIF for Ubiquitination and Degradation. *Proc. Natl. Acad. Sci. USA* **2013**, *110*, 5115–5120. [CrossRef] [PubMed]
17. Blair, S.S. Notch Signaling: Fringe Really Is a Glycosyltransferase. *Curr. Biol.* **2000**, *10*, R608–R612. [CrossRef] [PubMed]
18. Shao, L.; Haltiwanger, R.S. O-Fucose Modifications of Epidermal Growth Factor-like Repeats and Thrombospondin Type 1 Repeats: Unusual Modifications in Unusual Places. *Cell. Mol. Life Sci.* **2003**, *60*, 241–250. [CrossRef]
19. Wang, Z.; Li, Y.; Banerjee, S.; Sarkar, F.H. Emerging Role of Notch in Stem Cells and Cancer. *Cancer Lett.* **2009**, *279*, 8–12. [CrossRef]
20. Márton, M.; Tihanyi, N.; Gyulavári, P.; Bánhegyi, G.; Kapuy, O. NRF2-Regulated Cell Cycle Arrest at Early Stage of Oxidative Stress Response Mechanism. *PLoS ONE* **2018**, *13*, e0207549. [CrossRef]
21. Kopacz, A.; Kloska, D.; Forman, H.J.; Jozkowicz, A.; Grochot-Przeczek, A. Beyond Repression of Nrf2: An Update on Keap1. *Free Radic. Biol. Med.* **2020**, *157*, 63–74. [CrossRef] [PubMed]

22. Aristin Revilla, S.; Kranenburg, O.; Coffer, P.J. Colorectal Cancer-Infiltrating Regulatory T Cells: Functional Heterogeneity, Metabolic Adaptation, and Therapeutic Targeting. *Front. Immunol.* **2022**, *13*, 903564. [CrossRef]
23. Lewit-Bentley, A.; Réty, S. EF-Hand Calcium-Binding Proteins. *Curr. Opin. Struct. Biol.* **2000**, *10*, 637–643. [CrossRef]
24. Chazin, W.J. Relating Form and Function of EF-Hand Calcium Binding Proteins. *Acc. Chem. Res.* **2011**, *44*, 171–179. [CrossRef]
25. Yap, K.L.; Ames, J.B.; Swindells, M.B.; Ikura, M. Diversity of Conformational States and Changes within the EF-Hand Protein Superfamily. *Proteins Struct. Funct. Genet.* **1999**, *37*, 499–507. [CrossRef]
26. Miyara, M.; Sakaguchi, S. Natural Regulatory T Cells: Mechanisms of Suppression. *Trends Mol. Med.* **2007**, *13*, 108–116. [CrossRef] [PubMed]
27. Capozzi, F.; Casadei, F.; Luchinat, C. EF-Hand Protein Dynamics and Evolution of Calcium Signal Transduction: An NMR View. *JBIC J. Biol. Inorg. Chem.* **2006**, *11*, 949–962. [CrossRef]
28. Mickael, M.E.; Rajput, A.; Steyn, J.; Wiemerslage, L.; Bürglin, T. An Optimised Phylogenetic Method Sheds More Light on the Main Branching Events of Rhodopsin-like Superfamily. *Comp. Biochem. Physiol. Part D Genom. Proteom.* **2016**, *20*, 85–94. [CrossRef] [PubMed]
29. Kubick, N.; Klimovich, P.; Bieńkowska, I.; Poznanski, P.; Łazarczyk, M.; Sacharczuk, M.; Mickael, M.-E. Investigation of Evolutionary History and Origin of the Tre1 Family Suggests a Role in Regulating Hemocytes Cells Infiltration of the Blood–Brain Barrier. *Insects* **2021**, *12*, 882. [CrossRef] [PubMed]
30. Mickael, M.E.; Bieńkowska, I.; Sacharczuk, M. An Update on the Evolutionary History of Bregs. *Genes* **2022**, *13*, 890. [CrossRef] [PubMed]
31. Kanehisa, M.; Sato, Y.; Kawashima, M. KEGG Mapping Tools for Uncovering Hidden Features in Biological Data. *Protein Sci.* **2022**, *31*, 47–53. [CrossRef]
32. Yang, Z. PAML 4: Phylogenetic Analysis by Maximum Likelihood. *Mol. Biol. Evol.* **2007**, *24*, 1586–1591. [CrossRef] [PubMed]
33. Mickael, M.-E.; Kubick, N.; Gurba, A.; Klimovich, P.; Bieńkowska, I.; Kocki, T.; Sacharczuk, M. Fezf2 and Aire1 Evolutionary Trade-off in Negative Selection of T Cells in the Thymus. *bioRxiv* **2022**. [CrossRef]

Disclaimer/Publisher's Note: The statements, opinions and data contained in all publications are solely those of the individual author(s) and contributor(s) and not of MDPI and/or the editor(s). MDPI and/or the editor(s) disclaim responsibility for any injury to people or property resulting from any ideas, methods, instructions or products referred to in the content.

Review

Immune Checkpoint Inhibitors, Small-Molecule Immunotherapies and the Emerging Role of Neutrophil Extracellular Traps in Therapeutic Strategies for Head and Neck Cancer

Connor H. O'Meara [1,*], Zuhayr Jafri [2] and Levon M. Khachigian [2]

[1] Department of Otorhinolaryngology, Head and Neck Surgery, Prince of Wales Hospital, Randwick, NSW 2031, Australia
[2] Vascular Biology and Translational Research, School of Biomedical Sciences, UNSW Faculty of Medicine and Health, University of New South Wales, Sydney, NSW 2052, Australia
* Correspondence: drch_omeara@outlook.com

Abstract: Immune checkpoint inhibitor (ICI) therapy has revolutionized the treatment of many cancer types, including head and neck cancers (HNC). When checkpoint and partner proteins bind, these send an "off" signal to T cells, which prevents the immune system from destroying tumor cells. However, in HNC, and indeed many other cancers, more people do not respond and/or suffer from toxic effects than those who do respond. Hence, newer, more effective approaches are needed. The challenge to durable therapy lies in a deeper understanding of the complex interactions between immune cells, tumor cells and the tumor microenvironment. This will help develop therapies that promote lasting tumorlysis by overcoming T-cell exhaustion. Here we explore the strengths and limitations of current ICI therapy in head and neck squamous cell carcinoma (HNSCC). We also review emerging small-molecule immunotherapies and the growing promise of neutrophil extracellular traps in controlling tumor progression and metastasis.

Keywords: immune checkpoint inhibitor; head and neck cancers; head and neck squamous cell carcinoma; neutrophil extracellular traps

1. Introduction

The immune system is a dynamic and equipped mechanism, an intricate system of "recognition" and "on-off" switches. Unfortunately, cancers utilize this system to enable growth and escape. The role of the immune system in tumor regulation is particularly evident in the immunocompromised. Iatrogenic solid organ transplant, diabetes, autoimmunity requiring immunosuppressive therapy, HIV and hemoproliferative malignant disease or disorders and aging, are all associated with an increased risk of developing head and neck cancer (HNC) and worse outcomes [1–11]. Proliferating tumors utilize many forms of immunosuppression to tip the balance of immunoediting toward tumor progression [12]. Identifying therapies capable of shifting this balance back toward immunosurveillance should play an integral role in reducing morbidity- and mortality-associated HNC.

HNC, the sixth most common group of malignancies worldwide, results in 680,000 new cases annually, with squamous cell carcinoma (SCC) being the most common [13–15]. The incidence of HNC is increasing due to a range of factors including smoking, alcohol, human papillomavirus (HPV) infection and extended life expectancy [16].

Despite the vital role played by traditional therapies for HNSCC, namely surgery, radiotherapy and chemotherapy, prognosis remains poor and survival remains correlated to stage, with a 5-year survival rate of 50–60% and more than 60% presenting in the advanced stage [17,18]. More than 50% of HNSCCs have tumor recurrence and metastasis in less than 3 years [19]. Targeting the epidermal growth factor receptor (EGFR) was hailed a paradigm shift in personalizing HNSCC treatment, with the monoclonal antibody cetuximab demonstrating promise [20,21]; however, this has since demonstrated limited efficacy [22].

Compared with traditional therapies, new immunotherapy agents, namely antibodies targeting the PD-1/PD-L1 system, so-called immune checkpoint inhibitors (ICI) provide improved efficacy and comparatively lower toxicity for patients with advanced HNSCC [23–26]. KEYNOTE-048 (NCT02358031), a randomized open-label phase 3 study comparing the humanized monoclonal antibody pembrolizumab (Keytruda) targeting PD-1 alone or in conjunction with chemotherapy (platinum and 5-fluorouracil) against cetuximab with chemotherapy, demonstrated overall survival improvement in both treatment arms over standard-of-care therapy in recurrent or metastatic HNSCC [27]. Pembrolizumab was subsequently approved as a first-line therapeutic drug for patients with metastatic, unresectable and recurrent HNSCC. Unfortunately, the objective response rate (ORR) of pembrolizumab (or nivolumab/Optivo) in HNSCC is only 15%, with short-term durability [28,29]. In addition, immune-related adverse events (irAEs) secondary to immunotherapy treatment were identified in over 50% of patients, impacting clinical outcomes [30], with adverse-event-associated mortality evident in 0.3–1.3% of patients [31]. Common irAEs include gastrointestinal, dermatologic and endocrine toxicities, more specifically dermatitis, rash, nausea/vomiting, fever, headache, myalgia, hypothyroidism and fatigue [32]. Rarely, irAEs can be severe, resulting in carditis, nephritis, hepatitis, pneumonitis, gastrointestinal perforation and severe hematological dysfunction [33]. irAEs in ICI therapy have been associated with benefits, namely improvements in PFS, OS and ORR [34–37]. Consequently, balancing immunotherapy de-escalation or commencement of immunosuppressive therapy against a sub-optimal oncological outcome can be difficult.

Predictive biomarkers may be the key to identifying patients at risk of irAEs. To date, circulating blood counts and ratios, autoantibodies and autoantigens, microRNAs, gastrointestinal microbiome, T-cell diversification and expansion and cytokines are all being investigated; however, they remain to be validated for clinical use [38].

Biological, etiological, phenotypic and clinical heterogeneities characterize HNSCC and challenge the development of personalized medicine. However, poor survival, significant morbidity and compromised quality of life emphasize the requirement for innovative therapy. Immunoediting is the process through which the immune system can promote and constrain tumor development [39]. This article explores current and developing therapies in immunomodulation and the developing role of neutrophil extracellular traps (NETs), net-like structures comprised of DNA-histone complexes and proteins in immune-mediated tumorigenesis.

2. Immune Checkpoint Inhibitor Targets and Therapies

A successful objective ICI response revitalizes the immune system to recognize and target cancer cells. The roles of known key immune checkpoints CTLA-4, PD-1 and LAG-3 are summarized in Figure 1.

2.1. CTLA-4 and PD-1/PD-L1

CTLA-4 (cytotoxic T-lymphocyte associated protein 4, also known as cluster of differentiation 152, CD152) and programmed cell death protein 1 (PD-1) (and its ligands PD-L1 and PD L2) are immune checkpoints targeted by humanized antibodies for the treatment of HNSCC. CTLA-4 is bound by ipilimumab (Yervoy), whereas PD-1 is targeted by pembrolizumab and nivolumab [32]. The antibodies atezolizumab (Tecentriq), durvalumab (Imfinzi) and avelumab (Bavencio) have also been approved as inhibitors of PD-L1 [33]. Both checkpoints regulate different stages of the immune response. CTLA-4 is considered the "leader" of the immune response and prevents the stimulation of autoreactive T-cells in the initial stage of naïve T-cell activation, whereas PD-1 is thought to regulate previously activated T-cells at the later stages of the immune response [32].

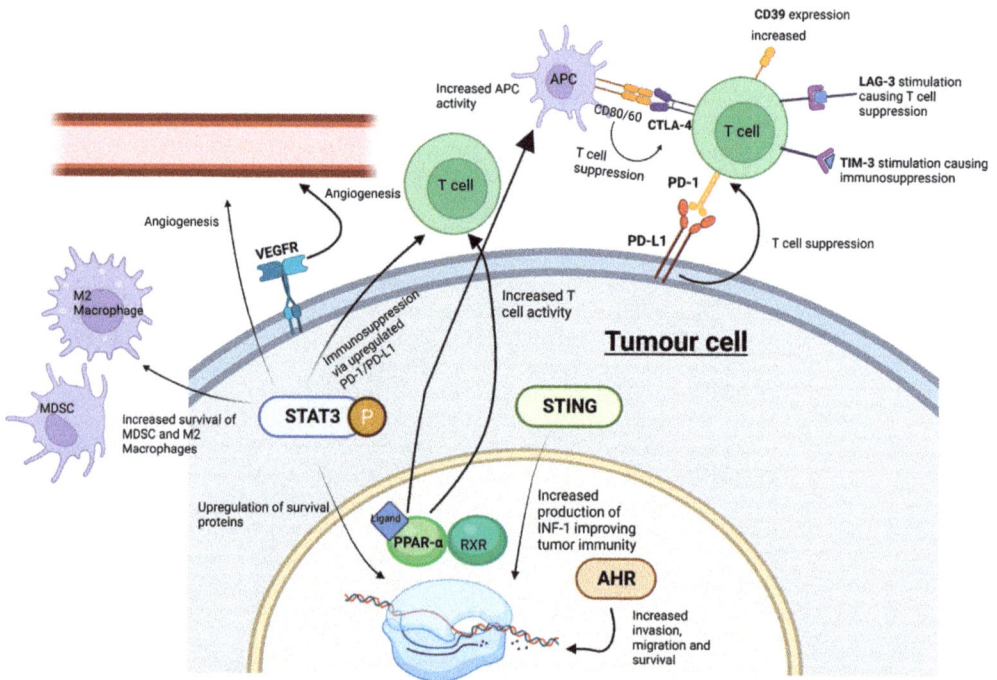

Figure 1. ICI and SMI actions within the tumor microenvironment. Whilst ICIs influence cell signaling at cell surface receptors, SMIs can interact with "upstream" intracellular signaling pathways-potentially playing a more effective role in abrogating tumor cell progression. MDSC, myeloid-derived suppressor cells; M2 macrophages, pro-tumorigenic macrophages; STING, stimulator of interferon genes; PPAR-α, peroxisome proliferator-activated receptor-α; AHR, aryl hydrocarbon receptor; STAT3, signal transducer and activator of transcription 3; P, phosphorylation of STAT3. Created with BioRender.com.

CTLA-4 is a homolog of CD28, but unlike CD28, CTLA-4 activation has an immunosuppressive effect opposite to the stimulatory effect of CD28 and the T-cell receptor (TCR) [40]. The binding of CD80/CD86 on antigen-presenting cells to CTLA-4 on T-cells in the tumor microenvironment suppresses the immune system, enabling tumor proliferation [41]. PD-1's interaction with PD-L1 and PD-L2 has an immunosuppressive effect [41]. PD-L1 and PD-L2 are expressed by a range of tumors including HNSCC [42]. Critically, increased PD-1 levels serve as a biomarker for T cell exhaustion; this state of exhaustion is linked to T-cell dysfunction, which can facilitate tumor proliferation [43]. PD-L1's interaction with PD-1 has an immunosuppressive effect, thus protecting cancer cells from lysis by activated T-cells [44].

Despite ICI therapy demonstrating survival advantage, comparatively few patients develop an effective response, the durability of which attenuates with acquired tumor resistance. Acquired resistance leads to tumor progression, and both arms of the immune system, innate and adaptive, can play a critical role in this change. Mechanisms of resistance to immunotherapy can be either intrinsic (tumor cell-mediated) or extrinsic (processes associated with T-cell activation) and shift the balance of immunomodulation towards tumor proliferation. Intrinsic resistance can include the downregulation of antigen-presenting machinery (APM) [45], the up-regulation of signaling pathways promoting T-cell exhaustion [46], the expression of multiple checkpoint inhibitors to mitigate T-cell activation [47], changes in tumor cell DNA repair, damage and genomic instability [48] and altered kinase signaling pathways [49]. Extrinsic resistance involves the complex interplay between tumor cells and the tumor microenvironment and its ability to regulate phenotypical characteristics of immune cells, especially TANs, TAMs, Tregs, MDSCs, T-cells, their associated regulatory cytokines and signaling pathways and a newly identified player, NETs [50–56].

Despite the clear improvements in overall survival due to immune checkpoint therapy, such treatments have limitations. For example, since CTLA-4 prevents the stimulation of autoreactive T-cells, inhibiting CTLA-4 can lead to grade 3 or 4 autoimmune-related adverse effects in 10–15% of patients [57]. Immune checkpoint immunotherapies are also associated with low response rates. For example, pembrolizumab has a response rate of only 15% in HNSCC [58].

To improve therapeutic failure and overcome immunotherapy resistance, significant energy is being invested in exploring biomarkers to predict clinical response and combinational therapies or changes in adjuvant delivery of immunotherapy to increase success rates. Biomarkers that have shown potential to determine improved clinical response in HNSCC include the tumor mutational burden, CCND1 amplification (CCND1 encodes cyclin D1, which regulates the retinoblastoma protein activity and cell-cycle progression), PD-1, IFN-γ, tumor-infiltrating lymphocytes (TILs) and cancer-associated fibroblasts (CAFs), CTLA-4, exosomes, CXCL, MTAP and SFR4/CPXM1/COL5A1 molecules [25,59–69].

Clinical trials exploring combinational immunotherapy in HNSCC are underway. The phase 3 randomized trial CheckMate 651 NCT02741570), which compared nivolumab and ipilimumab against EXTREME (platinum/5-fluorouracil/cetuximab) for R/M HNSCC, was unsuccessful in demonstrating OS improvement, although there was an association between elevated CPS and OS and durable response [70]. Other combination ICI therapy clinical trials have been largely unsuccessful (Table 1).

Table 1. Combination ICI Therapy Clinical Trials in HNSCC.

Target	Combination	Phase	Trial	Intent	Outcome
PD-1, CTLA-4	Nivolumab, Ipilimumab	3	NCT027441570 (CheckMate 651) [71]	Combination nivolumab + ipilimumab vs. EXTREME Regime (platinum/5-fluorouracil/cetuximab) for R/M HNSCC	Failed endpoint (OS). No difference between dual ICI blockade and EXTREME arm. Improvement in dual ICI arm if CPS > 20 (ns)
PD-L1, CTLA-4	Durvalumab, Tremelimumab	3	NCT02551159 (KESTRAL) [72]	Combination durvalumab + tremelimumab vs. duravalumab monotherapy vs. SOC CT in R/M HNSCC	Results pending
PD-1, CTLA-4	Nivolumab, Ipilimumab	2	NCT02823574 (CheckMate 714) [73]	Combination nivolumab + ipilimumab vs. nivolumab + ipilimumab placebo in R/M HNSCC	Failed ORR and OS endpoints. Subpopulation assessment ongoing
PD-L1, CTLA-4	Durvalumab, Tremelimumab	3	NCT02369874 (EAGLE) [74,75]	Combination durvalumab + tremelimumab vs. durvalumab monotherapy vs. SOC in R/M HNSCC	Failed to meet primary OS improvement endpoint

Concurrent neoadjuvant and adjuvant delivery of ICIs has recently demonstrated benefits in surgically resectable advanced melanoma (Stage IIIB to IVC). In a recently completed Phase 2 randomized study (NCT03698019), neoadjuvant-adjuvant delivery of pembrolizumab was compared to an adjuvant alone in demonstrating an event-free survival of 72% in the neoadjuvant-adjuvant group compared to 49% in the adjuvant group after 2 years [76].

2.2. LAG-3

LAG-3 is expressed on activated human T-cells and natural killer cells and plays a similar role in T-cell regulation to CTLA-4 and PD-1 [77]. LAG-3 may represent an intrinsic resistance mechanism to PD-1 inhibitors due to its synergistic co-expression with

PD-1 on exhausted T-cells [77]. To combat resistance, the FDA-approved drug opdualag® (combined LAG-3 and PD-1 inhibitor) became a first-line treatment for unresectable or metastatic melanoma in March 2022 [77]. Opdualag has shown success in clinical trials, more than doubling progression-free survival compared to melanoma patients treated with nivolumab alone [78].

2.3. Tim-3 and CD39

T cell immunoglobulin and mucin domain-containing protein 3 (Tim-3) is a co-inhibitory receptor expressed on IFN-γ-producing T-cells Tim-3. Studies by Liu et al. showed that Tim-3 is linked to immunosuppression in HNSCC and that targeting Tim-3 (with monoclonal antibodies) can enhance the anti-tumor immune response by reducing Tregs in HNSCC [79]. Similarly, the expression of the cell-surface ectonucleosidase CD39 in HNSCC positively correlates with tumor stage and predicts poor prognosis [80]. There are no approved inhibitors of Tim-3 or CD39, and opdualag has not yet been approved for HNSCC.

3. Emerging Immunotherapeutic Targets and Strategies

Small-molecule immunotherapy (SMI) may represent the paradigm shift required to improve quality of life (QOL) and survival in HNSCC. Unlike current ICI therapies, small molecules can be delivered orally, are potentially less expensive than antibodies and utilized to target intracellular signaling and transcriptional pathways upstream of receptors expressed on the cell surface. Several promising SMIs in various phases of development are listed in Table 2 and described below.

Table 2. SMI Targets and Clinical Trials in HNSCC.

Target	Drug	Phase	Trial	Intent	Outcome
STAT3	AZD9150	1b/2	NCT02499328 [81]	Combination ASD9150 + MED14736 (duravalumab) vs. MED14736 alone; in platinum refractory recurrent metastatic HNSCC	Acceptable toxicity profile. Combination therapy more effective than PD-L1 monotherapy
STING	MK-1454	1	NCT03010176 [82]	Combination MK-1454 (ulveostinag) + pembrilizumab vs. MK-1454 monotherapy; in advanced HNSCC	Acceptable toxicity profile. Combination therapy more effective (DCR 48%) than monotherapy (DCR 20%)
PPAR-α	TPST-1120	1/1b	03829436 [83]	Combination TPST-1120 + nivolumab vs. TPST-1120 monotherapy; in advanced solid tumors; including HNSCC	Acceptable toxicity (several patients suffered Grade 3 Adverse reactions. Optimal disease response in combination therapy (38%)
RTKs	AL3818	2	NCT04999800 [84]	Combination AL3818 (analotinib) + pembrolizumab; as a first line therapy for platinum refractory recurrent or metastatic HNSCC	Manageable side effects. Encouraging anti-tumor activity. ORR: 46.7% (7/15) & DCR: 100% Median PFS & OS not reached (median follow-up: 8.2 months.
RTKs	Afatinib	3	NCT01345682 [85]	LUX-Head & Neck 1: second-line afatinib therapy vs. methotrexate for platinum refractory recurrent/metastatic HNSCC	n = 483 patients. Median PFS: afatinib over methotrexate (2.7 months vs. 1.6 months) Afatinib more effective in all tumor subsets except HPV + OPSCC
AHR	BAY2416964	1	NCT04069026 [86]	AHR antagonist: safety and tumor response study in advanced HNSCC & nSCLC	Well tolerated at all dose regimes. Initial evaluation of biomarkers shows inhibition of AHR and modulation of immune functions. Encouraging preliminary anti-tumor activity in heavily pretreated patients.

3.1. STAT3

Among the intracellular signal transducer and activator of transcription (STAT) proteins, STAT3 plays an important hemostatic role in normal cells by helping to regulate cell growth, survival, differentiation, angiogenesis, immune response and cellular respiration [87,88]. STAT3 can be activated by both JAK and EGFR (via Src) signaling, subsequently binding target DNA to regulate gene expression [89,90]. Importantly, STAT3 activation can upregulate multiple survival proteins, namely Bcl-xL, survivin and Bcl-2 [91]. STAT3 is also an upstream regulator of PD-1 expression [92,93]. STAT3 can behave as an oncogene and is expressed in approximately 70% of human cancers [19]. Overexpression of STAT3 regulates tumor progression and is associated with poor prognosis in various malignancies [94]. Elevated levels of IL-6, released by tumor-infiltrating lymphocytes, M2 phenotype macrophages [95], cancer-associated fibroblasts [96] and tumor cells [97,98], can activate a pro-tumorigenic IL-6/STAT3 pathway [99], inhibiting dendritic cell maturation, suppressing $CD8^+$ T-cell and NK cell activation [100–102] and promoting $CD4^+$ T cell differentiation to a T regulatory phenotype [103]. This pathway supports the survival of immunosuppressive MDSCs and M2 phenotype macrophages [104,105] and supports tumor survival, invasiveness and proliferation [106]. STAT3 can also regulate metabolism-related genes that favor cancer progression [107] and promote angiogenesis via the upregulation of VEGF [108]. STAT3 activation was identified to be 10.6- and 8.8-fold higher in tumors and normal mucosa, respectively, of HNSCC patients compared to the mucosa of noncancer patients, supporting the concept of "field cancerization" [109]. Furthermore, there is a strong association between downstream proteins transcribed by STAT3 and locoregional metastasis, stage, recurrence and mortality in OCSCC [110]. There is also evidence that constitutive STAT3 activation plays a prominent role in mediating drug resistance to many targeted cancer therapies and chemotherapies, including the poor response to the EGFR blockade, with less than 15% of patients benefiting from cetuximab as a single agent or 36% when combined with chemotherapy [111]. Certainly, STAT3 inhibition has improved radiotherapy and cetuximab responsiveness in HNSCC cell lines [91,112]. Hence preclinical and early clinical trial data suggest that targeting STAT3 is a promising therapeutic strategy [113].

Flubendazole is a benzimidazole, long utilized as a macrofilaricide in humans and animals. Recently, it has been repurposed and recognized as a promising anti-cancer agent, effective in breast cancer, melanoma, neuroblastoma, colorectal, liver and squamous cell carcinoma [114–124]. In melanoma, flubendazole reduced the expression of phosphorylated STAT3 in tumor tissue and the expression of PD-1 expression, while also decreasing MDSC levels in tumors [125]. Immunological signature gene sets, including those associated with T cell differentiation, proliferation and function correlated with FLU treatment [126]. Flubendazole has also been found to have synergistic antiproliferative effects in vitro with 5-fluorouracil [127], which raised the potential benefit of use topically in conjunction with 5-FU for the treatment of premalignant and malignant non-melanoma skin cancers [128–130]. Flubendazole has not entered clinical trials for HNC.

An alternative approach is provided by danvatirsen (AZD9150), an antisense oligonucleotide inhibitor of STAT3 comprised of 16 nucleotides, which has demonstrated antiproliferative effects in xenograft models showing reduced STAT3 expression, paving its way to clinical trial [131]. Combining durvalumab (MEDI4736, PD-L1 inhibitor) with danvatirsen or AZD5069 (CXCR2 inhibitor) (NCT02499328) in patients with advanced solid malignancies and HNSCC improved anticancer activity as compared to PD-L1 monotherapy [81].

3.2. STING

STING (stimulator of interferon genes) is a cytosolic pattern-recognition receptor (PRR) that recognizes non-self-dsDNA, upregulating type 1 interferon [132]. Recent evidence indicates that type 1 IFN plays an important role in many anticancer modalities, including immunotherapy, helping to promote dendritic cell activation and prime and recruit cytotoxic CD8$^+$ T-cells against tumor-associated antigens [133–135]. Evidence suggests that the STING pathway may help potentiate checkpoint blockade therapy [136,137].

3.3. PPAR

Peroxisome proliferator-activated receptors (PPARs) regulate a multitude of cellular functions. They are a family of ligand-inducible nuclear hormone receptors belonging to the steroid receptor superfamily. PPAR-α is commonly expressed in skeletal muscle, liver, heart and brown adipose tissue. Its activation suppresses NF-κB signaling, which decreases the inflammatory cytokine production by different cells and modulates the proliferation, differentiation and survival of macrophages, B-cells and T-cells, whilst also playing a role in angiogenesis, homeostasis and glucose and lipid metabolism [138–141]. Notwithstanding a pleiotropic role in cancer, which appears type and tumor microenvironment (TME)-dependent, increasing evidence is demonstrating that PPAR-γ can modulate carcinogenesis, showing promise as a focus for cancer therapies. PPAR-γ agonists have been shown to inhibit cancer cell proliferation and Warburg effects.

3.4. RTKs

There are over 50 known RTKs in humans. These are transmembrane receptors integral to cell-to-cell communication and the regulation of cell growth, metabolism, motility and cell differentiation. These mediate the activation of a variety of signaling pathways, including JAK/STAT, PI-3K/AKT/mTOR, PLC/PKC and RAS/MAPK. Their dysregulation plays a role in multiple human disease processes, including carcinogenesis, which can confer constitutive activation by genomic amplification, chromosomal rearrangements, autocrine activation, gain-of-function mutations or kinase domain duplication [142,143].

There is evidence that IL-33, although a pleiotropic cytokine in HNSCC [144], can regulate immune cells in the TME, namely CD4$^+$ T-helper cells, CD8$^+$ T-cells, NK cells, DCs and macrophages [144]. Developing evidence suggests that IL-33 may regulate the immune response through a signaling complex between IL-33R and EGFR in gastrointestinal helminth infections [145]; however, we are not aware of current research supporting this pathway in cancer.

EGFR is a prototypic RTK and is well recognized to be susceptible to gain-of-function mutations and is commonly overexpressed in HNSCC. These mutations can hyperactivate the kinase and its downstream signaling, conferring oncogenic properties [146]. Eighty to ninety percent of HNSCCs overexpress or demonstrate EGFR mutation, with these changes detrimentally affecting both PFS and OS [147,148]. Certainly, EGFR status has been identified as a survival predictor and guide to the effectiveness of chemoradiation [149]. Although known mutations in EGFR are rare in HNC, its overexpression with TGF-α is common, and auto or paracrine activation is important in HNC EGFR function. Unfortunately, mutation commonly alters drug binding dynamics, leading to resistance, a similar phenomenon leading to a reduced radiation response and overall survival in HNSCC. Mutation status is also associated with the tumor stage [150]. Chromosomal rearrangements have been identified in the RET kinase in thyroid cancer [151] and TRKA, TRKB and TRKC tyrosine kinases in thyroid and HNC.

The primary site of action of TKIs on EGFR is the intracellular tyrosine kinase domain, inhibiting downstream signaling. Geftinib and erlotinib have been ineffective in HNC, comparatively lapatinib, afatinib and dacomitinib have demonstrated benefits in clinical trials and can target VEGFR to reduce tumor angiogenesis. Cetuximab has shown an ORR of 13% as a monotherapy, and gefitinib and erlotinib demonstrated ORR of 1.4% and 10.6% with a median OS of 5.5 and 8.1 months, respectively [152–154]. Anlotinib (AL3818) is a novel multi-target RTK antagonist against PDGFR, FGFR, VEGFR and c-Kit. In human OCSCC cell lines, it effectively reduced tumor cell proliferation and promoted apoptosis [155].

3.5. AHR

AHR (aryl hydrocarbon receptor) is a ligand-activated transcription factor activated by both anthropogenic and natural agonists, with recent studies reporting a key role in regulating host immunity [156,157]. As a transcription factor, AHR can regulate the expression of cytochrome P450 family genes. Chronically active AHR is capable of driving cancer cell invasion, migration, cancer stem cell characteristics and survival [158]. Tumor expression of AHR can result in an autocrine AHR-IL-6/STAT3 signaling loop via kynurenine, an immunosuppressive AHR agonist ligand produced by the metabolism of the essential amino acid tryptophan [159].

4. Neutrophils Extracellular Traps (NETs)

Neutrophils are the largest group of leukocytes within the blood and play an integral role in immune-mediated host defense mechanisms. As activated phagocytes, these secrete neutrophil elastase (NE), reactive oxygen species (ROS), nicotinamide adenine dinucleotide phosphate oxidase (NADPH) and myeloperoxidase (MPO) to digest pathogens [160]. Chemotactically attracted to the TME, tumor-associated neutrophils (TANs) phenotypically polarize to either N1 or N2 sub-types. Similar to M1 phenotype tumor-associated macrophages (TAMs), N1 TANs are anti-tumorigenic, whilst N2 TANs are pro-tumorigenic and regulate immunosuppression, tumor cell proliferation, angiogenesis and metastasis [161–163].

Derived from neutrophils undergoing a signal-mediated cell death program known as NETosis, NETs are extracellular "spider webs" of unwound chromatin, comprising histones, neutrophil elastase and granular antimicrobial enzymes [164,165] (Figure 2). These are key antimicrobial components of the innate defense system that sequester and contribute to bacterial cytotoxicity and phagocytosis [166,167]. Their antimicrobial purview includes inhibiting replication, containing and eliminating viral infections via the activation of TLR4, 7 and 8 pathways, PKC pathway blockade or aggregation and neutralizing effects of cationic histones, particularly arginine-rich H3 and H4 [168–172]. NETs are important in clearing large pathogens and are activated by β-glucan on fungal hyphae [173–175]. NETs recognize the activation of platelets and monocytes and limit the dissemination of parasites by trapping and killing these with histones, neutrophil elastase, MPO and collagenase mediating cytotoxicity [176–179]. Unfortunately, NET dysregulation is pivotal to the pathogenesis of numerous diseases, including sepsis [180,181], acute respiratory disease syndromes [182,183], ischemia-reperfusion injury [184], diabetes [185], venous thromboembolism [186] and chemotherapy-induced peripheral neuropathy [187].

New evidence indicates that NETs potentiate pro-tumorigenic effects, with neutrophils attracted to the tumor microenvironment being reprogrammed by tumor-associated factors to undergo NETosis and potentiate tumor activity. In this regard, NETs are capable of suppressing tumor cell apoptosis and promoting tumor cell invasion [188–195]. Factors involved in tumor-mediated NET formation include tumor-derived inflammatory and chemoattractant cytokines (IL-8, IL-6, TNF-α, G-CSF and IL-1β) [189,190,196–199], tumor extracellular vesicles [200], tumor-activated platelets [201,202], tumor-derived HMGB1 [203–206], KRAS oncogene mutation [207] and hypoxia [208].

Recent evidence has highlighted the ability of NETs to actively drive tumor growth and metastasis. NET-associated HMGB1 promotes tumor cell proliferation involving interaction with tumor RAGE, activating and NF-κB signaling [209]. Additionally, neutrophil elastase, via the PI3K signaling pathway, promotes the proliferation of adenocarcinoma cells [210]. NETs play a key role in shielding tumor cells from tumor-recognizing NK- and cytotoxic CD8+ T-cells [211,212], whilst promoting cytotoxic CD8+ T-cell exhaustion through the upregulation of PD-L1 [213]. Their immunosuppressive role also extends to the programming of T regulatory phenotype cells, which inhibit macrophage, dendritic, cytotoxic CD8+ T cell anti-tumor effects via a TLR4-dependent mechanism [214–216], which may be histone dependent, while they also play a key role in thrombosis [217–221]. Further compromising immunorecognition, NET-activated platelets may facilitate plasma membrane transfer, enabling tumor cell expression of platelet markers and MHC receptors to camouflage their presence within the platelet aggregate attached to a NET scaffold [222–224].

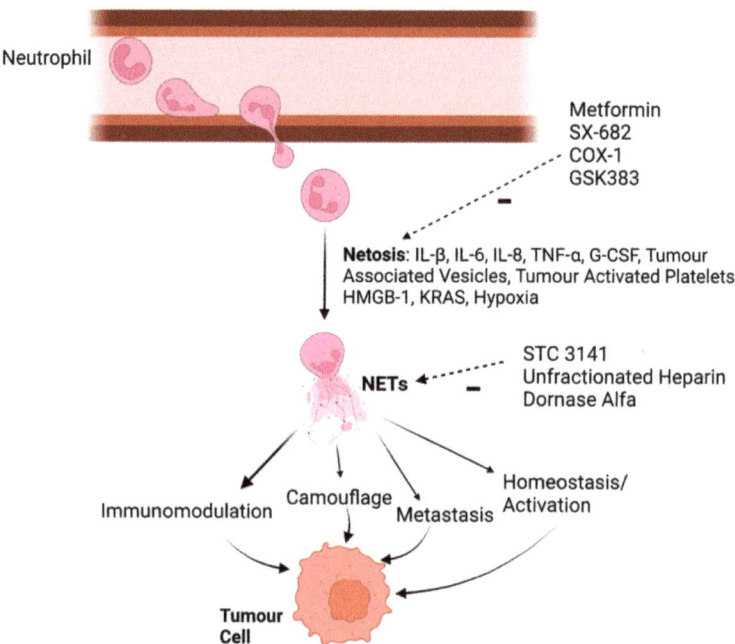

Figure 2. Pro-tumorigenic effects of NETs and developing therapeutic strategies. NETs have been demonstrated to play a role in tumor immunomodulation, camouflage, homeostasis/activation, and metastasis. Therapies under development can be grouped into those that prevent or "modulate" NETosis (the generation of NETs) or "dissolve" the chromatin "web" of the neutrophil extracellular trap, mitigating the ability for NETs to play a pro-tumorigenic role. HMGB-1, high mobility group box-1; KRAS oncogene mutation. Created with BioRender.com.

NETs can affect all arms of Virchow's triad [225], activating platelets and endothelial cells, aggregating erythrocytes [226] and promoting tissue factor release from both platelets and endothelial cells. Specifically, by way of histone expression, NETs induced an endothelial cell shift toward a pro-coagulant phenotype [227–229]. NETs also promote thrombosis by acting as a scaffold that can capture and activate platelets, facilitate fibrin deposition and express TF for coagulation [230]. There is likely significant crosstalk between NETs and platelets, working in concert to enhance tumor cell survival and metastasis. Platelet TLR4 can trigger NET production, and NET-expressed histone H3 and H4 can activate platelets in a positive feedback loop [217,231].

The ability of NETs to promote thrombosis and capture circulating tumor cells enhances the ability of tumor cells to metastasize. However, their further ability to facilitate distant tumor growth is much more extensive. NETs can promote tumor cell migration and enhance invasiveness, with the rearrangement of the cytoskeleton elements, via their CCDC25 receptor, which may aid tumor cell transmigration across the endothelium [232]. Although the mechanism remains unclear, NETs can reprogram the epithelial-to-mesenchymal transition in tumor cells, essentially allowing them to disconnect cell-to-cell and cell-to-extracellular matrix chelation and commence migration and invasion [233–235]. NETs can also revive dormant tumor cells that have metastasized to an unfavorable microenvironment. Evidence suggests that extracellular matrix

NET remodeled laminin binds dormant tumor a3B1 integrin, driving their activation via FAK/ERK/MLCK/YAP signalling, in turn supporting proliferation [236].

Angiogenesis is also a priority for hypoxic tumor cells. NETs have been demonstrated to promote angiogenesis [237–239] and this may well be via a histone-dependent mechanism [237]. Aldabbous and colleagues showed that NETs increased the vascularization of Matrigel plugs and release of MMP-9, TGF-β1 latency-associated peptide, HB-EGFGF and uPA, also promoting endothelial permeability and cell motility [238].

Interactions between NETs and cancer cells are also thought to drive resistance to various cancer therapies, including chemotherapy, immunotherapy and radiation therapies. Therefore, the development of therapies to mitigate the pro-tumorigenic role of NETs is an absolute necessity but should minimize interference with immunity, wound healing, and host defense mechanisms. There are several novel therapies being developed for the management of NETs in sterile systemic inflammatory response syndromes and several agents that may be repurposed to mitigate the pro-tumorigenic role of NETs [240–243].

5. Potential NET-Based Therapies
5.1. Novel Compounds Facilitating NET Prevention or Modulation
5.1.1. Conceptual

STC3141 (methyl β-cellobioside per-O-sulfate) is a small polyanion (SPA) that interacts electrostatically with histones, neutralizing their pathological effects preclinically in several pathologies [240]. This agent was developed specifically to target histones on NETs, to help preserve the host defense benefit of the protease-labelled chromatin web and facilitate microbe cytotoxicity in sepsis (histone neutralization with NET stabilization). Phase 1b results demonstrate favorable safety profiles and clinical benefits in ARDS. STC3141 is likely to interfere with the pro-tumorigenic functions of NETs, including tumor cell camouflage, migration and dormant cell reactivation, and may represent an effective small-molecule NET modulator. STC3141 may inhibit histone-dependent pathways, including TLR4/histone-dependent TME immunosuppression, histone-dependent endothelial and platelet activation and thrombosis, conferring survival and metastatic ability.

5.1.2. Preclinical

Sivelestat is an inhibitor of the NET-expressed serine protease neutrophil elastase, competitively inhibiting it with high specificity. NE plays a key role in NETosis and NET formation, and the pro-tumorigenic role of NE has been confirmed in breast, lung and colon cancers [244–246]. In keeping with the role of NETs in metastasis, Okamoto and colleagues demonstrated that sivelestat reduced NET formation and liver metastasis in a murine model of colorectal cancer (CRC) but had no effect on primary tumor growth or the suppression of liver metastasis if the CRC cells had already metastasized [247].

GSK383 chloramidine is a PAD4 inhibitor. PAD4 is a peptidyl arginine deiminase type IV enzyme, critical to the formation of NETs [248]. In 4T1 murine breast cancer cells, co-culture with GSK383 significantly attenuated NET production and inhibited NET-mediated tumor cell invasion.

5.1.3. Clinical Trials

Given the importance of the CXCR1/2:IL-8 axis in neutrophil/NET-mediated carcinogenesis, there are currently several CXCR1/2 inhibitors undergoing clinical trials in combination with ICIs. SX-682 (a CXCR1/2 antagonist) in combination with pembrolizumab entered Phase 1 trials in metastatic melanoma (NCT03161431), while a combination of avarixin (CXCR1/2 antagonist) and pembrolizumab is being trialed in advanced/metastatic solid tumors in a Phase 2 study (NCT03473925). CXCR1/2 inhibitors are not currently being trialed in HNC.

5.2. Repurposed Compounds Facilitating NET Prevention or Modulation
Preclinical/Clinical Cohort Studies

Aspirin is a COX-1 inhibitor commonly utilized as an antagonist to the primary hemostatic role of platelets. In a murine model of lipopolysaccharide-induced lung injury, aspirin reduced target tissue invasion by neutrophils and NET production. This is thought to be mediated by the amelioration of platelet-dependent release of CXCL4 (PF4) and CCL5 (RANTES), both of which increase neutrophil recruitment. Low-dose aspirin can have an anti-metastatic effect via a COX-1 inhibition-mediated reduction in NET production [249–251].

Metformin is a PKC inhibitor that attenuates NETosis. Previous studies identified that circulating NE, citrillinated histone, ds-DNA and proteinase-3 levels were reduced in the presence of metformin, and furthermore, metformin decreased expected NETosis in the presence of stimuli [252]. New evidence in hepatocellular carcinoma (HCC) and pancreatic cancer (PC) demonstrated that metformin attenuated NET production and reduced the metastatic potential of HCC and PC cells [253,254]. The role of metformin in attenuating NET-mediated carcinogenesis was corroborated in further murine models [255]. Interestingly, a recent presentation at AACR's Annual Meeting revealed that patients with type 2 diabetes mellitus suffering from colorectal cancer had significantly improved DFS in the presence of metformin, with tissue analysis identifying a significant reduction in tumor-associated NETs and a significant increase in CD8$^+$ T-cells. The authors concluded that metformin inhibits neutrophil infiltration and NET expression whilst promoting the infiltration of cytotoxic CD8$^+$ T-cells in the TME [256].

5.3. Compounds Facilitating NET Destruction
Preclinical

Unfractionated heparin (UH) is a glycosaminoglycan that potentiates the enzyme antithrombin III, inactivating thrombin, factor Xa and other proteases. It has a high affinity for extracellular histones and has been shown to promote the degradation of NETs [257]. UH is readily available and no longer under patent but has an off-target side effect in increasing the risk of bleeding.

Dornase alfa (rhDNase 1) is a recombinant human deoxyribonuclease, which selectively cleaves DNA. A co-culture of triple-negative breast cancer cells with neutrophils formed significant NETs, with DNase 1 blocking both NET-mediated cancer cell migration and invasion. In vivo, DNase 1 has demonstrated some ability to attenuate metastasis in a murine lung cancer model [258]. Park and colleagues concluded that this effect could be improved by increasing the half-life of DNase 1 and developing DNase-1-coated nanoparticles, which reduced the metastatic burden in a 4T1 breast cancer murine model [259]. Wang and colleagues also demonstrated that rhDNase 1 mitigated the pro-tumorigenic effects of NETs in murine pancreatic cancer [253]. rhDNase 1 is an FDA-approved therapy to degrade chromatin in cystic fibrosis and is currently undergoing a clinical trial to determine similar benefits in COVID-19-associated ARDs (NCT04402944) [260].

6. Conclusions

While tumor development may be controlled by cytotoxic adaptive and innate immune cells, the challenge to durable therapy lies in understanding the complex interaction between the immune cells, the tumor and its microenvironment and delivering therapies that re-program immunomodulation toward tumorlysis. In this article, we explored the strengths and limitations of current ICI therapies and outlined emerging SMI targets and modalities including our growing understanding of the important role NETs play in immunomodulation and their promise in ameliorating tumor progression and metastasis.

Author Contributions: Conceptualization, C.H.O. and L.M.K.; writing—original draft preparation, C.H.O.; writing—review and editing, C.H.O., Z.J. and L.M.K.; figure preparation, Z.J. All authors have read and agreed to the published version of the manuscript.

Funding: This research received no external funding.

Institutional Review Board Statement: Not applicable.

Informed Consent Statement: Not applicable.

Data Availability Statement: Not applicable.

Acknowledgments: The figures were created using BioRender under a licence acquired through the School of Biomedical Sciences, UNSW. CHO was supported by a Prince of Wales Hospital Foundation Grant to complete this review.

Conflicts of Interest: There are no conflict of interest.

References

1. Tam, S.; Yao, C.M.K.L.; Amit, M.; Gajera, M.; Luo, X.; Treistman, R.; Khanna, A.; Aashiq, M.; Nagarajan, P.; Bell, D.; et al. Association of Immunosuppression with Outcomes of Patients with Cutaneous Squamous Cell Carcinoma of the Head and Neck. *JAMA Otolaryngol. Head Neck Surg.* **2020**, *146*, 128–135. [CrossRef] [PubMed]
2. Szturz, P.; Vermorken, J.B. Treatment of Elderly Patients with Squamous Cell Carcinoma of the Head and Neck. *Front. Oncol.* **2016**, *6*, 199. [CrossRef] [PubMed]
3. Pritchett, E.N.; Doyle, A.; Shaver, C.M.; Miller, B.; Abdelmalek, M.; Cusack, C.A.; Malat, G.E.; Chung, C.L. Nonmelanoma Skin Cancer in Nonwhite Organ Transplant Recipients. *JAMA Dermatol.* **2016**, *152*, 1348–1353. [CrossRef] [PubMed]
4. Gilbert, M.; Liang, E.; Li, P.; Salgia, R.; Abouljoud, M.; Siddiqui, F. Outcomes of Primary Mucosal Head and Neck Squamous Cell Carcinoma in Solid Organ Transplant Recipients. *Cureus* **2022**, *14*, e24305. [CrossRef]
5. Mowery, A.J.; Conlin, M.J.; Clayburgh, D.R. Elevated incidence of head and neck cancer in solid organ transplant recipients. *Head Neck* **2019**, *41*, 4009–4017. [CrossRef]
6. Mowery, A.; Conlin, M.; Clayburgh, D. Risk of Head and Neck Cancer in Patients with Prior Hematologic Malignant Tumors. *JAMA Otolaryngol. Head Neck Surg.* **2019**, *145*, 1121–1127. [CrossRef]
7. Wang, X.; Wang, H.; Zhang, T.; Cai, L.; Dai, E.; He, J. Diabetes and its Potential Impact on Head and Neck Oncogenesis. *J. Cancer* **2020**, *11*, 583–591. [CrossRef]
8. Zhong, W.; Mao, Y. Daily Insulin Dose and Cancer Risk among Patients with Type 1 Diabetes. *JAMA Oncol.* **2022**, *8*, 1356–1358. [CrossRef]
9. Batista, N.V.R.; Valdez, R.M.A.; Silva, E.; Melo, T.S.; Pereira, J.R.D.; Warnakulasuriya, S.; Santos-Silva, A.R.; Duarte, A.; Mariz, H.A.; Gueiros, L.A. Association between autoimmune rheumatic diseases and head and neck cancer: Systematic review and meta-analysis. *J. Oral Pathol. Med.* **2022**, *52*, 357–364. [CrossRef]
10. Li, C.M.; Chen, Z. Autoimmunity as an Etiological Factor of Cancer: The Transformative Potential of Chronic Type 2 Inflammation. *Front. Cell Dev. Biol.* **2021**, *9*, 664305. [CrossRef]
11. D'souza, G.; Carey, T.E.; William, W.N., Jr.; Nguyen, M.L.; Ko, E.C.; Riddell, J., IV; Pai, S.I.; Gupta, V.; Walline, H.M.; Lee, J.J.; et al. Epidemiology of head and neck squamous cell cancer among HIV-infected patients. *J. Acquir. Immune Defic. Syndr.* **2014**, *65*, 603–610. [CrossRef] [PubMed]
12. Dunn, G.P.; Old, L.J.; Schreiber, R.D. The three Es of cancer immunoediting. *Annu. Rev. Immunol.* **2004**, *22*, 329–360. [CrossRef] [PubMed]
13. Dorsey, K.; Agulnik, M. Promising new molecular targeted therapies in head and neck cancer. *Drugs* **2013**, *73*, 315–325. [CrossRef] [PubMed]
14. Torre, L.A.; Bray, F.; Siegel, R.L.; Ferlay, J.; Lortet-Tieulent, J.; Jemal, A. Global cancer statistics, 2012. *CA Cancer J. Clin.* **2015**, *65*, 87–108. [CrossRef]
15. Haddad, R.I.; Shin, D.M. Recent advances in head and neck cancer. *N. Engl. J. Med.* **2008**, *359*, 1143–1154. [CrossRef]
16. Qiao, X.W.; Jiang, J.; Pang, X.; Huang, M.C.; Tang, Y.J.; Liang, X.H.; Tang, Y.L. The Evolving Landscape of PD-1/PD-L1 Pathway in Head and Neck Cancer. *Front. Immunol.* **2020**, *11*, 1721. [CrossRef]
17. Leemans, C.R.; Braakhuis, B.J.; Brakenhoff, R.H. The molecular biology of head and neck cancer. *Nat. Rev. Cancer* **2011**, *11*, 9–22. [CrossRef]
18. Chow, L.Q.M. Head and Neck Cancer. *N. Engl. J. Med.* **2020**, *382*, 60–72. [CrossRef]
19. Chi, A.C.; Day, T.A.; Neville, B.W. Oral cavity and oropharyngeal squamous cell carcinoma–An update. *CA Cancer J. Clin.* **2015**, *65*, 401–421. [CrossRef]
20. Mehra, R.; Cohen, R.B.; Burtness, B.A. The role of cetuximab for the treatment of squamous cell carcinoma of the head and neck. *Clin. Adv. Hematol. Oncol.* **2008**, *6*, 742–750.
21. Sundvall, M.; Karrila, A.; Nordberg, J.; Grénman, R.; Elenius, K. EGFR targeting drugs in the treatment of head and neck squamous cell carcinoma. *Expert Opin. Emerg. Drugs* **2010**, *15*, 185–201. [CrossRef] [PubMed]
22. Sacco, A.G.; Worden, F.P. Molecularly targeted therapy for the treatment of head and neck cancer: A review of the ErbB family inhibitors. *Onco Targets Ther.* **2016**, *9*, 1927–1943. [CrossRef] [PubMed]

23. Addeo, R.; Ghiani, M.; Merlino, F.; Ricciardiello, F.; Caraglia, M. CheckMate 141 trial: All that glitters is not gold. *Expert Opin. Biol. Ther.* **2019**, *19*, 169–171. [CrossRef] [PubMed]
24. Ferris, R.L.; Blumenschein, G., Jr.; Fayette, J.; Guigay, J.; Colevas, A.D.; Licitra, L.; Harrington, K.; Kasper, S.; Vokes, E.E.; Even, C.; et al. Nivolumab for Recurrent Squamous-Cell Carcinoma of the Head and Neck. *N. Engl. J. Med.* **2016**, *375*, 1856–1867. [CrossRef]
25. Seiwert, T.Y.; Burtness, B.; Mehra, R.; Weiss, J.; Berger, R.; Eder, J.P.; Heath, K.; McClanahan, T.; Lunceford, J.; Gause, C.; et al. Safety and clinical activity of pembrolizumab for treatment of recurrent or metastatic squamous cell carcinoma of the head and neck (KEYNOTE-012): An open-label, multicentre, phase 1b trial. *Lancet Oncol.* **2016**, *17*, 956–965. [CrossRef]
26. Cohen, E.E.W.; Soulières, D.; Le Tourneau, C.; Dinis, J.; Licitra, L.; Ahn, M.J.; Soria, A.; Machiels, J.P.; Mach, N.; Mehra, R.; et al. Pembrolizumab versus methotrexate, docetaxel, or cetuximab for recurrent or metastatic head-and-neck squamous cell carcinoma (KEYNOTE-040): A randomised, open-label, phase 3 study. *Lancet* **2019**, *393*, 156–167. [CrossRef]
27. Burtness, B.; Harrington, K.J.; Greil, R.; Soulières, D.; Tahara, M.; de Castro, G., Jr.; Psyrri, A.; Basté, N.; Neupane, P.; Bratland, Å.; et al. Pembrolizumab alone or with chemotherapy versus cetuximab with chemotherapy for recurrent or metastatic squamous cell carcinoma of the head and neck (KEYNOTE-048): A randomised, open-label, phase 3 study. *Lancet* **2019**, *394*, 1915–1928. [CrossRef]
28. Ribas, A.; Wolchok, J.D. Cancer immunotherapy using checkpoint blockade. *Science* **2018**, *359*, 1350–1355. [CrossRef]
29. Zandberg, D.P.; Algazi, A.P.; Jimeno, A.; Good, J.S.; Fayette, J.; Bouganim, N.; Ready, N.E.; Clement, P.M.; Even, C.; Jang, R.W.; et al. Durvalumab for recurrent or metastatic head and neck squamous cell carcinoma: Results from a single-arm, phase II study in patients with ≥25% tumour cell PD-L1 expression who have progressed on platinum-based chemotherapy. *Eur. J. Cancer* **2019**, *107*, 142–152. [CrossRef]
30. Jamieson, L.; Forster, M.D.; Zaki, K.; Mithra, S.; Alli, H.; O'Connor, A.; Patel, A.; Wong, I.C.K.; Chambers, P. Immunotherapy and associated immune-related adverse events at a large UK centre: A mixed methods study. *BMC Cancer* **2020**, *20*, 743. [CrossRef]
31. Wang, D.Y.; Salem, J.E.; Cohen, J.V.; Chandra, S.; Menzer, C.; Ye, F.; Zhao, S.; Das, S.; Beckermann, K.E.; Ha, L.; et al. Fatal Toxic Effects Associated with Immune Checkpoint Inhibitors: A Systematic Review and Meta-analysis. *JAMA Oncol.* **2018**, *4*, 1721–1728. [CrossRef] [PubMed]
32. Kessler, R.; Pandruvada, S. Immune-related adverse events following checkpoint inhibitor treatment in head and neck cancers: A comprehensive review. *Oral Oncol. Rep.* **2023**, *6*, 100036. [CrossRef]
33. Wu, X.; Gu, Z.; Chen, Y.; Chen, B.; Chen, W.; Weng, L.; Liu, X. Application of PD-1 blockade in cancer immunotherapy. *Comput. Struct. Biotechnol. J.* **2019**, *17*, 661–674. [CrossRef] [PubMed]
34. Park, R.; Lopes, L.; Saeed, A. Anti-PD-1/L1-associated immune-related adverse events as harbinger of favorable clinical outcome: Systematic review and meta-analysis. *Clin. Transl. Oncol.* **2021**, *23*, 100–109. [CrossRef] [PubMed]
35. Griewing, L.M.; Schweizer, C.; Schubert, P.; Rutzner, S.; Eckstein, M.; Frey, B.; Haderlein, M.; Weissmann, T.; Semrau, S.; Gostian, A.-O. Questionnaire-based detection of immune-related adverse events in cancer patients treated with PD-1/PD-L1 immune checkpoint inhibitors. *BMC Cancer* **2021**, *21*, 314. [CrossRef]
36. Schweizer, C.; Schubert, P.; Rutzner, S.; Eckstein, M.; Haderlein, M.; Lettmaier, S.; Semrau, S.; Gostian, A.-O.; Frey, B.; Gaipl, U.S. Prospective evaluation of the prognostic value of immune-related adverse events in patients with non-melanoma solid tumour treated with PD-1/PD-L1 inhibitors alone and in combination with radiotherapy. *Eur. J. Cancer* **2020**, *140*, 55–62. [CrossRef]
37. Matsuki, T.; Okamoto, I.; Fushimi, C.; Takahashi, H.; Okada, T.; Kondo, T.; Sato, H.; Ito, T.; Tokashiki, K.; Tsukahara, K. Real-world, long-term outcomes of nivolumab therapy for recurrent or metastatic squamous cell carcinoma of the head and neck and impact of the magnitude of best overall response: A retrospective multicenter study of 88 patients. *Cancers* **2020**, *12*, 3427. [CrossRef]
38. Les, I.; Martínez, M.; Pérez-Francisco, I.; Cabero, M.; Teijeira, L.; Arrazubi, V.; Torrego, N.; Campillo-Calatayud, A.; Elejalde, I.; Kochan, G.; et al. Predictive Biomarkers for Checkpoint Inhibitor Immune-Related Adverse Events. *Cancers* **2023**, *15*, 1629. [CrossRef]
39. O'Donnell, J.S.; Teng, M.W.L.; Smyth, M.J. Cancer immunoediting and resistance to T cell-based immunotherapy. *Nat. Rev. Clin. Oncol.* **2019**, *16*, 151–167. [CrossRef]
40. Masteller, E.L.; Chuang, E.; Mullen, A.C.; Reiner, S.L.; Thompson, C.B. Structural analysis of CTLA-4 function in vivo. *J. Immunol.* **2000**, *164*, 5319–5327. [CrossRef]
41. Buchbinder, E.I.; Desai, A. CTLA-4 and PD-1 Pathways: Similarities, Differences, and Implications of Their Inhibition. *Am. J. Clin. Oncol.* **2016**, *39*, 98–106. [CrossRef] [PubMed]
42. Qiao, Y.; Liu, C.; Zhang, X.; Zhou, Q.; Li, Y.; Xu, Y.; Gao, Z.; Xu, Y.; Kong, L.; Yang, A.; et al. PD-L2 based immune signature confers poor prognosis in HNSCC. *Oncoimmunology* **2021**, *10*, 1947569. [CrossRef] [PubMed]
43. Wherry, E.J. T cell exhaustion. *Nat. Immunol.* **2011**, *12*, 492–499. [CrossRef] [PubMed]
44. Hino, R.; Kabashima, K.; Kato, Y.; Yagi, H.; Nakamura, M.; Honjo, T.; Okazaki, T.; Tokura, Y. Tumor cell expression of programmed cell death-1 ligand 1 is a prognostic factor for malignant melanoma. *Cancer* **2010**, *116*, 1757–1766. [CrossRef]
45. Rasmussen, M.; Durhuus, J.A.; Nilbert, M.; Andersen, O.; Therkildsen, C. Response to Immune Checkpoint Inhibitors Is Affected by Deregulations in the Antigen Presentation Machinery: A Systematic Review and Meta-Analysis. *J. Clin. Med.* **2023**, *12*, 329. [CrossRef]
46. De Sousa Linhares, A.; Leitner, J.; Grabmeier-Pfistershammer, K.; Steinberger, P. Not All Immune Checkpoints Are Created Equal. *Front. Immunol.* **2018**, *9*, 1909. [CrossRef]

47. Haibe, Y.; El Husseini, Z.; El Sayed, R.; Shamseddine, A. Resisting Resistance to Immune Checkpoint Therapy: A Systematic Review. *Int. J. Mol. Sci.* **2020**, *21*, 6176. [CrossRef]
48. Mouw, K.W.; Goldberg, M.S.; Konstantinopoulos, P.A.; D'Andrea, A.D. DNA Damage and Repair Biomarkers of Immunotherapy Response. *Cancer Discov.* **2017**, *7*, 675–693. [CrossRef]
49. García-Aranda, M.; Redondo, M. Targeting Protein Kinases to Enhance the Response to anti-PD-1/PD-L1 Immunotherapy. *Int. J. Mol. Sci.* **2019**, *20*, 2296. [CrossRef]
50. Ohue, Y.; Nishikawa, H. Regulatory T (Treg) cells in cancer: Can Treg cells be a new therapeutic target? *Cancer Sci.* **2019**, *110*, 2080–2089. [CrossRef]
51. Yan, M.; Zheng, M.; Niu, R.; Yang, X.; Tian, S.; Fan, L.; Li, Y.; Zhang, S. Roles of tumor-associated neutrophils in tumor metastasis and its clinical applications. *Front. Cell Dev. Biol.* **2022**, *10*, 938289. [CrossRef] [PubMed]
52. Xiao, M.; He, J.; Yin, L.; Chen, X.; Zu, X.; Shen, Y. Tumor-Associated Macrophages: Critical Players in Drug Resistance of Breast Cancer. *Front. Immunol.* **2021**, *12*, 799428. [CrossRef] [PubMed]
53. Ma, T.; Renz, B.W.; Ilmer, M.; Koch, D.; Yang, Y.; Werner, J.; Bazhin, A.V. Myeloid-Derived Suppressor Cells in Solid Tumors. *Cells* **2022**, *11*, 310. [CrossRef] [PubMed]
54. Martin-Orozco, E.; Sanchez-Fernandez, A.; Ortiz-Parra, I.; Ayala-San Nicolas, M. WNT Signaling in Tumors: The Way to Evade Drugs and Immunity. *Front. Immunol.* **2019**, *10*, 2854. [CrossRef]
55. Berraondo, P.; Sanmamed, M.F.; Ochoa, M.C.; Etxeberria, I.; Aznar, M.A.; Pérez-Gracia, J.L.; Rodríguez-Ruiz, M.E.; Ponz-Sarvise, M.; Castañón, E.; Melero, I. Cytokines in clinical cancer immunotherapy. *Br. J. Cancer* **2019**, *120*, 6–15. [CrossRef]
56. Zhong, W.; Wang, Q.; Shen, X.; Du, J. The emerging role of neutrophil extracellular traps in cancer: From lab to ward. *Front. Oncol.* **2023**, *13*, 1163802. [CrossRef]
57. Hodi, F.S.; O'Day, S.J.; McDermott, D.F.; Weber, R.W.; Sosman, J.A.; Haanen, J.B.; Gonzalez, R.; Robert, C.; Schadendorf, D.; Hassel, J.C.; et al. Improved Survival with Ipilimumab in Patients with Metastatic Melanoma. *N. Engl. J. Med.* **2010**, *363*, 711–723. [CrossRef]
58. Mehra, R.; Seiwert, T.Y.; Gupta, S.; Weiss, J.; Gluck, I.; Eder, J.P.; Burtness, B.; Tahara, M.; Keam, B.; Kang, H.; et al. Efficacy and safety of pembrolizumab in recurrent/metastatic head and neck squamous cell carcinoma: Pooled analyses after long-term follow-up in KEYNOTE-012. *Br. J. Cancer* **2018**, *119*, 153–159. [CrossRef]
59. Noji, R.; Tohyama, K.; Kugimoto, T.; Kuroshima, T.; Hirai, H.; Tomioka, H.; Michi, Y.; Tasaki, A.; Ohno, K.; Ariizumi, Y.; et al. Comprehensive Genomic Profiling Reveals Clinical Associations in Response to Immune Therapy in Head and Neck Cancer. *Cancers* **2022**, *14*, 3476. [CrossRef]
60. Ayers, M.; Lunceford, J.; Nebozhyn, M.; Murphy, E.; Loboda, A.; Kaufman, D.R.; Albright, A.; Cheng, J.D.; Kang, S.P.; Shankaran, V.; et al. IFN-γ-related mRNA profile predicts clinical response to PD-1 blockade. *J. Clin. Investig.* **2017**, *127*, 2930–2940. [CrossRef]
61. Jing, F.; Wang, J.; Zhou, L.; Ning, Y.; Xu, S.; Zhu, Y. Bioinformatics analysis of the role of CXC ligands in the microenvironment of head and neck tumor. *Aging* **2021**, *13*, 17789–17817. [CrossRef] [PubMed]
62. de Ruiter, E.J.; Ooft, M.L.; Devriese, L.A.; Willems, S.M. The prognostic role of tumor infiltrating T-lymphocytes in squamous cell carcinoma of the head and neck: A systematic review and meta-analysis. *Oncoimmunology* **2017**, *6*, e1356148. [CrossRef] [PubMed]
63. Xu, Q.; Wang, C.; Yuan, X.; Feng, Z.; Han, Z. Prognostic Value of Tumor-Infiltrating Lymphocytes for Patients with Head and Neck Squamous Cell Carcinoma. *Transl. Oncol.* **2017**, *10*, 10–16. [CrossRef] [PubMed]
64. Daud, A.I.; Loo, K.; Pauli, M.L.; Sanchez-Rodriguez, R.; Sandoval, P.M.; Taravati, K.; Tsai, K.; Nosrati, A.; Nardo, L.; Alvarado, M.D.; et al. Tumor immune profiling predicts response to anti-PD-1 therapy in human melanoma. *J. Clin. Investig.* **2016**, *126*, 3447–3452. [CrossRef]
65. Kao, H.F.; Liao, B.C.; Huang, Y.L.; Huang, H.C.; Chen, C.N.; Chen, T.C.; Hong, Y.J.; Chan, C.Y.; Chia, J.S.; Hong, R.L. Afatinib and Pembrolizumab for Recurrent or Metastatic Head and Neck Squamous Cell Carcinoma (ALPHA Study): A Phase II Study with Biomarker Analysis. *Clin. Cancer Res.* **2022**, *28*, 1560–1571. [CrossRef] [PubMed]
66. Mascarella, M.A.; Mannard, E.; Silva, S.D.; Zeitouni, A. Neutrophil-to-lymphocyte ratio in head and neck cancer prognosis: A systematic review and meta-analysis. *Head Neck* **2018**, *40*, 1091–1100. [CrossRef]
67. Zhang, L.; Yu, D. Exosomes in cancer development, metastasis, and immunity. *Biochim. Biophys. Acta Rev. Cancer* **2019**, *1871*, 455–468. [CrossRef]
68. Chen, Y.; Li, Z.Y.; Zhou, G.Q.; Sun, Y. An Immune-Related Gene Prognostic Index for Head and Neck Squamous Cell Carcinoma. *Clin. Cancer Res.* **2021**, *27*, 330–341. [CrossRef]
69. Litchfield, K.; Reading, J.L.; Puttick, C.; Thakkar, K.; Abbosh, C.; Bentham, R.; Watkins, T.B.K.; Rosenthal, R.; Biswas, D.; Rowan, A.; et al. Meta-analysis of tumor- and T cell-intrinsic mechanisms of sensitization to checkpoint inhibition. *Cell* **2021**, *184*, 596–614.e14. [CrossRef]
70. Argiris, A.; Harrington, K.; Tahara, M.; Ferris, R.; Gillison, M.; Fayette, J.; Daste, A.; Koralewski, P.; Nin, R.M.; Saba, N. LBA36 Nivolumab (N)+ ipilimumab (I) vs EXTREME as first-line (1L) treatment (tx) for recurrent/metastatic squamous cell carcinoma of the head and neck (R/M SCCHN): Final results of CheckMate 651. *Ann. Oncol.* **2021**, *32*, S1310–S1311. [CrossRef]
71. Argiris, A.; Gillison, M.; Ferris, R.; Harrington, K.; Sanchez, T.; Baudelet, C.; Geese, W.; Shaw, J.; Haddad, R. A randomized, open-label, phase 3 study of nivolumab in combination with ipilimumab vs extreme regimen (cetuximab + cisplatin/carboplatin + fluorouracil) as first-line therapy in patients with recurrent or metastatic squamous cell carcinoma of the head and neck-CheckMate 651. *Ann. Oncol.* **2016**, *27*, vi350.

72. Seiwert, T.Y.; Weiss, J.; Baxi, S.S.; Ahn, M.-J.; Fayette, J.; Gillison, M.L.; Machiels, J.-P.H.; Takahashi, S.; Melillo, G.; Franks, A. A phase 3, randomized, open-label study of first-line durvalumab (MEDI4736) ± tremelimumab versus standard of care (SoC; EXTREME regimen) in recurrent/metastatic (R/M) SCCHN: KESTREL. *J. Clin. Oncol.* **2016**, *34*, TPS6101. [CrossRef]
73. Harrington, K.J.; Ferris, R.L.; Gillison, M.; Tahara, M.; Argiris, A.; Fayette, J.; Schenker, M.; Bratland, Å.; Walker, J.W.T.; Grell, P.; et al. Efficacy and Safety of Nivolumab Plus Ipilimumab vs Nivolumab Alone for Treatment of Recurrent or Metastatic Squamous Cell Carcinoma of the Head and Neck: The Phase 2 CheckMate 714 Randomized Clinical Trial. *JAMA Oncol.* **2023**, *9*, 779–789. [CrossRef] [PubMed]
74. Ferris, R.L.; Haddad, R.; Even, C.; Tahara, M.; Dvorkin, M.; Ciuleanu, T.E.; Clement, P.M.; Mesia, R.; Kutukova, S.; Zholudeva, L.; et al. Durvalumab with or without tremelimumab in patients with recurrent or metastatic head and neck squamous cell carcinoma: EAGLE, a randomized, open-label phase III study. *Ann. Oncol.* **2020**, *31*, 942–950. [CrossRef] [PubMed]
75. Licitra, L.F.; Haddad, R.I.; Even, C.; Tahara, M.; Dvorkin, M.; Ciuleanu, T.-E.; Clement, P.M.; Mesia, R.; Kutukova, S.I.; Zholudeva, L. EAGLE: A phase 3, randomized, open-label study of durvalumab (D) with or without tremelimumab (T) in patients (pts) with recurrent or metastatic head and neck squamous cell carcinoma (R/M HNSCC). *J. Clin. Oncol.* **2019**, *37*, 6012. [CrossRef]
76. Patel, S.P.; Othus, M.; Chen, Y.; Wright, G.P., Jr.; Yost, K.J.; Hyngstrom, J.R.; Hu-Lieskovan, S.; Lao, C.D.; Fecher, L.A.; Truong, T.G.; et al. Neoadjuvant-Adjuvant or Adjuvant-Only Pembrolizumab in Advanced Melanoma. *N. Engl. J. Med.* **2023**, *388*, 813–823. [CrossRef] [PubMed]
77. Chocarro, L.; Bocanegra, A.; Blanco, E.; Fernández-Rubio, L.; Arasanz, H.; Echaide, M.; Garnica, M.; Ramos, P.; Piñeiro-Hermida, S.; Vera, R.; et al. Cutting-Edge: Preclinical and Clinical Development of the First Approved Lag-3 Inhibitor. *Cells* **2022**, *11*, 2351. [CrossRef]
78. Albrecht, L.J.; Livingstone, E.; Zimmer, L.; Schadendorf, D. The Latest Option: Nivolumab and Relatlimab in Advanced Melanoma. *Curr. Oncol. Rep.* **2023**, *25*, 647–657. [CrossRef]
79. Liu, J.F.; Wu, L.; Yang, L.L.; Deng, W.W.; Mao, L.; Wu, H.; Zhang, W.F.; Sun, Z.J. Blockade of TIM3 relieves immunosuppression through reducing regulatory T cells in head and neck cancer. *J. Exp. Clin. Cancer Res.* **2018**, *37*, 44. [CrossRef]
80. Mandapathil, M.; Boduc, M.; Roessler, M.; Güldner, C.; Walliczek-Dworschak, U.; Mandic, R. Ectonucleotidase CD39 expression in regional metastases in head and neck cancer. *Acta Oto-Laryngol.* **2018**, *138*, 428–432. [CrossRef]
81. Cohen, E.E.; Hong, D.S.; Wise Draper, T.; Nassib William, W.; Schrijvers, D.; Mesia Nin, R.; Scott, M.L.; Lyne, P.; Mugundu, G.; McCoon, P.; et al. Phase 1b/2 Study (SCORES) assessing safety, tolerability, and preliminary anti-tumor activity of durvalumab plus AZD9150 or AZD5069 in patients with advanced solid malignancies and squamous cell carcinoma of the head and neck (SCCHN). *Ann. Oncol.* **2017**, *28*, V403. [CrossRef]
82. Harrington, K.J.; Brody, J.; Ingham, M.; Strauss, J.; Cemerski, S.; Wang, M.; Tse, A.; Khilnani, A.; Marabelle, A.; Golan, T. Preliminary results of the first-in-human (FIH) study of MK-1454, an agonist of stimulator of interferon genes (STING), as monotherapy or in combination with pembrolizumab (pembro) in patients with advanced solid tumours of lymphomas. *Ann. Oncol.* **2018**, *29*, viii712. [CrossRef]
83. Yarchoan, M.; Powderly, J.D.; Bastos, B.R.; Karasic, T.B.; Crysler, O.V.; Munster, P.N.; McKean, M.; Emens, L.A.; Saenger, Y.M.; Ged, Y.; et al. A phase 1 study of TPST-1120 as a single agent and in combination with nivolumab in subjects with advanced solid tumors. *J. Clin. Oncol.* **2022**, *40*, 3005. [CrossRef]
84. Gui, L.; He, X.; Yang, J.; Liu, P.; Qin, Y.; Shi, Y.K. 230MO Pembrolizumab plus anlotinib as first-line treatment in patients of CPS > 1 with recurrent or metastatic head and neck squamous-cell carcinoma: A prospective phase II study. *Ann. Oncol.* **2022**, *33*, S1524. [CrossRef]
85. Cohen, E.E.W.; Licitra, L.F.; Burtness, B.; Fayette, J.; Gauler, T.; Clement, P.M.; Grau, J.J.; Del Campo, J.M.; Mailliez, A.; Haddad, R.I.; et al. Biomarkers predict enhanced clinical outcomes with afatinib versus methotrexate in patients with second-line recurrent and/or metastatic head and neck cancer. *Ann. Oncol.* **2017**, *28*, 2526–2532. [CrossRef]
86. Cecchini, M.; Zugazagoitia, J.; Lopez, J.S.; Jäger, D.; Oliva, M.; Ochsenreither, S.; Gambardella, V.; Chung, K.Y.; Longo, F.; Razak, A.R.A.; et al. Initial results from a first-in-human, phase I study of immunomodulatory aryl hydrocarbon receptor (AhR) inhibitor BAY2416964 in patients with advanced solid tumors. *J. Clin. Oncol.* **2023**, *41*, 2502. [CrossRef]
87. Zhou, Y.; Chen, J.J. STAT3 plays an important role in DNA replication by turning on WDHD1. *Cell Biosci.* **2021**, *11*, 10. [CrossRef]
88. Xiong, A.; Yang, Z.; Shen, Y.; Zhou, J.; Shen, Q. Transcription Factor STAT3 as a Novel Molecular Target for Cancer Prevention. *Cancers* **2014**, *6*, 926–957. [CrossRef]
89. Greenhill, C.J.; Rose-John, S.; Lissilaa, R.; Ferlin, W.; Ernst, M.; Hertzog, P.J.; Mansell, A.; Jenkins, B.J. IL-6 trans-signaling modulates TLR4-dependent inflammatory responses via STAT3. *J. Immunol.* **2011**, *186*, 1199–1208. [CrossRef]
90. Jones, G.W.; McLoughlin, R.M.; Hammond, V.J.; Parker, C.R.; Williams, J.D.; Malhotra, R.; Scheller, J.; Williams, A.S.; Rose-John, S.; Topley, N.; et al. Loss of CD4+ T cell IL-6R expression during inflammation underlines a role for IL-6 trans signaling in the local maintenance of Th17 cells. *J. Immunol.* **2010**, *184*, 2130–2139. [CrossRef]
91. Bonner, J.A.; Yang, E.S.; Trummell, H.Q.; Nowsheen, S.; Willey, C.D.; Raisch, K.P. Inhibition of STAT-3 results in greater cetuximab sensitivity in head and neck squamous cell carcinoma. *Radiother. Oncol.* **2011**, *99*, 339–343. [CrossRef] [PubMed]
92. Loeuillard, E.; Yang, J.; Buckarma, E.; Wang, J.; Liu, Y.; Conboy, C.; Pavelko, K.D.; Li, Y.; O'Brien, D.; Wang, C.; et al. Targeting tumor-associated macrophages and granulocytic myeloid-derived suppressor cells augments PD-1 blockade in cholangiocarcinoma. *J. Clin. Investig.* **2020**, *130*, 5380–5396. [CrossRef] [PubMed]

93. Grauers Wiktorin, H.; Nilsson, M.S.; Kiffin, R.; Sander, F.E.; Lenox, B.; Rydström, A.; Hellstrand, K.; Martner, A. Histamine targets myeloid-derived suppressor cells and improves the anti-tumor efficacy of PD-1/PD-L1 checkpoint blockade. *Cancer Immunol. Immunother.* **2019**, *68*, 163–174. [CrossRef] [PubMed]
94. Yu, H.; Lee, H.; Herrmann, A.; Buettner, R.; Jove, R. Revisiting STAT3 signalling in cancer: New and unexpected biological functions. *Nat. Rev. Cancer* **2014**, *14*, 736–746. [CrossRef]
95. Radharani, N.N.V.; Yadav, A.S.; Nimma, R.; Kumar, T.V.S.; Bulbule, A.; Chanukuppa, V.; Kumar, D.; Patnaik, S.; Rapole, S.; Kundu, G.C. Tumor-associated macrophage derived IL-6 enriches cancer stem cell population and promotes breast tumor progression via Stat-3 pathway. *Cancer Cell Int.* **2022**, *22*, 122. [CrossRef]
96. Fisher, D.T.; Appenheimer, M.M.; Evans, S.S. The two faces of IL-6 in the tumor microenvironment. *Semin. Immunol.* **2014**, *26*, 38–47. [CrossRef]
97. Eldesoky, A.; Shouma, A.; Mosaad, Y.; Elhawary, A. Clinical relevance of serum vascular endothelial growth factor and interleukin-6 in patients with colorectal cancer. *Saudi J. Gastroenterol.* **2011**, *17*, 170–173. [CrossRef]
98. Riegel, K.; Yurugi, H.; Schlöder, J.; Jonuleit, H.; Kaulich, M.; Kirschner, F.; Arnold-Schild, D.; Tenzer, S.; Schild, H.; Rajalingam, K. ERK5 modulates IL-6 secretion and contributes to tumor-induced immune suppression. *Cell Death Dis.* **2021**, *12*, 969. [CrossRef]
99. Chaudhry, A.; Rudra, D.; Treuting, P.; Samstein, R.M.; Liang, Y.; Kas, A.; Rudensky, A.Y. CD4+ regulatory T cells control TH17 responses in a Stat3-dependent manner. *Science* **2009**, *326*, 986–991. [CrossRef]
100. Wang, T.; Niu, G.; Kortylewski, M.; Burdelya, L.; Shain, K.; Zhang, S.; Bhattacharya, R.; Gabrilovich, D.; Heller, R.; Coppola, D.; et al. Regulation of the innate and adaptive immune responses by Stat-3 signaling in tumor cells. *Nat. Med.* **2004**, *10*, 48–54. [CrossRef]
101. Iwata-Kajihara, T.; Sumimoto, H.; Kawamura, N.; Ueda, R.; Takahashi, T.; Mizuguchi, H.; Miyagishi, M.; Takeda, K.; Kawakami, Y. Enhanced cancer immunotherapy using STAT3-depleted dendritic cells with high Th1-inducing ability and resistance to cancer cell-derived inhibitory factors. *J. Immunol.* **2011**, *187*, 27–36. [CrossRef] [PubMed]
102. Herrmann, A.; Kortylewski, M.; Kujawski, M.; Zhang, C.; Reckamp, K.; Armstrong, B.; Wang, L.; Kowolik, C.; Deng, J.; Figlin, R.; et al. Targeting Stat3 in the myeloid compartment drastically improves the in vivo antitumor functions of adoptively transferred T cells. *Cancer Res.* **2010**, *70*, 7455–7464. [CrossRef] [PubMed]
103. Yu, H.; Pardoll, D.; Jove, R. STATs in cancer inflammation and immunity: A leading role for STAT3. *Nat. Rev. Cancer* **2009**, *9*, 798–809. [CrossRef] [PubMed]
104. Kopf, M.; Baumann, H.; Freer, G.; Freudenberg, M.; Lamers, M.; Kishimoto, T.; Zinkernagel, R.; Bluethmann, H.; Köhler, G. Impaired immune and acute-phase responses in interleukin-6-deficient mice. *Nature* **1994**, *368*, 339–342. [CrossRef]
105. Zhao, C.; Zeng, N.; Zhou, X.; Tan, Y.; Wang, Y.; Zhang, J.; Wu, Y.; Zhang, Q. CAA-derived IL-6 induced M2 macrophage polarization by activating STAT3. *BMC Cancer* **2023**, *23*, 392. [CrossRef]
106. Zou, S.; Tong, Q.; Liu, B.; Huang, W.; Tian, Y.; Fu, X. Targeting STAT3 in Cancer Immunotherapy. *Mol. Cancer* **2020**, *19*, 145. [CrossRef]
107. Valle-Mendiola, A.; Soto-Cruz, I. Energy Metabolism in Cancer: The Roles of STAT3 and STAT5 in the Regulation of Metabolism-Related Genes. *Cancers* **2020**, *12*, 124. [CrossRef]
108. Masuda, M.; Ruan, H.Y.; Ito, A.; Nakashima, T.; Toh, S.; Wakasaki, T.; Yasumatsu, R.; Kutratomi, Y.; Komune, S.; Weinstein, I.B. Signal transducers and activators of transcription 3 up-regulates vascular endothelial growth factor production and tumor angiogenesis in head and neck squamous cell carcinoma. *Oral Oncol.* **2007**, *43*, 785–790. [CrossRef]
109. Grandis, J.R.; Drenning, S.D.; Zeng, Q.; Watkins, S.C.; Melhem, M.F.; Endo, S.; Johnson, D.E.; Huang, L.; He, Y.; Kim, J.D. Constitutive activation of Stat3 signaling abrogates apoptosis in squamous cell carcinogenesis in vivo. *Proc. Natl. Acad. Sci. USA* **2000**, *97*, 4227–4232. [CrossRef]
110. Zhao, Y.; Zhang, J.; Xia, H.; Zhang, B.; Jiang, T.; Wang, J.; Chen, X.; Wang, Y. Stat3 is involved in the motility, metastasis and prognosis in lingual squamous cell carcinoma. *Cell Biochem. Funct.* **2012**, *30*, 340–346. [CrossRef]
111. Vermorken, J.B.; Mesia, R.; Rivera, F.; Remenar, E.; Kawecki, A.; Rottey, S.; Erfan, J.; Zabolotnyy, D.; Kienzer, H.R.; Cupissol, D.; et al. Platinum-based chemotherapy plus cetuximab in head and neck cancer. *N. Engl. J. Med.* **2008**, *359*, 1116–1127. [CrossRef] [PubMed]
112. Li, X.; Wang, H.; Lu, X.; Di, B. STAT3 blockade with shRNA enhances radiosensitivity in Hep-2 human laryngeal squamous carcinoma cells. *Oncol. Rep.* **2010**, *23*, 345–353. [CrossRef] [PubMed]
113. Zhao, C.; Li, H.; Lin, H.J.; Yang, S.; Lin, J.; Liang, G. Feedback Activation of STAT3 as a Cancer Drug Resistance Mechanism. *Trends Pharmacol. Sci.* **2016**, *37*, 47–61. [CrossRef] [PubMed]
114. Lin, S.; Yang, L.; Yao, Y.; Xu, L.; Xiang, Y.; Zhao, H.; Wang, L.; Zuo, Z.; Huang, X.; Zhao, C. Flubendazole demonstrates valid antitumor effects by inhibiting STAT3 and activating autophagy. *J. Exp. Clin. Cancer Res.* **2019**, *38*, 293. [CrossRef] [PubMed]
115. Spagnuolo, P.A.; Hu, J.; Hurren, R.; Wang, X.; Gronda, M.; Sukhai, M.A.; Di Meo, A.; Boss, J.; Ashali, I.; Beheshti Zavareh, R.; et al. The antihelmintic flubendazole inhibits microtubule function through a mechanism distinct from Vinca alkaloids and displays preclinical activity in leukemia and myeloma. *Blood* **2010**, *115*, 4824–4833. [CrossRef]
116. Čáňová, K.; Rozkydalová, L.; Vokurková, D.; Rudolf, E. Flubendazole induces mitotic catastrophe and apoptosis in melanoma cells. *Toxicol. Vitr.* **2018**, *46*, 313–322. [CrossRef]
117. Tao, J.; Zhao, H.; Xie, X.; Luo, M.; Gao, Z.; Sun, H.; Huang, Z. The anthelmintic drug flubendazole induces cell apoptosis and inhibits NF-κB signaling in esophageal squamous cell carcinoma. *OncoTargets Ther.* **2019**, *12*, 471–478. [CrossRef]

118. Michaelis, M.; Agha, B.; Rothweiler, F.; Löschmann, N.; Voges, Y.; Mittelbronn, M.; Starzetz, T.; Harter, P.N.; Abhari, B.A.; Fulda, S.; et al. Identification of flubendazole as potential anti-neuroblastoma compound in a large cell line screen. *Sci. Rep.* **2015**, *5*, 8202. [CrossRef]
119. Hanušová, V.; Skálová, L.; Králová, V.; Matoušková, P. The Effect of Flubendazole on Adhesion and Migration in SW480 and SW620 Colon Cancer Cells. *Anticancer Agents Med. Chem.* **2018**, *18*, 837–846. [CrossRef]
120. Králová, V.; Hanušová, V.; Rudolf, E.; Čáňová, K.; Skálová, L. Flubendazole induces mitotic catastrophe and senescence in colon cancer cells in vitro. *J. Pharm. Pharmacol.* **2016**, *68*, 208–218. [CrossRef]
121. Raisová Stuchlíková, L.; Králová, V.; Lněničková, K.; Zárybnický, T.; Matoušková, P.; Hanušová, V.; Ambrož, M.; Šubrt, Z.; Skálová, L. The metabolism of flubendazole in human liver and cancer cell lines. *Drug Test. Anal.* **2018**, *10*, 1139–1146. [CrossRef] [PubMed]
122. Kim, Y.J.; Sung, D.; Oh, E.; Cho, Y.; Cho, T.M.; Farrand, L.; Seo, J.H.; Kim, J.Y. Flubendazole overcomes trastuzumab resistance by targeting cancer stem-like properties and HER2 signaling in HER2-positive breast cancer. *Cancer Lett.* **2018**, *412*, 118–130. [CrossRef] [PubMed]
123. Oh, E.; Kim, Y.J.; An, H.; Sung, D.; Cho, T.M.; Farrand, L.; Jang, S.; Seo, J.H.; Kim, J.Y. Flubendazole elicits anti-metastatic effects in triple-negative breast cancer via STAT3 inhibition. *Int. J. Cancer* **2018**, *143*, 1978–1993. [CrossRef]
124. Hou, Z.J.; Luo, X.; Zhang, W.; Peng, F.; Cui, B.; Wu, S.J.; Zheng, F.M.; Xu, J.; Xu, L.Z.; Long, Z.J.; et al. Flubendazole, FDA-approved anthelmintic, targets breast cancer stem-like cells. *Oncotarget* **2015**, *6*, 6326–6340. [CrossRef] [PubMed]
125. Li, Y.; Acharya, G.; Elahy, M.; Xin, H.; Khachigian, L.M. The anthelmintic flubendazole blocks human melanoma growth and metastasis and suppresses programmed cell death protein-1 and myeloid-derived suppressor cell accumulation. *Cancer Lett.* **2019**, *459*, 268–276. [CrossRef] [PubMed]
126. Li, Y.; Wu, B.; Hossain, M.J.; Quagliata, L.; O'Meara, C.; Wilkins, M.R.; Corley, S.; Khachigian, L.M. Flubendazole inhibits PD-1 and suppresses melanoma growth in immunocompetent mice. *J. Transl. Med.* **2023**, *21*, 467. Available online: https://translational-medicine.biomedcentral.com/articles/10.1186/s12967-023-04289-y (accessed on 16 July 2023). [CrossRef]
127. Pillaiyar, T.; Meenakshisundaram, S.; Manickam, M.; Sankaranarayanan, M. A medicinal chemistry perspective of drug repositioning: Recent advances and challenges in drug discovery. *Eur. J. Med. Chem.* **2020**, *195*, 112275. [CrossRef]
128. Tutrone, W.D.; Saini, R.; Caglar, S.; Weinberg, J.M.; Crespo, J. Topical therapy for actinic keratoses, I: 5-Fluorouracil and imiquimod. *Cutis* **2003**, *71*, 365–370.
129. Hamouda, B.; Jamila, Z.; Najet, R.; Slim, T.; Rafiaa, N.; Noureddine, B.; Ahmed, E.M.; Mohamed, F.; Ridha, K.M.; Abderrahman, L. Topical 5-fluorouracil to treat multiple or unresectable facial squamous cell carcinomas in xeroderma pigmentosum. *J. Am. Acad. Dermatol.* **2001**, *44*, 1054. [CrossRef]
130. van Ruth, S.; Jansman, F.G.; Sanders, C.J. Total body topical 5-fluorouracil for extensive non-melanoma skin cancer. *Pharm. World Sci.* **2006**, *28*, 159–162. [CrossRef]
131. Nishina, T.; Fujita, T.; Yoshizuka, N.; Sugibayashi, K.; Murayama, K.; Kuboki, Y. Safety, tolerability, pharmacokinetics and preliminary antitumour activity of an antisense oligonucleotide targeting STAT3 (danvatirsen) as monotherapy and in combination with durvalumab in Japanese patients with advanced solid malignancies: A phase 1 study. *BMJ Open* **2022**, *12*, e055718. [CrossRef] [PubMed]
132. Dhanwani, R.; Takahashi, M.; Sharma, S. Cytosolic sensing of immuno-stimulatory DNA, the enemy within. *Curr. Opin. Immunol.* **2018**, *50*, 82–87. [CrossRef] [PubMed]
133. Zhou, S.; Zhang, Y.; Wang, Y.; Zhang, M.; Sun, W.; Dai, T.; Wang, A.; Wu, X.; Zhang, S.; Wang, S.; et al. A Dual Role of Type I Interferons in Antitumor Immunity. *Adv. Biosyst.* **2020**, *4*, e1900237. [CrossRef] [PubMed]
134. Zitvogel, L.; Galluzzi, L.; Kepp, O.; Smyth, M.J.; Kroemer, G. Type I interferons in anticancer immunity. *Nat. Rev. Immunol.* **2015**, *15*, 405–414. [CrossRef] [PubMed]
135. Flood, B.A.; Higgs, E.F.; Li, S.; Luke, J.J.; Gajewski, T.F. STING pathway agonism as a cancer therapeutic. *Immunol. Rev.* **2019**, *290*, 24–38. [CrossRef]
136. Kinkead, H.L.; Hopkins, A.; Lutz, E.; Wu, A.A.; Yarchoan, M.; Cruz, K.; Woolman, S.; Vithayathil, T.; Glickman, L.H.; Ndubaku, C.O.; et al. Combining STING-based neoantigen-targeted vaccine with checkpoint modulators enhances antitumor immunity in murine pancreatic cancer. *JCI Insight* **2018**, *3*, e122857. [CrossRef]
137. Ager, C.R.; Reilley, M.J.; Nicholas, C.; Bartkowiak, T.; Jaiswal, A.R.; Curran, M.A. Intratumoral STING Activation with T-cell Checkpoint Modulation Generates Systemic Antitumor Immunity. *Cancer Immunol. Res.* **2017**, *5*, 676–684. [CrossRef]
138. Kostadinova, R.; Wahli, W.; Michalik, L. PPARs in diseases: Control mechanisms of inflammation. *Curr. Med. Chem.* **2005**, *12*, 2995–3009. [CrossRef]
139. Decara, J.; Rivera, P.; López-Gambero, A.J.; Serrano, A.; Pavón, F.J.; Baixeras, E.; Rodríguez de Fonseca, F.; Suárez, J. Peroxisome Proliferator-Activated Receptors: Experimental Targeting for the Treatment of Inflammatory Bowel Diseases. *Front. Pharmacol.* **2020**, *11*, 730. [CrossRef]
140. Yousefnia, S.; Momenzadeh, S.; Seyed Forootan, F.; Ghaedi, K.; Nasr Esfahani, M.H. The influence of peroxisome proliferator-activated receptor γ (PPARγ) ligands on cancer cell tumorigenicity. *Gene* **2018**, *649*, 14–22. [CrossRef]
141. Han, L.; Shen, W.J.; Bittner, S.; Kraemer, F.B.; Azhar, S. PPARs: Regulators of metabolism and as therapeutic targets in cardiovascular disease. Part II: PPAR-β/δ and PPAR-γ. *Future Cardiol.* **2017**, *13*, 279–296. [CrossRef] [PubMed]
142. Lemmon, M.A.; Schlessinger, J. Cell signaling by receptor tyrosine kinases. *Cell* **2010**, *141*, 1117–1134. [CrossRef] [PubMed]

143. Du, Z.; Lovly, C.M. Mechanisms of receptor tyrosine kinase activation in cancer. *Mol. Cancer* **2018**, *17*, 58. [CrossRef] [PubMed]
144. Peng, L.; Sun, W.; Chen, L.; Wen, W.P. The Role of Interleukin-33 in Head and Neck Squamous Cell Carcinoma Is Determined by Its Cellular Sources in the Tumor Microenvironment. *Front. Oncol.* **2020**, *10*, 588454. [CrossRef]
145. Minutti, C.M.; Drube, S.; Blair, N.; Schwartz, C.; McCrae, J.C.; McKenzie, A.N.; Kamradt, T.; Mokry, M.; Coffer, P.J.; Sibilia, M.; et al. Epidermal Growth Factor Receptor Expression Licenses Type-2 Helper T Cells to Function in a T Cell Receptor-Independent Fashion. *Immunity* **2017**, *47*, 710–722.e6. [CrossRef]
146. Wang, Z.; Longo, P.A.; Tarrant, M.K.; Kim, K.; Head, S.; Leahy, D.J.; Cole, P.A. Mechanistic insights into the activation of oncogenic forms of EGF receptor. *Nat. Struct. Mol. Biol.* **2011**, *18*, 1388–1393. [CrossRef]
147. Comprehensive genomic characterization of head and neck squamous cell carcinomas. *Nature* **2015**, *517*, 576–582. [CrossRef]
148. Johnson, D.E.; Burtness, B.; Leemans, C.R.; Lui, V.W.Y.; Bauman, J.E.; Grandis, J.R. Head and neck squamous cell carcinoma. *Nat. Rev. Dis. Primers* **2020**, *6*, 92. [CrossRef]
149. Byeon, H.K.; Ku, M.; Yang, J. Beyond EGFR inhibition: Multilateral combat strategies to stop the progression of head and neck cancer. *Exp. Mol. Med.* **2019**, *51*, 1–14. [CrossRef]
150. Vatte, C.; Al Amri, A.M.; Cyrus, C.; Chathoth, S.; Acharya, S.; Hashim, T.M.; Al Ali, Z.; Alshreadah, S.T.; Alsayyah, A.; Al-Ali, A.K. Tyrosine kinase domain mutations of EGFR gene in head and neck squamous cell carcinoma. *OncoTargets Ther.* **2017**, *10*, 1527–1533. [CrossRef]
151. Stransky, N.; Cerami, E.; Schalm, S.; Kim, J.L.; Lengauer, C. The landscape of kinase fusions in cancer. *Nat. Commun.* **2014**, *5*, 4846. [CrossRef] [PubMed]
152. Soulieres, D.; Senzer, N.N.; Vokes, E.E.; Hidalgo, M.; Agarwala, S.S.; Siu, L.L. Multicenter phase II study of erlotinib, an oral epidermal growth factor receptor tyrosine kinase inhibitor, in patients with recurrent or metastatic squamous cell cancer of the head and neck. *J. Clin. Oncol.* **2004**, *22*, 77–85. [CrossRef] [PubMed]
153. Stewart, J.S.; Cohen, E.E.; Licitra, L.; Van Herpen, C.M.; Khorprasert, C.; Soulieres, D.; Vodvarka, P.; Rischin, D.; Garin, A.M.; Hirsch, F.R.; et al. Phase III study of gefitinib compared with intravenous methotrexate for recurrent squamous cell carcinoma of the head and neck [corrected]. *J. Clin. Oncol.* **2009**, *27*, 1864–1871. [CrossRef] [PubMed]
154. Vermorken, J.B.; Trigo, J.; Hitt, R.; Koralewski, P.; Diaz-Rubio, E.; Rolland, F.; Knecht, R.; Amellal, N.; Schueler, A.; Baselga, J. Open-label, uncontrolled, multicenter phase II study to evaluate the efficacy and toxicity of cetuximab as a single agent in patients with recurrent and/or metastatic squamous cell carcinoma of the head and neck who failed to respond to platinum-based therapy. *J. Clin. Oncol.* **2007**, *25*, 2171–2177. [CrossRef]
155. Deng, Z.; Liao, W.; Wei, W.; Zhong, G.; He, C.; Zhang, H.; Liu, Q.; Xu, X.; Liang, J.; Liu, Z. Anlotinib as a promising inhibitor on tumor growth of oral squamous cell carcinoma through cell apoptosis and mitotic catastrophe. *Cancer Cell Int.* **2021**, *21*, 37. [CrossRef]
156. Boule, L.A.; Burke, C.G.; Jin, G.B.; Lawrence, B.P. Aryl hydrocarbon receptor signaling modulates antiviral immune responses: Ligand metabolism rather than chemical source is the stronger predictor of outcome. *Sci. Rep.* **2018**, *8*, 1826. [CrossRef]
157. Gutiérrez-Vázquez, C.; Quintana, F.J. Regulation of the Immune Response by the Aryl Hydrocarbon Receptor. *Immunity* **2018**, *48*, 19–33. [CrossRef]
158. Wang, Z.; Snyder, M.; Kenison, J.E.; Yang, K.; Lara, B.; Lydell, E.; Bennani, K.; Novikov, O.; Federico, A.; Monti, S.; et al. How the AHR Became Important in Cancer: The Role of Chronically Active AHR in Cancer Aggression. *Int. J. Mol. Sci.* **2021**, *22*, 387. [CrossRef]
159. Dean, J.W.; Zhou, L. Cell-intrinsic view of the aryl hydrocarbon receptor in tumor immunity. *Trends Immunol.* **2022**, *43*, 245–258. [CrossRef]
160. Zeng, W.; Song, Y.; Wang, R.; He, R.; Wang, T. Neutrophil elastase: From mechanisms to therapeutic potential. *J. Pharm. Anal.* **2023**, *13*, 355–366. [CrossRef]
161. Fridlender, Z.G.; Sun, J.; Kim, S.; Kapoor, V.; Cheng, G.; Ling, L.; Worthen, G.S.; Albelda, S.M. Polarization of tumor-associated neutrophil phenotype by TGF-beta: "N1" versus "N2" TAN. *Cancer Cell* **2009**, *16*, 183–194. [CrossRef] [PubMed]
162. Swierczak, A.; Mouchemore, K.A.; Hamilton, J.A.; Anderson, R.L. Neutrophils: Important contributors to tumor progression and metastasis. *Cancer Metastasis Rev.* **2015**, *34*, 735–751. [CrossRef] [PubMed]
163. Mantovani, A. The yin-yang of tumor-associated neutrophils. *Cancer Cell* **2009**, *16*, 173–174. [CrossRef] [PubMed]
164. Brinkmann, V.; Reichard, U.; Goosmann, C.; Fauler, B.; Uhlemann, Y.; Weiss, D.S.; Weinrauch, Y.; Zychlinsky, A. Neutrophil extracellular traps kill bacteria. *Science* **2004**, *303*, 1532–1535. [CrossRef]
165. Neumann, A.; Berends, E.T.; Nerlich, A.; Molhoek, E.M.; Gallo, R.L.; Meerloo, T.; Nizet, V.; Naim, H.Y.; von Köckritz-Blickwede, M. The antimicrobial peptide LL-37 facilitates the formation of neutrophil extracellular traps. *Biochem. J.* **2014**, *464*, 3–11. [CrossRef]
166. Mutua, V.; Gershwin, L.J. A Review of Neutrophil Extracellular Traps (NETs) in Disease: Potential Anti-NETs Therapeutics. *Clin. Rev. Allergy Immunol.* **2021**, *61*, 194–211. [CrossRef]
167. Delgado-Rizo, V.; Martínez-Guzmán, M.A.; Iñiguez-Gutierrez, L.; García-Orozco, A.; Alvarado-Navarro, A.; Fafutis-Morris, M. Neutrophil Extracellular Traps and Its Implications in Inflammation: An Overview. *Front. Immunol.* **2017**, *8*, 81. [CrossRef]
168. Saitoh, T.; Komano, J.; Saitoh, Y.; Misawa, T.; Takahama, M.; Kozaki, T.; Uehata, T.; Iwasaki, H.; Omori, H.; Yamaoka, S.; et al. Neutrophil extracellular traps mediate a host defense response to human immunodeficiency virus-1. *Cell Host Microbe* **2012**, *12*, 109–116. [CrossRef]

169. Muraro, S.P.; De Souza, G.F.; Gallo, S.W.; Da Silva, B.K.; De Oliveira, S.D.; Vinolo, M.A.R.; Saraiva, E.M.; Porto, B.N. Respiratory Syncytial Virus induces the classical ROS-dependent NETosis through PAD-4 and necroptosis pathways activation. *Sci. Rep.* **2018**, *8*, 14166. [CrossRef]
170. Narasaraju, T.; Yang, E.; Samy, R.P.; Ng, H.H.; Poh, W.P.; Liew, A.A.; Phoon, M.C.; van Rooijen, N.; Chow, V.T. Excessive neutrophils and neutrophil extracellular traps contribute to acute lung injury of influenza pneumonitis. *Am. J. Pathol.* **2011**, *179*, 199–210. [CrossRef]
171. Gwyer Findlay, E.; Currie, S.M.; Davidson, D.J. Cationic host defence peptides: Potential as antiviral therapeutics. *BioDrugs* **2013**, *27*, 479–493. [CrossRef] [PubMed]
172. Hoeksema, M.; Tripathi, S.; White, M.; Qi, L.; Taubenberger, J.; van Eijk, M.; Haagsman, H.; Hartshorn, K.L. Arginine-rich histones have strong antiviral activity for influenza A viruses. *Innate Immun.* **2015**, *21*, 736–745. [CrossRef] [PubMed]
173. Branzk, N.; Lubojemska, A.; Hardison, S.E.; Wang, Q.; Gutierrez, M.G.; Brown, G.D.; Papayannopoulos, V. Neutrophils sense microbe size and selectively release neutrophil extracellular traps in response to large pathogens. *Nat. Immunol.* **2014**, *15*, 1017–1025. [CrossRef] [PubMed]
174. Bruns, S.; Kniemeyer, O.; Hasenberg, M.; Aimanianda, V.; Nietzsche, S.; Thywissen, A.; Jeron, A.; Latgé, J.P.; Brakhage, A.A.; Gunzer, M. Production of extracellular traps against Aspergillus fumigatus in vitro and in infected lung tissue is dependent on invading neutrophils and influenced by hydrophobin RodA. *PLoS Pathog.* **2010**, *6*, e1000873. [CrossRef]
175. McCormick, A.; Heesemann, L.; Wagener, J.; Marcos, V.; Hartl, D.; Loeffler, J.; Heesemann, J.; Ebel, F. NETs formed by human neutrophils inhibit growth of the pathogenic mold Aspergillus fumigatus. *Microbes Infect.* **2010**, *12*, 928–936. [CrossRef]
176. Baker, V.S.; Imade, G.E.; Molta, N.B.; Tawde, P.; Pam, S.D.; Obadofin, M.O.; Sagay, S.A.; Egah, D.Z.; Iya, D.; Afolabi, B.B.; et al. Cytokine-associated neutrophil extracellular traps and antinuclear antibodies in Plasmodium falciparum infected children under six years of age. *Malar. J.* **2008**, *7*, 41. [CrossRef]
177. Abi Abdallah, D.S.; Lin, C.; Ball, C.J.; King, M.R.; Duhamel, G.E.; Denkers, E.Y. Toxoplasma gondii triggers release of human and mouse neutrophil extracellular traps. *Infect. Immun.* **2012**, *80*, 768–777. [CrossRef]
178. Guimarães-Costa, A.B.; Nascimento, M.T.; Froment, G.S.; Soares, R.P.; Morgado, F.N.; Conceição-Silva, F.; Saraiva, E.M. Leishmania amazonensis promastigotes induce and are killed by neutrophil extracellular traps. *Proc. Natl. Acad. Sci. USA* **2009**, *106*, 6748–6753. [CrossRef]
179. Díaz-Godínez, C.; Carrero, J.C. The state of art of neutrophil extracellular traps in protozoan and helminthic infections. *Biosci. Rep.* **2019**, *39*, BSR20180916. [CrossRef]
180. Ekaney, M.L.; Otto, G.P.; Sossdorf, M.; Sponholz, C.; Boehringer, M.; Loesche, W.; Rittirsch, D.; Wilharm, A.; Kurzai, O.; Bauer, M.; et al. Impact of plasma histones in human sepsis and their contribution to cellular injury and inflammation. *Crit. Care* **2014**, *18*, 543. [CrossRef]
181. Xu, J.; Zhang, X.; Pelayo, R.; Monestier, M.; Ammollo, C.T.; Semeraro, F.; Taylor, F.B.; Esmon, N.L.; Lupu, F.; Esmon, C.T. Extracellular histones are major mediators of death in sepsis. *Nat. Med.* **2009**, *15*, 1318–1321. [CrossRef] [PubMed]
182. Abrams, S.T.; Zhang, N.; Manson, J.; Liu, T.; Dart, C.; Baluwa, F.; Wang, S.S.; Brohi, K.; Kipar, A.; Yu, W.; et al. Circulating histones are mediators of trauma-associated lung injury. *Am. J. Respir. Crit. Care Med.* **2013**, *187*, 160–169. [CrossRef] [PubMed]
183. Scozzi, D.; Liao, F.; Krupnick, A.S.; Kreisel, D.; Gelman, A.E. The role of neutrophil extracellular traps in acute lung injury. *Front. Immunol.* **2022**, *13*, 953195. [CrossRef] [PubMed]
184. Zhang, F.; Li, Y.; Wu, J.; Zhang, J.; Cao, P.; Sun, Z.; Wang, W. The role of extracellular traps in ischemia reperfusion injury. *Front. Immunol.* **2022**, *13*, 1022380. [CrossRef]
185. Njeim, R.; Azar, W.S.; Fares, A.H.; Azar, S.T.; Kfoury Kassouf, H.; Eid, A.A. NETosis contributes to the pathogenesis of diabetes and its complications. *J. Mol. Endocrinol.* **2020**, *65*, R65–R76. [CrossRef]
186. Middleton, E.A.; He, X.Y.; Denorme, F.; Campbell, R.A.; Ng, D.; Salvatore, S.P.; Mostyka, M.; Baxter-Stoltzfus, A.; Borczuk, A.C.; Loda, M.; et al. Neutrophil extracellular traps contribute to immunothrombosis in COVID-19 acute respiratory distress syndrome. *Blood* **2020**, *136*, 1169–1179. [CrossRef]
187. Wang, C.Y.; Lin, T.T.; Hu, L.; Xu, C.J.; Hu, F.; Wan, L.; Yang, X.; Wu, X.F.; Zhang, X.T.; Li, Y.; et al. Neutrophil extracellular traps as a unique target in the treatment of chemotherapy-induced peripheral neuropathy. *eBioMedicine* **2023**, *90*, 104499. [CrossRef]
188. Masucci, M.T.; Minopoli, M.; Del Vecchio, S.; Carriero, M.V. The Emerging Role of Neutrophil Extracellular Traps (NETs) in Tumor Progression and Metastasis. *Front. Immunol.* **2020**, *11*, 1749. [CrossRef]
189. Podaza, E.; Sabbione, F.; Risnik, D.; Borge, M.; Almejún, M.B.; Colado, A.; Fernández-Grecco, H.; Cabrejo, M.; Bezares, R.F.; Trevani, A.; et al. Neutrophils from chronic lymphocytic leukemia patients exhibit an increased capacity to release extracellular traps (NETs). *Cancer Immunol. Immunother.* **2017**, *66*, 77–89. [CrossRef]
190. Li, B.; Liu, Y.; Hu, T.; Zhang, Y.; Zhang, C.; Li, T.; Wang, C.; Dong, Z.; Novakovic, V.A.; Hu, T.; et al. Neutrophil extracellular traps enhance procoagulant activity in patients with oral squamous cell carcinoma. *J. Cancer Res. Clin. Oncol.* **2019**, *145*, 1695–1707. [CrossRef]
191. Li, R.; Zou, X.; Zhu, T.; Xu, H.; Li, X.; Zhu, L. Destruction of Neutrophil Extracellular Traps Promotes the Apoptosis and Inhibits the Invasion of Gastric Cancer Cells by Regulating the Expression of Bcl-2, Bax and NF-κB. *OncoTargets Ther.* **2020**, *13*, 5271–5281. [CrossRef]

192. Li, M.; Lin, C.; Deng, H.; Strnad, J.; Bernabei, L.; Vogl, D.T.; Burke, J.J.; Nefedova, Y. A Novel Peptidylarginine Deiminase 4 (PAD4) Inhibitor BMS-P5 Blocks Formation of Neutrophil Extracellular Traps and Delays Progression of Multiple Myeloma. *Mol. Cancer Ther.* **2020**, *19*, 1530–1538. [CrossRef] [PubMed]
193. Cristinziano, L.; Modestino, L.; Loffredo, S.; Varricchi, G.; Braile, M.; Ferrara, A.L.; de Paulis, A.; Antonelli, A.; Marone, G.; Galdiero, M.R. Anaplastic Thyroid Cancer Cells Induce the Release of Mitochondrial Extracellular DNA Traps by Viable Neutrophils. *J. Immunol.* **2020**, *204*, 1362–1372. [CrossRef] [PubMed]
194. Wu, L.; Saxena, S.; Goel, P.; Prajapati, D.R.; Wang, C.; Singh, R.K. Breast Cancer Cell-Neutrophil Interactions Enhance Neutrophil Survival and Pro-Tumorigenic Activities. *Cancers* **2020**, *12*, 2884. [CrossRef] [PubMed]
195. Garley, M.; Jabłońska, E.; Miltyk, W.; Grubczak, K.; Surażyński, A.; Ratajczak-Wrona, W.; Grudzińska, M.; Nowacka, K.H.; Moniuszko, M.; Pałka, J.A.; et al. Cancers Cells in Traps? The Pathways of NETs Formation in Response to OSCC in Humans-A Pilot Study. *Cancer Control* **2020**, *27*, 1073274820960473. [CrossRef] [PubMed]
196. Cai, Z.; Zhang, M.; Boafo Kwantwi, L.; Bi, X.; Zhang, C.; Cheng, Z.; Ding, X.; Su, T.; Wang, H.; Wu, Q. Breast cancer cells promote self-migration by secreting interleukin 8 to induce NET formation. *Gene* **2020**, *754*, 144902. [CrossRef]
197. Thålin, C.; Lundström, S.; Seignez, C.; Daleskog, M.; Lundström, A.; Henriksson, P.; Helleday, T.; Phillipson, M.; Wallén, H.; Demers, M. Citrullinated histone H3 as a novel prognostic blood marker in patients with advanced cancer. *PLoS ONE* **2018**, *13*, e0191231. [CrossRef]
198. Guan, X.; Lu, Y.; Zhu, H.; Yu, S.; Zhao, W.; Chi, X.; Xie, C.; Yin, Z. The Crosstalk Between Cancer Cells and Neutrophils Enhances Hepatocellular Carcinoma Metastasis via Neutrophil Extracellular Traps-Associated Cathepsin G Component: A Potential Therapeutic Target. *J. Hepatocell. Carcinoma* **2021**, *8*, 451–465. [CrossRef]
199. Gomes, T.; Várady, C.B.S.; Lourenço, A.L.; Mizurini, D.M.; Rondon, A.M.R.; Leal, A.C.; Gonçalves, B.S.; Bou-Habib, D.C.; Medei, E.; Monteiro, R.Q. IL-1β Blockade Attenuates Thrombosis in a Neutrophil Extracellular Trap-Dependent Breast Cancer Model. *Front. Immunol.* **2019**, *10*, 2088. [CrossRef]
200. Guimarães-Bastos, D.; Frony, A.C.; Barja-Fidalgo, C.; Moraes, J.A. Melanoma-derived extracellular vesicles skew neutrophils into a pro-tumor phenotype. *J. Leukoc. Biol.* **2022**, *111*, 585–596. [CrossRef]
201. Abdol Razak, N.; Elaskalani, O.; Metharom, P. Pancreatic Cancer-Induced Neutrophil Extracellular Traps: A Potential Contributor to Cancer-Associated Thrombosis. *Int. J. Mol. Sci.* **2017**, *18*, 487. [CrossRef] [PubMed]
202. Zhang, Y.; Wang, C.; Yu, M.; Zhao, X.; Du, J.; Li, Y.; Jing, H.; Dong, Z.; Kou, J.; Bi, Y.; et al. Neutrophil extracellular traps induced by activated platelets contribute to procoagulant activity in patients with colorectal cancer. *Thromb. Res.* **2019**, *180*, 87–97. [CrossRef] [PubMed]
203. Wang, Z.; Yang, C.; Li, L.; Jin, X.; Zhang, Z.; Zheng, H.; Pan, J.; Shi, L.; Jiang, Z.; Su, K.; et al. Tumor-derived HMGB1 induces CD62L^{dim} neutrophil polarization and promotes lung metastasis in triple-negative breast cancer. *Oncogenesis* **2020**, *9*, 82. [CrossRef] [PubMed]
204. Mittal, D.; Saccheri, F.; Vénéreau, E.; Pusterla, T.; Bianchi, M.E.; Rescigno, M. TLR4-mediated skin carcinogenesis is dependent on immune and radioresistant cells. *Embo J.* **2010**, *29*, 2242–2252. [CrossRef] [PubMed]
205. Shinde-Jadhav, S.; Mansure, J.J.; Rayes, R.F.; Marcq, G.; Ayoub, M.; Skowronski, R.; Kool, R.; Bourdeau, F.; Brimo, F.; Spicer, J.; et al. Role of neutrophil extracellular traps in radiation resistance of invasive bladder cancer. *Nat. Commun.* **2021**, *12*, 2776. [CrossRef]
206. Li, J.; Xia, Y.; Sun, B.; Zheng, N.; Li, Y.; Pang, X.; Yang, F.; Zhao, X.; Ji, Z.; Yu, H.; et al. Neutrophil extracellular traps induced by the hypoxic microenvironment in gastric cancer augment tumour growth. *Cell Commun. Signal.* **2023**, *21*, 86. [CrossRef]
207. Shang, A.; Gu, C.; Zhou, C.; Yang, Y.; Chen, C.; Zeng, B.; Wu, J.; Lu, W.; Wang, W.; Sun, Z.; et al. Exosomal KRAS mutation promotes the formation of tumor-associated neutrophil extracellular traps and causes deterioration of colorectal cancer by inducing IL-8 expression. *Cell Commun. Signal.* **2020**, *18*, 52. [CrossRef]
208. Tohme, S.; Yazdani, H.O.; Al-Khafaji, A.B.; Chidi, A.P.; Loughran, P.; Mowen, K.; Wang, Y.; Simmons, R.L.; Huang, H.; Tsung, A. Neutrophil Extracellular Traps Promote the Development and Progression of Liver Metastases after Surgical Stress. *Cancer Res.* **2016**, *76*, 1367–1380. [CrossRef]
209. Zha, C.; Meng, X.; Li, L.; Mi, S.; Qian, D.; Li, Z.; Wu, P.; Hu, S.; Zhao, S.; Cai, J.; et al. Neutrophil extracellular traps mediate the crosstalk between glioma progression and the tumor microenvironment via the HMGB1/RAGE/IL-8 axis. *Cancer Biol. Med.* **2020**, *17*, 154–168. [CrossRef]
210. Houghton, A.M.; Rzymkiewicz, D.M.; Ji, H.; Gregory, A.D.; Egea, E.E.; Metz, H.E.; Stolz, D.B.; Land, S.R.; Marconcini, L.A.; Kliment, C.R.; et al. Neutrophil elastase-mediated degradation of IRS-1 accelerates lung tumor growth. *Nat. Med.* **2010**, *16*, 219–223. [CrossRef]
211. Teijeira, Á.; Garasa, S.; Gato, M.; Alfaro, C.; Migueliz, I.; Cirella, A.; de Andrea, C.; Ochoa, M.C.; Otano, I.; Etxeberria, I.; et al. CXCR1 and CXCR2 Chemokine Receptor Agonists Produced by Tumors Induce Neutrophil Extracellular Traps that Interfere with Immune Cytotoxicity. *Immunity* **2020**, *52*, 856–871.e8. [CrossRef] [PubMed]
212. Ireland, A.S.; Oliver, T.G. Neutrophils Create an ImpeNETrable Shield between Tumor and Cytotoxic Immune Cells. *Immunity* **2020**, *52*, 729–731. [CrossRef] [PubMed]
213. Kaltenmeier, C.; Yazdani, H.O.; Morder, K.; Geller, D.A.; Simmons, R.L.; Tohme, S. Neutrophil Extracellular Traps Promote T Cell Exhaustion in the Tumor Microenvironment. *Front. Immunol.* **2021**, *12*, 785222. [CrossRef] [PubMed]
214. Okeke, E.B.; Uzonna, J.E. The Pivotal Role of Regulatory T Cells in the Regulation of Innate Immune Cells. *Front. Immunol.* **2019**, *10*, 680. [CrossRef] [PubMed]

215. Wang, H.; Zhang, H.; Wang, Y.; Brown, Z.J.; Xia, Y.; Huang, Z.; Shen, C.; Hu, Z.; Beane, J.; Ansa-Addo, E.A.; et al. Regulatory T-cell and neutrophil extracellular trap interaction contributes to carcinogenesis in non-alcoholic steatohepatitis. *J. Hepatol.* **2021**, *75*, 1271–1283. [CrossRef] [PubMed]
216. Velliou, R.I.; Mitroulis, I.; Chatzigeorgiou, A. Neutrophil extracellular traps contribute to the development of hepatocellular carcinoma in NASH by promoting Treg differentiation. *Hepatobiliary Surg. Nutr.* **2022**, *11*, 415–418. [CrossRef]
217. Kimball, A.S.; Obi, A.T.; Diaz, J.A.; Henke, P.K. The Emerging Role of NETs in Venous Thrombosis and Immunothrombosis. *Front. Immunol.* **2016**, *7*, 236. [CrossRef]
218. Diaz, J.A.; Fuchs, T.A.; Jackson, T.O.; Kremer Hovinga, J.A.; Lämmle, B.; Henke, P.K.; Myers, D.D., Jr.; Wagner, D.D.; Wakefield, T.W. Plasma DNA is Elevated in Patients with Deep Vein Thrombosis. *J. Vasc. Surg. Venous Lymphat. Disord.* **2013**, *1*, 341–348.e1. [CrossRef]
219. Fuchs, T.A.; Brill, A.; Duerschmied, D.; Schatzberg, D.; Monestier, M.; Myers, D.D., Jr.; Wrobleski, S.K.; Wakefield, T.W.; Hartwig, J.H.; Wagner, D.D. Extracellular DNA traps promote thrombosis. *Proc. Natl. Acad. Sci. USA* **2010**, *107*, 15880–15885. [CrossRef]
220. Brill, A.; Fuchs, T.A.; Savchenko, A.S.; Thomas, G.M.; Martinod, K.; De Meyer, S.F.; Bhandari, A.A.; Wagner, D.D. Neutrophil extracellular traps promote deep vein thrombosis in mice. *J. Thromb. Haemost.* **2012**, *10*, 136–144. [CrossRef]
221. Martinod, K.; Wagner, D.D. Thrombosis: Tangled up in NETs. *Blood* **2014**, *123*, 2768–2776. [CrossRef] [PubMed]
222. Wang, S.; Li, Z.; Xu, R. Human Cancer and Platelet Interaction, a Potential Therapeutic Target. *Int. J. Mol. Sci.* **2018**, *19*, 1246. [CrossRef] [PubMed]
223. Waldmann, T.A. Cytokines in Cancer Immunotherapy. *Cold Spring Harb. Perspect. Biol.* **2018**, *10*, a028472. [CrossRef] [PubMed]
224. Placke, T.; Örgel, M.; Schaller, M.; Jung, G.; Rammensee, H.G.; Kopp, H.G.; Salih, H.R. Platelet-derived MHC class I confers a pseudonormal phenotype to cancer cells that subverts the antitumor reactivity of natural killer immune cells. *Cancer Res.* **2012**, *72*, 440–448. [CrossRef] [PubMed]
225. Batu, E.D. Neutrophil-mediated Thrombosis and NETosis in Behçet's Disease: A Hypothesis. *J. Korean Med. Sci.* **2020**, *35*, e213. [CrossRef] [PubMed]
226. Kordbacheh, F.; O'Meara, C.H.; Coupland, L.A.; Lelliott, P.M.; Parish, C.R. Extracellular histones induce erythrocyte fragility and anemia. *Blood* **2017**, *130*, 2884–2888. [CrossRef] [PubMed]
227. Zhang, H.; Wang, Y.; Qu, M.; Li, W.; Wu, D.; Cata, J.P.; Miao, C. Neutrophil, neutrophil extracellular traps and endothelial cell dysfunction in sepsis. *Clin. Transl. Med.* **2023**, *13*, e1170. [CrossRef]
228. Yang, X.; Li, L.; Liu, J.; Lv, B.; Chen, F. Extracellular histones induce tissue factor expression in vascular endothelial cells via TLR and activation of NF-κB and AP-1. *Thromb. Res.* **2016**, *137*, 211–218. [CrossRef]
229. Saffarzadeh, M.; Juenemann, C.; Queisser, M.A.; Lochnit, G.; Barreto, G.; Galuska, S.P.; Lohmeyer, J.; Preissner, K.T. Neutrophil extracellular traps directly induce epithelial and endothelial cell death: A predominant role of histones. *PLoS ONE* **2012**, *7*, e32366. [CrossRef]
230. Folco, E.J.; Mawson, T.L.; Vromman, A.; Bernardes-Souza, B.; Franck, G.; Persson, O.; Nakamura, M.; Newton, G.; Luscinskas, F.W.; Libby, P. Neutrophil Extracellular Traps Induce Endothelial Cell Activation and Tissue Factor Production Through Interleukin-1α and Cathepsin G. *Arterioscler. Thromb. Vasc. Biol.* **2018**, *38*, 1901–1912. [CrossRef]
231. Cools-Lartigue, J.; Spicer, J.; Najmeh, S.; Ferri, L. Neutrophil extracellular traps in cancer progression. *Cell. Mol. Life Sci.* **2014**, *71*, 4179–4194. [CrossRef] [PubMed]
232. Yang, L.; Liu, Q.; Zhang, X.; Liu, X.; Zhou, B.; Chen, J.; Huang, D.; Li, J.; Li, H.; Chen, F.; et al. DNA of neutrophil extracellular traps promotes cancer metastasis via CCDC25. *Nature* **2020**, *583*, 133–138. [CrossRef] [PubMed]
233. Dongre, A.; Weinberg, R.A. New insights into the mechanisms of epithelial-mesenchymal transition and implications for cancer. *Nat. Rev. Mol. Cell Biol.* **2019**, *20*, 69–84. [CrossRef]
234. Martins-Cardoso, K.; Almeida, V.H.; Bagri, K.M.; Rossi, M.I.D.; Mermelstein, C.S.; König, S.; Monteiro, R.Q. Neutrophil Extracellular Traps (NETs) Promote Pro-Metastatic Phenotype in Human Breast Cancer Cells through Epithelial-Mesenchymal Transition. *Cancers* **2020**, *12*, 1542. [CrossRef]
235. Zhu, T.; Zou, X.; Yang, C.; Li, L.; Wang, B.; Li, R.; Li, H.; Xu, Z.; Huang, D.; Wu, Q. Neutrophil extracellular traps promote gastric cancer metastasis by inducing epithelial-mesenchymal transition. *Int. J. Mol. Med.* **2021**, *48*, 127. [CrossRef] [PubMed]
236. Albrengues, J.; Shields, M.A.; Ng, D.; Park, C.G.; Ambrico, A.; Poindexter, M.E.; Upadhyay, P.; Uyeminami, D.L.; Pommier, A.; Küttner, V.; et al. Neutrophil extracellular traps produced during inflammation awaken dormant cancer cells in mice. *Science* **2018**, *361*, eaao4227. [CrossRef] [PubMed]
237. Jung, H.S.; Gu, J.; Kim, J.E.; Nam, Y.; Song, J.W.; Kim, H.K. Cancer cell-induced neutrophil extracellular traps promote both hypercoagulability and cancer progression. *PLoS ONE* **2019**, *14*, e0216055. [CrossRef]
238. Aldabbous, L.; Abdul-Salam, V.; McKinnon, T.; Duluc, L.; Pepke-Zaba, J.; Southwood, M.; Ainscough, A.J.; Hadinnapola, C.; Wilkins, M.R.; Toshner, M.; et al. Neutrophil Extracellular Traps Promote Angiogenesis: Evidence From Vascular Pathology in Pulmonary Hypertension. *Arterioscler. Thromb. Vasc. Biol.* **2016**, *36*, 2078–2087. [CrossRef]
239. Yuan, K.; Zheng, J.; Huang, X.; Zhang, Y.; Han, Y.; Hu, R.; Jin, X. Neutrophil extracellular traps promote corneal neovascularization-induced by alkali burn. *Int. Immunopharmacol.* **2020**, *88*, 106902. [CrossRef]
240. Meara, C.H.O.; Coupland, L.A.; Kordbacheh, F.; Quah, B.J.C.; Chang, C.W.; Simon Davis, D.A.; Bezos, A.; Browne, A.M.; Freeman, C.; Hammill, D.J.; et al. Neutralizing the pathological effects of extracellular histones with small polyanions. *Nat. Commun.* **2020**, *11*, 6408. [CrossRef]

241. Volkov, D.V.; Tetz, G.V.; Rubtsov, Y.P.; Stepanov, A.V.; Gabibov, A.G. Neutrophil Extracellular Traps (NETs): Opportunities for Targeted Therapy. *Acta Naturae* **2021**, *13*, 15–23. [CrossRef] [PubMed]
242. Clark, S.R.; Ma, A.C.; Tavener, S.A.; McDonald, B.; Goodarzi, Z.; Kelly, M.M.; Patel, K.D.; Chakrabarti, S.; McAvoy, E.; Sinclair, G.D.; et al. Platelet TLR4 activates neutrophil extracellular traps to ensnare bacteria in septic blood. *Nat. Med.* **2007**, *13*, 463–469. [CrossRef] [PubMed]
243. Tilgner, J.; von Trotha, K.T.; Gombert, A.; Jacobs, M.J.; Drechsler, M.; Döring, Y.; Soehnlein, O.; Grommes, J. Aspirin, but Not Tirofiban Displays Protective Effects in Endotoxin Induced Lung Injury. *PLoS ONE* **2016**, *11*, e0161218. [CrossRef]
244. Vaguliene, N.; Zemaitis, M.; Lavinskiene, S.; Miliauskas, S.; Sakalauskas, R. Local and systemic neutrophilic inflammation in patients with lung cancer and chronic obstructive pulmonary disease. *BMC Immunol.* **2013**, *14*, 36. [CrossRef]
245. Kistowski, M.; Dębski, J.; Karczmarski, J.; Paziewska, A.; Olędzki, J.; Mikula, M.; Ostrowski, J.; Dadlez, M. A Strong Neutrophil Elastase Proteolytic Fingerprint Marks the Carcinoma Tumor Proteome. *Mol. Cell. Proteom.* **2017**, *16*, 213–227. [CrossRef] [PubMed]
246. Akizuki, M.; Fukutomi, T.; Takasugi, M.; Takahashi, S.; Sato, T.; Harao, M.; Mizumoto, T.; Yamashita, J. Prognostic significance of immunoreactive neutrophil elastase in human breast cancer: Long-term follow-up results in 313 patients. *Neoplasia* **2007**, *9*, 260–264. [CrossRef]
247. Okamoto, M.; Mizuno, R.; Kawada, K.; Itatani, Y.; Kiyasu, Y.; Hanada, K.; Hirata, W.; Nishikawa, Y.; Masui, H.; Sugimoto, N.; et al. Neutrophil Extracellular Traps Promote Metastases of Colorectal Cancers through Activation of ERK Signaling by Releasing Neutrophil Elastase. *Int. J. Mol. Sci.* **2023**, *24*, 1118. [CrossRef]
248. Hemmers, S.; Teijaro, J.R.; Arandjelovic, S.; Mowen, K.A. PAD4-mediated neutrophil extracellular trap formation is not required for immunity against influenza infection. *PLoS ONE* **2011**, *6*, e22043. [CrossRef]
249. Bruno, A.; Dovizio, M.; Tacconelli, S.; Patrignani, P. Mechanisms of the antitumoural effects of aspirin in the gastrointestinal tract. *Best Pract. Res. Clin. Gastroenterol.* **2012**, *26*, e1–e13. [CrossRef]
250. Guillem-Llobat, P.; Dovizio, M.; Bruno, A.; Ricciotti, E.; Cufino, V.; Sacco, A.; Grande, R.; Alberti, S.; Arena, V.; Cirillo, M.; et al. Aspirin prevents colorectal cancer metastasis in mice by splitting the crosstalk between platelets and tumor cells. *Oncotarget* **2016**, *7*, 32462–32477. [CrossRef]
251. Patrignani, P.; Patrono, C. Aspirin and Cancer. *J. Am. Coll. Cardiol.* **2016**, *68*, 967–976. [CrossRef] [PubMed]
252. Menegazzo, L.; Scattolini, V.; Cappellari, R.; Bonora, B.M.; Albiero, M.; Bortolozzi, M.; Romanato, F.; Ceolotto, G.; Vigili de Kreutzeberg, S.; Avogaro, A.; et al. The antidiabetic drug metformin blunts NETosis in vitro and reduces circulating NETosis biomarkers in vivo. *Acta Diabetol.* **2018**, *55*, 593–601. [CrossRef]
253. Wang, G.; Gao, H.; Dai, S.; Li, M.; Gao, Y.; Yin, L.; Zhang, K.; Zhang, J.; Jiang, K.; Miao, Y.; et al. Metformin inhibits neutrophil extracellular traps-promoted pancreatic carcinogenesis in obese mice. *Cancer Lett.* **2023**, *562*, 216155. [CrossRef] [PubMed]
254. Yang, L.Y.; Shen, X.T.; Sun, H.T.; Zhu, W.W.; Zhang, J.B.; Lu, L. Neutrophil extracellular traps in hepatocellular carcinoma are enriched in oxidized mitochondrial DNA which is highly pro-inflammatory and pro-metastatic. *J. Cancer* **2022**, *13*, 1261–1271. [CrossRef]
255. Chen, C.-J.; Wu, C.-C.; Chang, C.-Y.; Li, J.-R.; Ou, Y.-C.; Chen, W.-Y.; Liao, S.-L.; Wang, J.-D. Metformin Mitigated Obesity-Driven Cancer Aggressiveness in Tumor-Bearing Mice. *Int. J. Mol. Sci.* **2022**, *23*, 9134. [CrossRef] [PubMed]
256. Saito, A.; Ohzawa, H.; Kaneko, Y.; Tamura, K.; Futoh, Y.; Takahashi, K.; Kimura, Y.; Kawashima, R.; Miyato, H.; Kitayama, J.; et al. Abstract 1268: The effect of metformin for neutrophil extracellular traps (NETs) on tumor immune microenvironment of colorectal cancer with type 2 diabetes mellitus. *Cancer Res.* **2023**, *83*, 1268. [CrossRef]
257. Leppkes, M.; Knopf, J.; Naschberger, E.; Lindemann, A.; Singh, J.; Herrmann, I.; Stürzl, M.; Staats, L.; Mahajan, A.; Schauer, C.; et al. Vascular occlusion by neutrophil extracellular traps in COVID-19. *eBioMedicine* **2020**, *58*, 102925. [CrossRef]
258. Cools-Lartigue, J.; Spicer, J.; McDonald, B.; Gowing, S.; Chow, S.; Giannias, B.; Bourdeau, F.; Kubes, P.; Ferri, L. Neutrophil extracellular traps sequester circulating tumor cells and promote metastasis. *J. Clin. Investig.* **2013**, *123*, 3446–3458. [CrossRef]
259. Park, J.; Wysocki, R.W.; Amoozgar, Z.; Maiorino, L.; Fein, M.R.; Jorns, J.; Schott, A.F.; Kinugasa-Katayama, Y.; Lee, Y.; Won, N.H.; et al. Cancer cells induce metastasis-supporting neutrophil extracellular DNA traps. *Sci. Transl. Med.* **2016**, *8*, 361ra138. [CrossRef]
260. Law, S.M.; Gray, R.D. Neutrophil extracellular traps and the dysfunctional innate immune response of cystic fibrosis lung disease: A review. *J. Inflamm.* **2017**, *14*, 29. [CrossRef]

Disclaimer/Publisher's Note: The statements, opinions and data contained in all publications are solely those of the individual author(s) and contributor(s) and not of MDPI and/or the editor(s). MDPI and/or the editor(s) disclaim responsibility for any injury to people or property resulting from any ideas, methods, instructions or products referred to in the content.

Review

Upgrading Treatment and Molecular Diagnosis in Endometrial Cancer—Driving New Tools for Endometrial Preservation?

Miriam Dellino [1,*], Marco Cerbone [1], Antonio Simone Laganà [2], Amerigo Vitagliano [1], Antonella Vimercati [1], Marco Marinaccio [1], Giorgio Maria Baldini [3], Antonio Malvasi [1], Ettore Cicinelli [1], Gianluca Raffaello Damiani [1], Gerardo Cazzato [4,†] and Eliano Cascardi [5,6,†]

1. Obstetrics and Gynaecology Unit, Department of Biomedical Sciences and Human Oncology, University of Bari "Aldo Moro", 70124 Bari, Italy
2. Unit of Gynecologic Oncology, ARNAS "Civico—Di Cristina—Benfratelli", Department of Health Promotion, Mother and Child Care, Internal Medicine and Medical Specialties (PROMISE), University of Palermo, 90127 Palermo, Italy
3. MOMO' FertiLIFE IVF Center, 76011 Bisceglie, Italy
4. Department of Emergency and Organ Transplantation, Pathology Section, University of Bari "Aldo Moro", 70124 Bari, Italy
5. Department of Medical Sciences, University of Turin, 10124 Turin, Italy
6. Pathology Unit, FPO-IRCCS Candiolo Cancer Institute, 10060 Candiolo, Italy
* Correspondence: miriamdellino@hotmail.it
† These authors contributed equally to this work.

Abstract: One emerging problem for onco-gynecologists is the incidence of premenopausal patients under 40 years of age diagnosed with stage I Endometrial Cancer (EC) who want to preserve their fertility. Our review aims to define a primary risk assessment that can help fertility experts and onco-gynecologists tailor personalized treatment and fertility-preserving strategies for fertile patients wishing to have children. We confirm that risk factors such as myometrial invasion and The International Federation of Gynecology and Obstetrics (FIGO) staging should be integrated into the novel molecular classification provided by The Cancer Genome Atlas (TCGA). We also corroborate the influence of classical risk factors such as obesity, Polycystic ovarian syndrome (PCOS), and diabetes mellitus to assess fertility outcomes. The fertility preservation options are inadequately discussed with women with a diagnosis of gynecological cancer. A multidisciplinary team of gynecologists, oncologists, and fertility specialists could increase patient satisfaction and improve fertility outcomes. The incidence and death rates of endometrial cancer are rising globally. International guidelines recommend radical hysterectomy and bilateral salpingo-oophorectomy as the standard of care for this cancer; however, fertility-sparing alternatives should be tailored to motivated women of reproductive age, establishing an appropriate cost–benefit balance between childbearing desire and cancer risk. New molecular classifications such as that of TCGA provide a robust supplementary risk assessment tool that can tailor the treatment options to the patient's needs, curtail over- and under-treatment, and contribute to the spread of fertility-preserving strategies.

Keywords: endometrial cancer; fertility preservation; molecular characterization; reproductive outcome

1. Epidemiology and Risk Factors

The term "gynecologic cancer" defines any cancer that arises in a woman's reproductive system. During the period 2012–2016, 94,000 women were diagnosed with gynecologic cancer annually [1,2]. With an incidence of more than 3.6 million per year and mortality exceeding 1.3 million per year, these cancers constitute a public health issue, accounting for 40% of all cancer incidences and more than 30% of all cancer deaths in women worldwide [3,4]. Between 1990 and 2017, the age-standardized prevalence and incidence rate of EC increased globally by 0.89 percent and 0.58 percent a year, respectively, with a median age of diagnosis of 65 years [5–7]. According to recent statistics, incidence of EC is estimated to be 15% for patients $\leq$ 50 years old and 4–14% for those $\leq$40 years old [8]. For the year 2020, the World Health Organization estimated an EC incidence of 35,915 cases in women $\leq$ 44 years, 14,203 in women $\leq$ 39 years, and 2232 in women $\leq$ 29 years of age [9]. The Western lifestyle, with the associated spread of clustered risk factors such as excess weight, diabetes mellitus, hypertension, and high serum triglycerides, contributes to the emergence of EC [10]. PCOS, dysfunctional uterine bleeding/anovulation, and hypermenorrhea are other emerging risk factors linked to EC [9]. EC should be a trending topic for public health professionals: the IARC estimated that EC cases will increase by more than 50% worldwide by 2040 [5,11]. Different risk factors are associated with Type I and II EC, and partially mirror those of women's risk for other cancers [2,12–14]. The main risk factor for type I EC is prolonged exposure to excess estrogen without adequate balance by progestin [15,16]. Sources of exogenous estrogen include hormone replacement therapy and Tamoxifen, while endogenous estrogen exposure may result from high body weight, dysfunctional menstrual cycles, or, in rare cases, tumors that secrete estrogen. Unopposed systemic estrogen therapy results in a high risk of endometrial hyperplasia (20–50% of women) [17] with a relative risk of EC ranging from 1.1 to 15 [18]. The use of the selective estrogen receptor modulator Tamoxifen increases the risk of EC in women after menopause with an effect that is duration- and dose-dependent with a RR of 3.32, and 95% CI 1.95–5.67 [19]. The dietary supplementation of phytoestrogens (non-steroidal chemical products with estrogenic and anti-estrogenic properties) for a time longer than 12 months may potentially increase the risk of EC (3.8%) [20]. The most common disorder associated with anovulation is polycystic ovary syndrome. Anovulatory women have an imbalance of sex hormones that leads to irregular uterine bleeding and continued proliferation of the endometrium that can lead to endometrial hyperplasia. Obese women have high endogenous estrogen levels and endocrine abnormalities such as altered levels of insulin-like growth factor and insulin resistance. Early menarche and late menopause increase the risk of the disease. Some ovarian cancers that produce estrogen, such as granulosa cell tumors, are most likely associated with endometrial neoplasia (25–50 per cent of women affected) and carcinoma (5–10 per cent of the affected) [21]. Some genetic syndromes are decisive risk factors for EC; Lynch syndrome, for example, accounts for 2–5% of all EC [22], and BRCA gene mutation significantly increases uterine cancer (RR 2.65, 95% CI 1.69–4.16) [23]. The risk of EC is significantly elevated, especially for BRCA mutation carriers taking the drug Tamoxifen [24]. Some risk factors that are associated with EC include nulliparity, infertility, hypertension, and diabetes [25]. Protective factors for type I EC include combined estrogen–progestin oral contraceptive use (decreases endometrial carcinoma risk by 30 per cent or higher [26]), childbearing at an older age, and breastfeeding. Cigarette smoking is associated with a diminished risk of EC in postmenopausal women relative risk (RR 0.71, 95% CI 0.65–0.78) [27]. Increased physical activity appears to reduce the risk of EC (RR 0.80, 95% CI 0.75–0.85) [28]. The habit of drinking coffee decreases the risk of EC with a dose-dependent rate. The reductions in risk for low/moderate drinkers were RR 0.87 (95% CI 0.78–0.97), and for heavy coffee drinkers were RR 0.64 (95% CI 0.48–0.86) [29]. Additionally, tea consumers have a decreased risk of EC proportional to the quantity consumed, especially for green tea (RR 0.8, 95% CI 0.7–0.9) [30]. Type II endometrial neoplasms have different risk factors than type I EC and are less well known because of their rarity. Obesity is less strongly correlated [31]. Pluriparity is a risk factor [32]. Type II tumors have a different racial distribution; type II EC is more common in Black women than in White women [33].

2. Classification and Molecular Aspects

Jan V. Bokhman, in a paper published in 1983, analyzed 336 patients affected by EC, and proposed a dualistic model based on pathogenetic and prognostic features. The model differentiates EC into two classes: Type I EC and Type II EC [34,35]. Type I EC represents more than 70% of cases and develops into a hyperestrogenism condition. These tumors are generally low-grade, estrogen-receptor-positive, endometrioid adenocarcinomas (80–90%) that arise from endometrial hyperplasia with atypia (atypical hyperplasia, AH/endometrial intraepithelial hyperplasia, EIN) and with positive prognostic outcomes. Type II tumors, accounting for 10% of EC, are clinically and histologically more aggressive and account for more than 40% of deaths from EC [10]. These neoplasms are predominantly non-estrogen-associated serous carcinomas, mainly receptor-negative, and usually arise in an atrophic endometrium from serous-type endometrial intraepithelial carcinoma. Type II tumors are more likely to be clear cell, papillary, serous, and undifferentiated carcinomas (10–20%), and carry a poor prognosis because of their high histological grade and high invasion and relapse rate [36]. TCGA project carried out a genomic analysis of 373 cancers of the endometrium, stratifying them into four distinct prognostic groups: polymerase and (POLE) ultra mutated, mismatch repair-deficient (MMRd), p53 mutant/abnormal (p53abn), and NSMP (non-specific molecular profile) [37,38]. The genes implied in the development of Type I and Type II EC are PTEN, mismatch repair proteins, β-catenin, KRAS and TP53-, E-cadherin, and PIK3CA, combined with additional molecular and pathologic biomarkers such as the expression of p53, L1CAM, estrogen receptor, progesterone receptor, and the presence of invasion of lymphovascular space (LVSI) to create an integrated risk profile. For example, the abnormal expression of the product of the gene TPp53 strongly correlates to the tumors' high genomic instability and the consequent aggressive behavior of the tumor in progression and invasion. LVSI strongly correlates to lymphatic and capillary tumor spread. L1CAM expression is related to tumor cells with enhanced motility [39,40]. The new classification integrates molecular milieu with clinicopathological characteristics to define an accurate risk assessment [41].

3. The Risk Stratification Models

Type I endometrial cancer arises in cells exposed to high concentrations of estrogens. Many clinical conditions can lead to estrogen/progesterone imbalance, primarily high body weight (OR = 1.27; 95% CI, 1.17 to 1.38) and diabetes mellitus (OR = 1.20; 95% CI, 1.19 to 1.21; $p < 0.0001$). Obesity causes an increase in the risk of EC directly related to body mass index (BMI). For a BMI that ranges from 35 to 40, the risk of EC is increased 4-fold (OR = 4.45; 95% CI, 4.05 to 4.89; $p < 0.0001$), while for a BMI over 40 the risk of EC increases 7-fold (OR = 7.14; 95% CI 6.33 to 8.06; $p < 0.0001$). Other classically described risk factors include early menarche, failure to ovulate or infertility, PCOS, nulliparity, and late menopause [42]. Tamoxifen usage is associated with the insurgence of endometrial hyperplasia and dysplasia. A 10-year continuative therapy doubles the risk of EC (RR 2.29; $p < 0.001$) [43]. Protective factors are related to minor exposure to estrogens such as pluriparity, late menarche, and combined oral contraceptive and cigarette smoking [42]. Endometrioid EC type I arises based on atypical endometrial hyperplasia (AEH); concurrent AEH and EC is estimated in up to 29.5 percent of cases [43]. Autosomal dominant inherited syndromes such as Cowden, Lynch, and Peutz–Jeghers are associated with EC and account for approximately 2% to 5% of all cases. The most important is Lynch syndrome, which is caused by a deficit of the DNA mismatch repair proteins MSH.6, MLH.1, MSH.2, and PMS.2, and is associated with a lifetime risk of EC of 16–54% [44,45]. Cowden syndrome, caused by mutations of the protein PTEN tumor, has a lifetime risk for EC of up to 19–28% [45]. TCGA differentiates EC into four prognostically significant categories: POL-e (ultramutated), microsatellite instability (hyper-mutated), low copy number (CN-L) (endometrioid), and high copy number (serous-like) (CN-H) [46]. POLE is a DNA polymerase responsible for base excision repair. POLE-ultramutated tumors (4–12% of EC) have an extraordinarily high mutation rate and an excellent prognosis, regardless of tumor histotype and grade. Copy-number-high (MSI-H) tumors (23–36% of EC) are associated with PTEN, PIK3CA and PIK3R1 mutations [47], frequently with a low uterine location and Type I EC [47]. The TP53 gene encodes for the p53 protein, commonly mutated in cancer [48]. p53-abnormal tumors (8–24% of EC) have, typically, an aggressive behavior compared to other molecular subtypes; in fact, they are classically CN-H, serous, high-grade endometrioid/clear cell carcinomas and represent type II EC [48]. Low-grade endometrioid adenocarcinomas characterize CN-L tumors (30–60% of EC). These tumors are without a specific driver mutation, have no specific molecular profile (NSMP), and are also seen in type I EC. Regular expression of p53 tumors is common in endometrioid ER/PR-positive tumors, especially in obese patients [49,50]. TCGA subgroups are linked to the prognosis, progression-free survival, and risk profile of EC [51]. There is excellent prognosis in POLE-mutated tumors, and intermediate prognosis in MSI-H tumors and CN-L tumors, while CN-H tumors have poor outcomes [52,53]. The current joint guidelines of the European Societies of Gynaecological Oncology (ESGO), Radiotherapy and Oncology (ESTRO), and Pathology (ESP) proposed an integrated strategy which integrates the genome atlas molecular classification with EC characteristics such as myometrial invasion, histological type, and LVSI to define the correct treatment for these tumors [54]. This risk stratification model was summarized recently in 2022 by Crosbie et al. and is summarized in Table 1. [55]. The Pro-active Molecular Risk Classifier (ProMisE) proposed a novel molecular classification based on a combination of immunohistochemistry (IHC) for mismatch repair proteins, p53, and the molecular analysis of POLE [56]. The WHO proposed the algorithm shown in Figure 1 [55]. Raffone et al. calculated the accuracy of IHC for mismatch repair proteins in EC [57]. Assessing MMR status is crucial to propose humanized monoclonal antibody pembrolizumab and nivolumab to selected patients with a dramatic clinical improvement [58,59]. MMR-deficient tumors have increased resistance (2.1-fold) and recurrence (3.8-fold) compared to regular MMR expression. MMR specificity in recurrence rate of AEH/EC after initial regression is 100% [60].

Table 1. ESGO–ESP–ESTRO prognostic groups of risk modified and rearranged from Crosbie et al. [55]. Non-E.C = Non E.C. carcinoma: serous, clear cell, carcinosarcoma, mixed, undifferentiated carcinoma. E.C. = E.C. carcinoma. LVSI = lymphovascular space invasion. NSMP = non-specific molecular pattern. St. = Stage.

	Molecular Classification Known				Molecular Classification Unknown	
	Pole-Mutant	Mmr-Deficient	NSMP	P53 Abnormal		
Low risk	St. I-II, no residual disease	St. IA, E.C Low-grade, negative/focal LVSI			St. IA, E.C., low-grade, with negative or focal LVSI	
Intermediate risk		St. IB E.C Low-grade with negative/focal LVSI	St. IA without myometrial invasion		St. IB, E.C. low-grade, negative or focal LVSI	
		St. IA E.C., high-grade with negative/focal LVSI			St. IA, E.C. high-grade, negative or focal LVSI	
		St. IA non-E.C., without myometrial invasion			St. IA non-E.C. without myometrial invasion	
High-intermediate risk		St. I E.C., with substantial LVSI, regardless of grade or depth of invasion	St. I-IVA with myometrial invasion and no residual disease		St. I E.C. with substantial LVSI, regardless of grade or depth of invasion	
		St. IB high-grade E.C. with any LVSI			St. IB, E.C. high-grade, regardless of LVSI	
		St. II EC			St. II E.C.	

Table 1. *Cont.*

	Molecular Classification Known				Molecular Classification Unknown
	Pole-Mutant	Mmr-Deficient	NSMP	P53 Abnormal	
High risk		St. III-IVA E.C., with no residual disease		St. I–IVA, with myometrial invasion and no residual disease	St. III–IVA E.C. with no residual disease
		St. I-IVA non E.C. with no myometrial invasion and no residual disease			St. I–IVA non-E.C with myometrial invasion and no residual disease
Advanced/metastatic	St. II-IVA with residual disease				St. II-IVA with residual disease
	St. IV B				St. IVB

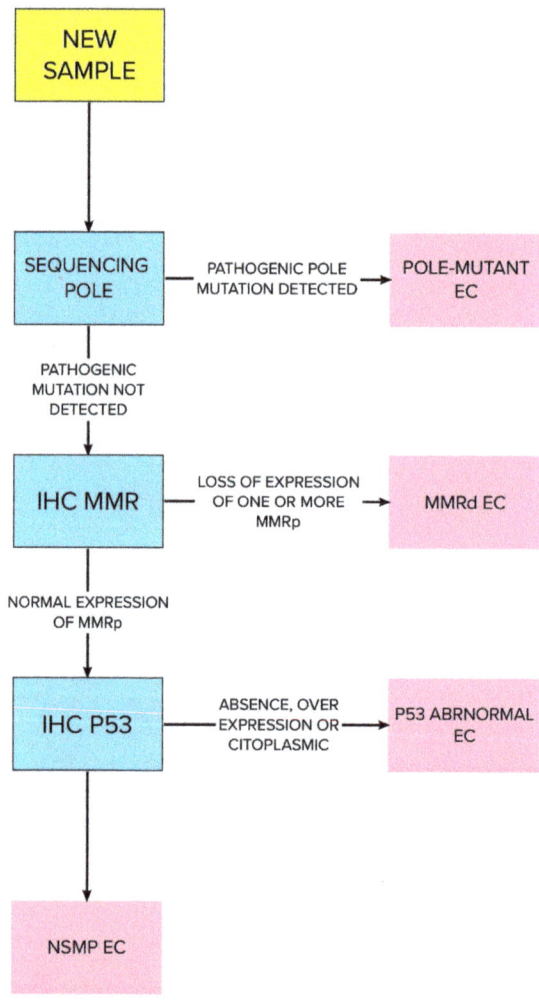

Figure 1. WHO-proposed algorithm for testing EC pathological specimens. Rearranged and modified from Crosbie et al. [55].

The novel DNA analysis tools such as "liquid biopsy" and next-generation sequencing are already in use for the diagnosis of hereditary cancer syndromes but have not yet been routinely established in the clinical diagnostics of EC [61]. Multiple gene panels have been proposed for assessment: Bolivar et al., for example, proposed sequencing PIK3-CA PTEN, K-RAS, and CTTNB-1 to assess endometrioid EC [61]. Prognostic biomarkers have been proposed, including CTNNB-1 mutation status, ER/PR expression, and LVSI or L1 cell-adhesion molecule (L1CAM). LVSI is an independent prognostic marker that increases the mortality and recurrence and/or progression of the disease by 1.5–2-fold [62]. Overexpression of L1CAM is found in most aggressive ECs, especially in p53-abnormal tumors (80%). L1CAM is associated with aggressive behavior of the tumor (cell migration, invasion, epithelial–mesenchymal transition, and chemoresistance), and predicts worse outcomes (recurrence, reduced survival) in p53wilde-type/NSMP tumors [63]. Molecular classification of ECs and endometrial intraepithelial neoplasia (EIN) prior to conservative management is able to differentiate aggressive tumors which should be treated with primary surgery [64]. The rate of progression in patients with p53-abnormal tumors is 50%, whereas in POLE-mutated tumors it is just 25%, independent of the pathological findings [64]. An open field of study is the significance of the ProMisE classifier in fertility-preserving strategies [65].

4. Diagnostic Work Up

Abnormal uterine bleeding (AUB) and heavy menstrual bleeding (HMB) are common conditions affecting 19.5% of women of reproductive age [66]. In 2011, FIGO provided definitions of AUB, HMB, and Chronic AUB that should be used in medical literature and current clinical practice to standardize language [67]. The acronym PALM-COEIN could facilitate accurate diagnosis and treatment of uterine bleeding: PALM stands for the causes that can be assessed by imaging and pathology, such as polyps, adenomyosis, leiomyoma and malignancy, while the word COEIN stands for non-structural causes such as coagulation disease, ovulation problems, endometrial causes, iatrogenic, and others causes [68]. Abnormal bleeding is one of the main symptoms of all types of uterine disease, but has low specificity for malignancy. There is no correlation between AUB and the FIGO stage of the tumor; in addition, women not presenting AUB at the diagnosis of EC showed significantly better prognosis [69]. Bleeding disorders such as abnormal premenopausal and postmenopausal bleeding are the main EC symptoms for which patients seek gynecological consultation. Transvaginal ultrasonography is a safe, straightforward, and easy way to examine double-layered endometrial thickness and to triage women for further investigations [70]. According to international guidelines, gynecologists should assess endometrial thickness with transvaginal ultrasonography (TVUS) in women with abnormal bleeding that arises after menopause [71,72]. A thin endometrium should reassure clinicians about EC and lead to expectation management with seriated TVUS. In cases with a thickened endometrium, endometrial biopsy is warranted. TVUS diagnostic accuracy for EC diagnosis depends on the cut-off in use. The British Gynaecological Cancer Society guidelines currently recommend an endometrial thickness cut-off of ≥ 4 mm that has shown 94.8% sensitivity, 46.7% specificity, and a 99% negative predictive value for EC detection [70,71]. There is not consensus on the best cut-off to use to select AUB patients requiring endometrial biopsy; there is a high prevalence of EC in symptomatic patients when TVUS showed a thickness < 4 mm (8.5%). Some authors have suggested new diagnostic tools for the assessment of EC [73]. One of the main prognostic factors of EC is the depth of myometrial invasion [74], which strongly correlates with the 5-year survival rate: 94% for EC confined to the endometrium, 91% for EC in the inner 1/3 of the myometrium, and 59% when EC is in the outer 1/3 of the myometrium [75]. In addition, myometrial invasion correlates with the risk of extrauterine extension of EC: tumors confined to the inner 1/3 of the myometrium have a 12% risk of extrauterine extension while tumors invading the outer 1/3 have a 46% risk [76]. Contrast-enhanced magnetic resonance imaging (MRI), prevalently T1-weighted imaging including DCE MRI, is the most accurate diagnostic tool for the deep myometrium, with a sensitivity of 72–94% and a specificity of 87–96% [77–80]. Because of the high cost and technical issues related to MRI, physical examination followed by office vaginal ultrasound is a more affordable and accessible diagnostic technique proposed for deep myometrial invasion assessment, with an estimated sensitivity of 75% and specificity of 86% [81]. Biopsy provides the definitive diagnosis: hysteroscopy-guided biopsy remains the gold standard for diagnostic EC, (sensitivity 99.2%, specificity of 86.4%) [82]. Tao brush cytology and Pipelle have a positive predictive value of 81.7% and negative predictive value of 99.1%, but have sampling issues [83]. Blind dilatation and curettage has the highest undiagnosed rate for EC. Hysteroscopy with directed biopsy/curettage is more effective in diagnosing cervical involvement (specificity 98.71% vs. 93.76% ($p < 0.01$)) and more accurate in diagnosis of EC histology type and tumor grade than blind D and C. In Figure 2 we summarized the main ultrasonographic and hysteroscopic findings in endometrial hyperplasia and endometrial cancer.

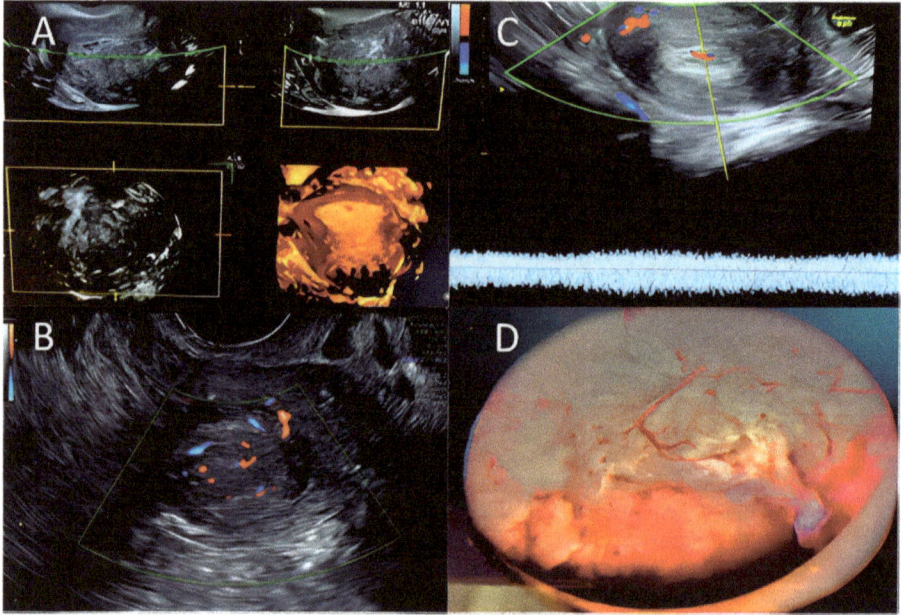

Figure 2. Hyperplasia/endometrial cancer: (**A**) 3D ultrasound exam; (**B,C**) Color–Doppler score examination—Courtesy of MD and VP; (**D**) Hysteroscopic typical pattern for endometrial hyperplasia/endometrial cancer. Courtesy of GRD.

5. Imaging: FIGO and TNM Staging

After the histological diagnosis, further investigations are performed to assess FIGO staging. A vaginal ultrasound may exclude concurrent cancers in the ovaries and presence of ascites, and can define myometrial invasion. A routine chest/abdomen/pelvis Computer Tomography (CT) scan should be performed on high-grade carcinomas to exclude metastatic disease. MRI should be used to differentiate EC from cervical cancer and assess soft tissue extension of EC [84]. MRI has shown low sensitivity (30.3%) for detecting metastatic lymph nodes; with PET/CT, this rate was 57.6% [85]. The low accuracy of imaging to assess lymph node involvement is the reason why accurate surgical staging remains the gold standard. A negative sentinel node evaluation confirms the pathologic absence of metastatic nodes (pN0) in patients with low/low–intermediate risk. Classical surgical lymph node staging remains a strategy of choice in patients with high–intermediate/high-risk disease [86]. Diagnostic laparoscopy is a minimally invasive surgical technique used for the assessment of intra-abdominal masses. It permits the direct inspection of intraabdominal organs and rules out endometrial cancer outside the myometrium or accompanying ovarian malignancies. Because of the relatively low incidence of synchronous endometrial and ovarian cancer (3–5%), a diagnostic laparoscopy is not mandatory in low-risk early EC, when there is no myometrium invasion, in grade 1 endometrial EC, in unsuspicious ovaries, and in normal cancer antigen 125 [87]. The final classification based on the operative staging of EC is that of the Tumour-Node-Metastasis (TNM)-Classification and FIGO, and is shown in Table 2 and Figure 1.

Table 2. FIGO staging of uterine corpus carcinoma and carcinosarcoma, TNM, and AJCC, modified and integrated from Koska et al. [88,89].

TNM STAGE	FIGO STAGE	Short Definition	Details
T1	I	Confined to the uterine corpus	EC inside the uterus, and/or the cervical glands. Not in the cervical stroma. No nearby lymph nodes (N0). No metastasis (M0).
N0			
M0			
T1a	IA	Involves < 50% of the myometrium	EC is in the endometrium and may have grown <50% the myometrium (T1a). No nearby lymph nodes (N0). No metastasis (M0).
N0			
M0			
T1b	IB	Invasion ≥ 50% of the myometrium	EC has grown ≥50% of myometrium, NOT beyond the uterus (T1b).
N0			
M0			
T2	II	Invasion of the cervical stroma but no extension outside the uterus	EC spread from the uterus body and growing into the cervical stroma. Not spread outside the uterus (T2).
N0			
M0			
T3	III	Local and/or regional spread of the tumor	EC is outside the uterus, but NOT in rectum or urinary bladder (T3). No nearby lymph nodes (N0). No metastasis (M0)
N0			
M0			
T3a	IIIA	Invasion of uterine serosa, adnexa, or both (direct extension or metastasis)	EC outside the serosa of the uterus and/or to the adnexa (T3a). No nearby lymph nodes (N0). No metastasis (M0).
N0			
M0			
T3b	IIIB	Metastases or direct spread to the vagina and/or spread to the parametria	EC in the vagina or in the parametrium (T3b). No nearby lymph nodes (N0). No metastasis (M0).
N0			
M0			
	IIIC	Metastases in pelvic or para aortic lymph nodes, or to both	
T1-T3	IIIC1	Metastases to pelvic lymph nodes	EC has extended to some nearby tissues, but NOT into the inside of the urinary bladder or rectum (T1 to T3). Spread to pelvic lymph nodes (N1, N1mi, or N1a), but NOT to aorta lymph nodes or distant sites (M0).
N1, N1mi or N1a			
M0			

Table 2. Cont.

TNM STAGE	FIGO STAGE	Short Definition	Details	
T1-T3	IIIC2	Metastases to para-aortic lymph nodes, with or without metastases to pelvic lymph nodes	Spread to lymph nodes around the aorta (para-aortic lymph nodes) (N2, N2mi, or N2a), but not to distant sites (M0).	
N2, N2mi or N2a				
M0				
	IV	Involvement of the bladder and/or intestinal mucosa and/or distant metastases		
T4	IVA	Invasion of the bladder, intestinal mucosa, or both	Extend to the inner lining of the rectum or urinary bladder (called the mucosa) (T4).	
Any N			M0	
M0				
Any T	IVB	Distant metastases, including metastases to the inguinal lymph nodes or intraperitoneal disease	Inguinal lymph nodes are positive, and/or distant metastases (lungs, liver, bones).	
Any N			Any size (Any N).	
M1				

6. Conventional Treatment: NCCN Guidelines

The standard management of EC involves surgery, chemotherapy, and/or radiation therapy. The gold standard staging procedure for EC is total hysterectomy with bilateral sal-pingo-ovariectomy (TH/BSO) with, if necessary, lymph node surgical assessment [90]. In some selected premenopausal patients, ovary preservation may be a safe choice in stage I endometrioid cancer [91]. Minimally invasive surgery does not compromise oncological outcomes and has a lower rate of complications, so should be proposed in patients with macroscopically uterine-confined cancer. A LAP2 trial compared oncological outcomes in laparoscopic vs. laparotomic surgery, showing recurrence rates of 11.4% for LPS versus 10.2% for LPT surgery and a 5-year overall survival rate of up to 84.8% [92,93]. A trial by Maurits et al. [93] showed a significant complication rate of 14.6% in laparoscopy versus 14.9% in laparotomy, and a minor complication rate of 13.0% in laparoscopy versus 11.7% in laparotomy. Laparotomy remains the gold standard for patients with old age, a large uterus, or metastatic presentations [94]. Robotic surgery may be the surgical choice for the severely obese and for patients at higher anesthesiologic risk [95]. During the surgery, suspicious intraperitoneal areas and enlarged lymph nodes should be biopsied and peritoneal cytology should be collected. Through surgical staging, an accurate diagnosis, extension of the disease, a prognostic assessment and patients who require further adjuvant therapy can be defined. Routine lymph node dissection identifies patients with nodal localization requiring adjuvant treatment with radio and/or chemotherapy [96–98]. Guidelines recommend sentinel lymph node biopsy in patients with low-risk and intermediate-risk diseases. Radiotherapy plus brachytherapy, external beam radiation, and the combination of both, or chemotherapy with carboplatin with a given area under the free carboplatin plasma concentration versus time curve of 5–6 plus paclitaxel 175 mg/m^2 are the standard adjuvant therapies that are proven to lower the risk of tumor recurrence. Adjuvant treatment recommendations for EC strongly depend on the prognostic risk group. For low-risk ECs, no adjuvant treatment is recommended [86]. In intermediate-risk populations, adjuvant brachytherapy should be proposed [86]. Adjuvant chemotherapy should be proposed in high-risk populations, especially for high grade and/or substantial LVSI. The omission of adjuvant treatment should be considered if a close follow-up is guaranteed. In stages III, IV, and recurrent EC, debulking surgery should be performed only if complete macroscopic resection is possible with acceptable morbidity. Primary chemotherapy should be used if debulking surgery is not feasible or acceptable [99].

Immunotherapy as a New Approach in EC

One of the fields of interest in gynecological carcinomas is immunotherapy. The basic principle is that cancer grows when the host's immune system is abnormal, and immunotherapy strengthens the patient's immune system, allowing it to act better against cancer cells, slowing the growth and inhibiting the spread of the cancer [100]. Therefore, as in other female solid carcinomas [101], the evaluation of tumor infiltrating lymphocytes (TILs) has a fundamental role in predicting the response to immunotherapy. Currently, there are several immunotherapy strategies, among which those related to programmed cell death protein 1 (PD-1) and its ligand (PDL-1) are very encouraging. PD-1 and PDL-1 are proteins that inhibit the T-lymphocyte-mediated inflammatory response and allow cancer to evade apoptosis. PD-1 is a transmembrane protein expressed on the surface of lymphocytes which acts as an immunological checkpoint, i.e., it prevents the excessive activation of immune system cells from which immune and autoimmune responses arise [102]. Using anti-PD1 or PDL-1 molecules, immunotherapy inhibits the immune inhibitory system and consequently activates the patient's immune system against cancer [103,104]. In a study of 437 ovarian and endometrial solid tumors, PD-1 expression was found in 80% to 90% of cases [105]. An example is Pembrolizumab, which is used as a promising therapy in carcinomas showing loss of MMR proteins such as melanoma and endometrial cancer [106–108]. Loss of function of the phosphatase tumor suppressor PTEN, which blocks the PI3K/AKT/mTOR pathway, is another area of research. The use of Temsirolimus, an mTOR inhibitor, was studied in a phase II study of 62 patients with recurrent metastatic endometrial cancer and demonstrated a remarkable response in patients who had not yet received any chemotherapy, regardless of PTEN status [109]. Unfortunately, in another phase II study of 42 patients with platinum-resistant ovarian cancer and advanced endometrial cancer, Temsirolimus treatment failed and the study was suspended [110]. Furthermore, patients with high microsatellite instability (MSI-high) also have a better response to immunotherapy. This is probably due to the fact that the excessive mutational load leads to an elevated expression of neo-antigens by each TILs-recalling cell, resulting in a response to immunotherapeutic drugs [111]. In May 2017, the US Food and Drugs Administration (FDA) accelerated the use of Pembrolizumab in patients with MSI-high or MMR protein loss in solid tumors. This was the first time the FDA has approved a treatment for patients with a specific molecular signature and not based on the location of the primary tumor. This decision was also supported by clinical studies in other cancer histotypes (NSCLC, melanoma, colon), which demonstrated that patients with high MSI molecular labeling or MMR protein deficiency had a very marked improvement in response outcomes [112–115]. These studies have been foundational and are leading to the validation of immunotherapy in endometrial cancer [116]. In 2017, the phase 1b study KEYNOTE-028 on the effect of Pembrolizumab on advanced or metastatic CE with PDL-1 positivity already treated with standard therapy showed a partial response in three patients, of which one had a mutation in POLE. The cumulative response was 13% with a six-month PFS of 19% and overall survival of 68.8%. Only mild adverse effects were found in 54.2% of patients [117]. Furthermore, in a recent phase Ib/II study, the combination of Pembrolizumab and the TKI Lenvatinib was tested in 23 patients with progressive metastatic EC, after standard chemotherapy. This study saw a 48% cumulative response with mild adverse effects [118]. Finally, there are numerous active clinical studies in the field of immunotherapy from which we expect promising answers in such a way as to be able to identify specific cohorts of patients also on the basis of molecular characteristics and genetic signatures.

7. Fertility-Sparing Options

Fertility-sparing treatments should be proposed to patients affected by AH/EIN/G1 with no myometrial invasion and who wish to have children. Continuous megestrol acetate, medroxyprogesterone, or levonorgestrel-IUD should be the medications of choice [119–121]. The NCCN defined patient selection criteria for fertility preservation: affected by well-differentiated endometrioid adenocarcinoma limited to the endometrium [122] on MRI/TVUS with the absence of suspicious/metastatic disease and no contraindications to progestin therapy and pregnancy with a close follow up with endometrial sampling every 3/6 months. We summarize the NCCN and ESGO guidelines for fertility-sparing management algorithm in Figure 3.

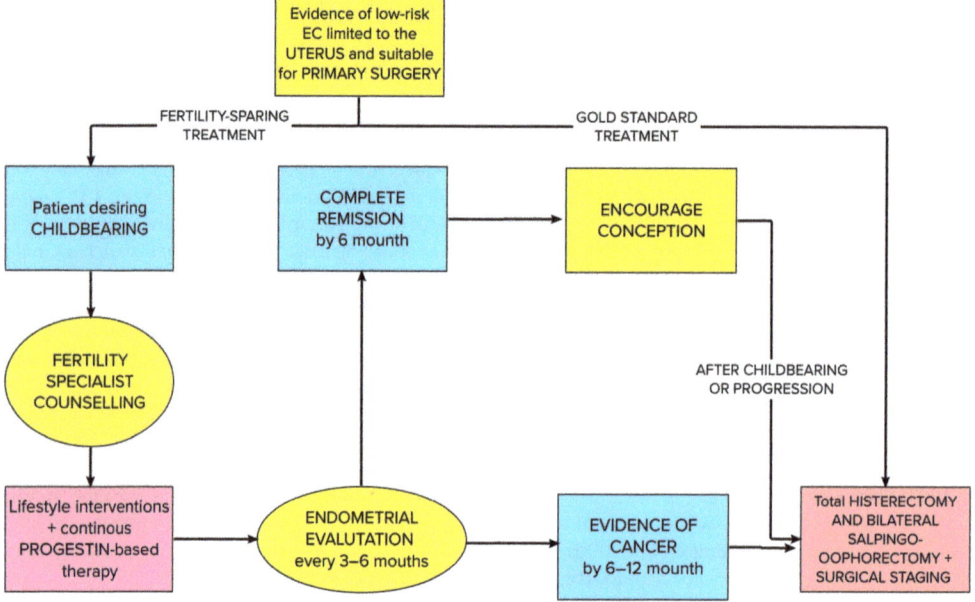

Figure 3. Integrated NCCN and ESGO guidelines flow chart for fertility-sparing management.

Hystopathology is the core strategy for stratification of the risk and the choice for fertility-sparing options for EC. The most common classification that may be useful to differentiate prognostic groups of EC implies the use of three IHC markers—TPp53, MSH6, PMS2—and the molecular test of POLE. In terms of prognostic values, EC should only be classified as POLE-mutated, which implies a low risk [123]. Some advance-stage ECs with POLEmut have excellent prognosis, while p53-abnormal (p53abn) tumors have poor prognosis [47,124]. MMRd or non-specific molecular profiles (NSMPs) have intermediate prognosis. New biomarkers, such as L1CAM or CTNNB1 mutation, may be useful to stratify low-grade endometrioid carcinomas [63]. An indicator of good response to EC is Dusp6, implicated in the MAP-Kinase cellular pathway, whereas MMR deficiency, mutations of PTEN, and overexpression of beta-catenin are indicators of therapy failure. Mutations of p53 and CTNNB1 are risk factors for recurrence. More studies are needed to assess the prognostic potential of new genes, such as KIF2C, CDK1, TPX2, and UBE2C [125]. Staging TAH/BSO is recommended after childbearing is completed or in case of progression. The reported complete response rate varies closely, depending on the stage and grade of EC. A durable CR occurs in about 50% of patients [119]. There is no uniform recommendation about progestin therapy timing, but maintenance treatment seems to lower the recurrence rate [126]. During follow up of patients who choose fertility-sparing therapy, clinicians should check symptom-oriented anamnesis and perform a complete clinical gynecological examination with a speculum, office echography, and rectovaginal palpation every 3/6 months during the first 3 years followed by every 6 months during the following 2 years. Cappelletti et al. performed a metanalysis of 42 studies that included 826 women about the possibility of achieving a pregnancy for EC patients treated with fertility-sparing progestin therapy (FSPT). In that study, the rate of live birth after FSPT was 20.5%. A complete response to fertility-sparing treatment was reported in 79.7% of EC patients, while the response rate of patients treated with FSPT was 79.9%. Recurrence was diagnosed in 35.3 per cent of a cohort of women with a previous complete response, and only one woman died during the follow-up.

Fertility-Sparing Treatment Outcomes

Of all women diagnosed with endometrial cancer, 6.5 percent are younger than 45 years old, and about 70 percent of them have not yet realized their wish to have children [127,128]. This new sociological milieu is responsible for the emergence of fertility-sparing treatments (FST) [129]. Young women should be informed by oncogynecologists about the strategies to improve fertility outcomes [130,131]. In a recent review [129] (*n* = 812), a complete or partial response to FST was found in about 83% of patients, and only a small percentage of patients were refractory to it. In 25.3% of cases a relapse occurred. Hormonal therapy plus hysteroscopy increased the pregnancy rate in EC-affected women to approximately 70% [132]. Some medications have been proposed for fertility-preservation strategies: Gonadotropin-RH agonists and aromatase inhibitors have had good outcomes in young and high-body-weight EC patients wishing to preserve their fertility [133–135]. In Table 3 we reported the results form Cappelletti et al. [136] about the mode of conception and pregnancy outcome after conservative management in early stage EC. Sub-fertility and worse obstetrical outcomes are multifactorial conditions linked to pre-existing metabolic disorders such as obesity, polycystic ovarian syndrome, insulin resistance, and a history of repeated curettage [134,135].

Table 3. Mode of conception and pregnancy outcome after conservative management of early stage endometrial cancer. (Modified and rearranged from Cappelletti et al. [136]).

Conception	Percentage of Pregnancy
Spontaneous	27.6
Fertility treatment	55.2
Outcome	**Percentage**
Miscarriage	26.9
Ongoing pregnancy	3.5
Delivery and live birth	68.9
Multiple birth	
No	71.1
Yes, twin	5.6
Yes, 3	1.5

The literature reports a high rate of complications in women who reach pregnancy by medically assisted reproductive techniques. The most common obstetric complications are gestational diabetes, hemolysis and HELLP syndrome, and hypertension. These patients should be followed up in a highly specialized hospital in which specialists are aware of obstetric complications. It could be challenging to consider emerging risk factors to tailor FST to young oncological patients. In Table 4 we reported the general characteristics and rate of response to FST for EC in the main studies in literature.

Table 4. General characteristics and rate of response of main studies on conservative treatment of EEC. (Modified from Cappelletti et al. [136]). (Ch: Cohort, Cs: case series, Os: oral, IU: intrauterine, Hr: Hysteroscopic resection).

Reference (First Author—Country)	Study Design	Subjects with EEC	Mean Age (Years)	Route of Progestin Administration	Additional Treatment	Subjects with Complete Response	Complete Response (%)	Mean Follow-Up (Months)
Andress et al. (2021)—de [137]	Ch	10	34	OS		5	50%	16.7
Ayhan et al. (2020)—tr [138]	Ch	30	32	OS AND/OR IU	Hr	22	73.3%	55.5
Cade et al. (2013)—Au [139]	Ch	10	32	OS AND/OR IU		10	100%	89.2
Casadio et al. (2020)—It [140]	Cs	36	33	OS	Hr	35	97.2%	30
Chen et al. (2016)—Cn [141]	Ch	37	32	OS		27	73%	54
Choi et al. (2013)—Kr [142]	Cs	11	31		Photodynamic therapy + IV photosensitizer	7	63.6%	82.7
Duska et al. (2001)—Us [143]	Ch	12	30	NR		10	83.3%	NR
Falcone et al. (2017)—It [144]	Ch	27	36	OS OR IU	Hr	26	96.3%	96
Giampaolino et al. (2019)—It [145]	Cs	14	35	IU	Hr	11	78.6%	NR
Kaku et al. (2001)—Jp [146]	Cs	10	30	OS		7	70%	33.6
Kim et al. (2013)—Kr [147]	Cs	16	34	OS AND IU		14	87.5%	31.1
Kudesia et al. (2014)—Us [148]	Ch	10	38	OS AND/OR IU		7	70%	21.3
Maggiore et al. (2019)—It [149]	Cs	16	33	IU		13	81.3%	85.3
Minaguchi et al. (2007)—Jp [150]	Ch	19	30	OS		15	78.9%	45.1
Minig et al. (2011)—It [151]	Ch	14	34	IU	GnRH agonists	8	57.1%	29
Niwa et al. (2005)—Jp [152]	Cs	10	30	OS		10	100%	52.2
Ohyagi-Hara et al. (2015)—Jp [153]	Cs	16	NR	OS		11	68.8%	NR
Ota et al. (2005)—Jp [154]	Cs	12	30	OS		5	41.7%	52.7
Park et al. (2013)—Kr [155]	Ch	177	NR	OS		14	79.7%	NR
Pashov et al. (2012)—Ru [156]	Cs	11	30	IU	GnRH agonists	11	100%	44.4
Perri et al. (2011)—Il [157]	Ch	25	NR	OS OR IM		22	88%	NR
Raffone et al. (2021)—It [158]	Ch	6	35	IU	Hr	2	33.3%	NR

Table 4. Cont.

Reference (First Author—Country)	Study Design	Subjects with EEC	Mean Age (Years)	Route of Progestin Administration	Additional Treatment	Subjects with Complete Response	Complete Response (%)	Mean Follow-Up (Months)
Shan et al. (2013)—Cn [159]	Ch	14	30	OS		11	78.6%	34.7
Shirali et al. (2012)—Ir [160]	Ch	16	33	OS		10	62.5%	NR
Ushijima et al. (2007)—Jp [161]	Ch	28	31	OS		14	50%	47.9
Wang et al. (2014)—Tw [162]	Ch	37	32	OS	Hr	30	81.1%	78.6
Wang et al. (2017)—Cn [163]	Cs	11	27	OS OR IM	Hr	9	81.8%	82.3
Yamagami et al. (2018)—Jp [164]	Ch	97	35	OS		88	90.7%	71.3
Yamazawa et al. (2007)—Jp [165]	Ch	9	36	OS		7	77.8%	38.9
Yu et al. (2009)—Cn [166]	Cs	8	25	OS OR IM		5	62.5%	31.8
Zhou et al. (2015)—Cn [167]	Ch	19	30	OS		15	78.9%	32.5

8. New Prospectives in Non-Invasive Diagnostic and Prognostic Biomarkers for EC: The Proteomic Landscape

There is an emergent need for clinical noninvasive biomarkers for EC that could triage a selected cohort of high-risk patients to more invasive tests. The emerging relevance of translational sciences—transcriptomic, proteomic, and metabolomic—draws attention to the cellular, subcellular, and intercellular environments which are significant for tumorigenesis and tumor progression [168–170]. At present, these omics sciences are of less importance in the field of EC, and there is a need for further studies to assess their clinical significance in assessing tumor microenvironment and fingerprint [171,172], in addition to adapting the precise therapeutic strategy. Some interesting biomarkers for EC diagnostics (CA125, CA 19-9, HE4) and prognostics (L1CAM, COX2, SURVIVIN, CERB2) have been proposed. Unfortunately, none of those are currently used in clinical practice due to their lack of sensibility and specificity [173]. The most crucial blood biomarker candidates for EC detection are prolactin, he4, cancer antigens, YKL-40, and adiponectin [168]. Prolactin has over 98% sensitivity and specificity for EC, but is also elevated in the pancreas, ovary, and lung cancers [174]. Human Epididymis Protein 4 has 90% specificity for detecting EC [175], but is also overexpressed in other tumors. However, these proteins are useful in association with other biomarkers: the association of HE4 with SERUM AMILOID A has 73 specificity and 64% specificity in EC diagnostics, respectively, and HE4 with CA125 has over 60% specificity and 90% specificity in EC diagnostics, respectively [176–179]. Antigens CA125, CA72.4, and CA15-3 have a sub-optimal diagnostic accuracy because they are not specific [180]. Other blood markers include Human chitinase-3-like protein1 (YKL-40) [181] and Adiponectin/Leptin. The most helpful protein candidates for EC detection in endometrial uterine lavage specimens are chaperonin 10, Pyruvate kinase, and serpina1. Chaperonin 10, a chaperone involved in protein folding, is upregulated in EC tissues but is elevated in many other conditions. There is evidence that Pyruvate Kinase, a protein involved in the glycolytic pathway, is upregulated in EC tissues. Like Chaperone 10, it is not specific because it is related to other malignant and physiologic metabolic conditions. There is limited evidence regarding Serpina 1, a serine protease inhibitor downregulated in EC tissues. However, combining the three biomarkers, CPN 10, PK, and SERPINA1, showed sensitivity, specificity, and positive predictive values of over 95% for EC diagnostic. In the field of prognostic markers, there is a growing attention to the tumor heterogenicity, i.e., the co-existence of distinct subpopulations of cancer cells with different metabolic pathways and signaling profiles in the same tissue. This field of study promises to explain why tumors with the same histology may differ in their response to target therapy [182–186]. Annexin 2, a phospholipid-binding protein that plays a crucial role in cell growth and signal pathway transduction, is one of the candidates for predicting the recurrence of EC in in vitro, experiments but there is a need for clinical studies to assess the use of this biomarker in clinical settings [187]. One field of interest in the assessment of therapeutic strategies for endometrial cancer is the prediction of response to adjuvant treatment. There are some genomic markers proposed by TCGA that can divide cancers by risk profile. The change in proteome profile during EC adjuvant therapy needs further studies to evaluate possible markers of response or recurrence to selected therapies and to propose fertility-sparing options to patients who will respond to conservative treatment [188].

9. Conclusions and Perspective

There have been recent advancements in the area of fertility-sparing treatments for early stage endometrial cancer, but more studies are needed to fully understand their safety and efficacy. Factors such as the patient's age, BMI, markers of ovarian reserve such as anti-Müllerian hormone and antral follicle count, and the stage and type of cancer can all impact the decision-making process. It is important for patients to have an open and honest discussion with their doctor and to be supported by a multidisciplinary team of medical specialists. The 2022 ESGO/ESHRE/ESGE Guidelines recommend referring patients with a pregnancy wish to a specialized center with a multidisciplinary team of gynecologic oncologists, fertility specialists, pathologists, and radiologists to aid in the decision-making process. Histological, molecular, and clinical features guide EC treatments. However, advanced research in translational science on carcinogenesis is revolutionizing the standard therapy of most cancers. EC is not yet benefiting from tailored therapies compared to other malignancies. Systemic traditional CHT is the only treatment international guidelines recommend for advanced and recurrent ECs. Targeted therapies should be one of the solutions in research settings for treating ECs. In endometrioid EC, the loss of function of PTEN, the activating mutation of PIK3CA, and the mutation of ARID1A are some of the common mutations. Intriguingly, ARID1A is a regulator of DNA damage checkpoint, leading the way to the use of Poly (ADP-Ribose) Polymerase (PARP) inhibitor. Dysregulation of the HER2 molecule tyrosine kinase is another attractive druggable molecular target in clear cell EC. Molecular aberrations that lead to carcinogenesis are still under study. Serous ECs are primarily independent of estrogen and are TP53-mutated. Common mutations in carcinosarcoma imply TP53, PTEN, PIK3CA, and PIK3R1, and are potentially targetable with an immune checkpoint inhibitor. Mismatch Repair Deficiency with mutations of the MMR genes MLH1, MSH2, MSH6, PMS2, or EPCAM leads to the accumulation of neoantigen loads, with promising outcomes with immune checkpoint inhibitors. Low-grade metastatic/recurrent endometrioid ECs with the expression of estrogen receptors have a reasonable response rate to hormonal therapy with drugs such as tamoxifen and megestrol/medroxyprogesterone acetate. Immunotherapies with checkpoint inhibitors could be effective in MSI-high/MMR-deficient or -high TILs. With the expression of PD-ligand1-2, activation of the PI3K/Akt/mTOR pathway leads to mTOR inhibitors that only have low activity. Additionally, overexpression of HER2, which is frequent in serous ECs, has a small clinical utility: treatment with trastuzumab lacks efficacy. ARID1A deficiency has a potential clinical utility for the use of PARP inhibitors. It is crucial for patients to be aware that pregnancy after endometrial cancer treatment may come with increased risks and to carefully consider the potential benefits and risks of fertility-sparing options. Achieving a pregnancy after conservative treatment for endometrial cancer could be followed by obstetrical complications. New predictive markers could assist reproduction specialists in helping these women conceive and predict and avoid complications.

Author Contributions: Conceptualization: M.D., G.C., M.C. and E.C. (Eliano Cascardi); investigation and literature review: G.M.B., A.S.L., A.V. (Antonella Vimercati), M.M., G.R.D., A.M. and E.C. (Ettore Cicinelli); writing—original draft preparation: M.C., M.D., G.C. and E.C. (Eliano Cascardi); writing—review and editing: G.M.B., A.S.L., A.V. (Amerigo Vitagliano), M.M., G.R.D., A.M. and E.C. (Ettore Cicinelli); data curation: M.C. and E.C. (Eliano Cascardi); supervision: A.S.L., G.C., M.D. and E.C. (Ettore Cicinelli). All authors have read and agreed to the published version of the manuscript.

Funding: This research received no external funding.

Institutional Review Board Statement: Not applicable.

Informed Consent Statement: Not applicable.

Data Availability Statement: Not applicable.

Conflicts of Interest: The authors declare no conflict of interest.

References

1. Corpus, C.U.; Vulva, U.O.V. *Gynecologic Cancer Incidence, United States—2012–2016*; Centers for Disease Control and Prevention, US Department of Health and Human Services: Atlanta, GA, USA, 2019.
2. Dellino, M.; Cascardi, E.; Laganà, A.S.; Di Vagno, G.; Malvasi, A.; Zaccaro, R.; Maggipinto, K.; Cazzato, G.; Scacco, S.; Tinelli, R.; et al. Lactobacillus crispatus M247 oral administration: Is it really an effective strategy in the management of papillomavirus-infected women? *Infect. Agents Cancer* **2022**, *17*, 53. [CrossRef] [PubMed]
3. Sung, H.; Ferlay, J.; Siegel, R.L.; Laversanne, M.; Soerjomataram, I.; Jemal, A.; Bray, F. Global Cancer Statistics 2020: GLOBOCAN Estimates of Incidence and Mortality Worldwide for 36 Cancers in 185 Countries. *CA Cancer J. Clin.* **2021**, *71*, 209–249. [CrossRef] [PubMed]
4. Viviani, S.; Dellino, M.; Ramadan, S.; Peracchio, C.; Marcheselli, L.; Minoia, C.; Guarini, A. Fertility preservation strategies for patients with lymphoma: A real-world practice survey among Fondazione Fertility preservation strategies for patients with lymphoma: A real-world practice survey among Fondazione Italiana Linfomi centerscenters. *Tumori J.* **2022**, *108*, 572–577. [CrossRef] [PubMed]
5. Zhang, S.; Gong, T.-T.; Liu, F.-H.; Jiang, Y.-T.; Sun, H.; Ma, X.-X.; Zhao, Y.-H.; Wu, Q.-J. Global, Regional, and National Burden of Endometrial Cancer, 1990–2017: Results from the Global Burden of Disease Study, 2017. *Front. Oncol.* **2019**, *9*, 1440. [CrossRef]
6. Dellino, M.; Cascardi, E.; Tomasone, V.; Zaccaro, R.; Maggipinto, K.; Giacomino, M.E.; De Nicolò, M.; De Summa, S.; Cazzato, G.; Scacco, S.; et al. Communications Is Time for Care: An Italian Monocentric Survey on Human Papillomavirus (HPV) Risk Information as Part of Cervical Cancer Screening. *J. Pers. Med.* **2022**, *12*, 1387. [CrossRef]
7. Tomao, F.; Peccatori, F.; del Pup, L.; Franchi, D.; Zanagnolo, V.; Panici, P.B.; Colombo, N. Special issues in fertility preservation for gynecologic malignancies. *Crit. Rev. Oncol.* **2016**, *97*, 206–219. [CrossRef]
8. Son, J.; Carr, C.; Yao, M.; Radeva, M.; Priyadarshini, A.; Marquard, J.; Michener, C.M.; AlHilli, M. Endometrial cancer in young women: Prognostic factors and treatment outcomes in women aged ≤ 40 years. *Int. J. Gynecol. Cancer* **2020**, *30*, 631–639. [CrossRef]
9. Roh, H.J.; Yoon, H.J.; Jeong, D.H.; Lee, T.H.; Kwon, B.S.; Suh, D.S.; Kim, K.H. Prognostic factors of oncologic outcomes after fertility-preservative management with progestin in early-stage of endometrial cancer. *J. Res. Med. Sci.* **2021**, *26*, 48. [CrossRef]
10. Vicky, M.; MacKay, H.; Ray-Coquard, I.; Levine, D.A.; Westin, S.N.; Daisuke, A.; Ana, O. Endometrial Cancer (Primer). *Nat. Rev. Dis. Prim.* **2021**, *7*, 88.
11. Dellino, M.; Cascardi, E.; Vinciguerra, M.; Lamanna, B.; Malvasi, A.; Scacco, S.; Acquaviva, S.; Pinto, V.; Di Vagno, G.; Cormio, G.; et al. Nutrition as Personalized Medicine against SARS-CoV-2 Infections: Clinical and Oncological Options with a Specific Female Groups Overview. *Int. J. Mol. Sci.* **2022**, *23*, 9136. [CrossRef]
12. Cascardi, E.; Cazzato, G.; Daniele, A.; Silvestris, E.; Cormio, G.; Di Vagno, G.; Malvasi, A.; Loizzi, V.; Scacco, S.; Pinto, V. Association between Cervical Microbiota and HPV: Could This Be the Key to Complete Cervical Cancer Eradication? *Biology* **2022**, *11*, 1114. [CrossRef] [PubMed]
13. Loizzi, V.; Dellino, M.; Cerbone, M.; Arezzo, F.; Chiariello, G.; Lepera, A.; Cazzato, G.; Cascardi, E.; Damiani, G.R.; Cicinelli, E.; et al. Hormone replacement therapy in BRCA mutation carriers: How shall we do no harm? *Hormones* **2023**, *22*, 19–23. [CrossRef]
14. Loizzi, V.; Dellino, M.; Cerbone, M.; Arezzo, F.; Cazzato, G.; Damiani, G.R.; Pinto, V.; Silvestris, E.; Kardhashi, A.; Cicinelli, E.; et al. The Role of Hormonal Replacement Therapy in BRCA Mutated Patients: Lights and Shadows. *Int. J. Mol. Sci.* **2023**, *24*, 764. [CrossRef] [PubMed]
15. Yamawaki, T.; Teshima, H.; Takeshima, N.; Yamauchi, K.; Hasumi, K. A clinicopathological study in clear cell adenocarcinoma of the endometrium. *Nihon Sanka Fujinka Gakkai Zasshi* **1996**, *48*, 328–334.
16. Zorn, K.K.; Bonome, T.; Gangi, L.; Chandramouli, G.V.; Awtrey, C.S.; Gardner, G.J.; Barrett, J.C.; Boyd, J.; Birrer, M.J. Gene Expression Profiles of Serous, Endometrioid, and Clear Cell Subtypes of Ovarian and Endometrial Cancer. *Clin. Cancer Res.* **2005**, *11*, 6422–6430. [CrossRef] [PubMed]

17. Woodruff, J.; Pickar, J.H.; The Menopause Study Group. Incidence of endometrial hyperplasia in postmenopausal women taking conjugated estrogens (Premarin) with medroxyprogesterone acetate or conjugated estrogens alone. *Am. J. Obstet. Gynecol.* **1994**, *170*, 1213–1223. [CrossRef]
18. Henderson, B.E. The cancer question: An overview of recent epidemiologic and retrospective data. *Am. J. Obstet. Gynecol.* **1989**, *161*, 1859–1864. [CrossRef]
19. Early Breast Cancer Trialists' Collaborative Group (EBCTCG); Davies, C.; Godwin, J.; Gray, R.; Clarke, M.; Cutter, D.; Darby, S.; McGale, P.; Pan, H.C.; Taylor, C.; et al. Relevance of breast cancer hormone receptors and other factors to the efficacy of adjuvant tamoxifen: Patient-level meta-analysis of randomised trials. *Lancet* **2011**, *378*, 771–784. [CrossRef]
20. Unfer, V.; Casini, M.L.; Costabile, L.; Mignosa, M.; Gerli, S.; Di Renzo, G.C. Endometrial effects of long-term treatment with phytoestrogens: A randomized, double-blind, placebo-controlled study. *Fertil. Steril.* **2004**, *82*, 145–148. [CrossRef]
21. Zanagnolo, V.; Pasinetti, B.; Sartori, E. Clinical review of 63 cases of sex cord stromal tumors. *Eur. J. Gynaecol. Oncol.* **2004**, *25*, 431–438.
22. Kwon, J.S.; Scott, J.L.; Gilks, C.B.; Daniels, M.; Sun, C.C.; Lu, K.H. Testing Women with Endometrial Cancer to Detect Lynch Syndrome. *J. Clin. Oncol.* **2011**, *29*, 2247–2252. [CrossRef] [PubMed]
23. Thompson, D.; Easton, D.F.; The Breast Cancer Linkage Consortium. Cancer Incidence in BRCA1 Mutation Carriers. *J. Natl. Cancer Inst.* **2002**, *94*, 1358–1365. [CrossRef] [PubMed]
24. Beiner, M.E.; Finch, A.; Rosen, B.; Lubinski, J.; Moller, P.; Ghadirian, P.; Lynch, H.T.; Friedman, E.; Sun, P.; Narod, S.A.; et al. The risk of endometrial cancer in women with BRCA1 and BRCA2 mutations. A prospective study. *Gynecol. Oncol.* **2007**, *104*, 7–10. [CrossRef] [PubMed]
25. Furberg, A.-S.; Thune, I. Metabolic abnormalities (hypertension, hyperglycemia and overweight), lifestyle (high energy intake and physical inactivity) and endometrial cancer risk in a Norwegian cohort. *Int. J. Cancer* **2003**, *104*, 669–676. [CrossRef]
26. Parslov, M.; Lidegaard, Ø.; Klintorp, S.; Pedersen, B.; Jønsson, L.; Eriksen, P.S.; Ottesen, B. Risk factors among young women with endometrial cancer: A Danish case-control study. *Am. J. Obstet. Gynecol.* **2000**, *182*, 23–29. [CrossRef]
27. Setiawan, V.W.; Pike, M.C.; Karageorgi, S.; Deming, S.L.; Anderson, K.; Bernstein, L.; Brinton, L.A.; Cai, H.; Cerhan, J.; Cozen, W.; et al. Age at Last Birth in Relation to Risk of Endometrial Cancer: Pooled Analysis in the Epidemiology of Endometrial Cancer Consortium. *Am. J. Epidemiol.* **2012**, *176*, 269–278. [CrossRef]
28. Schmid, D.; Behrens, G.; Keimling, M.; Jochem, C.; Ricci, C.; Leitzmann, M. A systematic review and meta-analysis of physical activity and endometrial cancer risk. *Eur. J. Epidemiol.* **2015**, *30*, 397–412. [CrossRef]
29. Yu, X.; Bao, Z.; Zou, J.; Dong, J. Coffee consumption and risk of cancers: A meta-analysis of cohort studies. *BMC Cancer* **2011**, *11*, 96. [CrossRef]
30. Zicha, J.; Kronauer, J.; Duhm, J. Effects of a chronic high salt intake on blood pressure and the kinetics of sodium and potassium transport in erythrocytes of young and adult subtotally nephrectomized Sprague-Dawley rats. *J. Hypertens.* **1990**, *8*, 207–217. [CrossRef]
31. Moore, K.N.; Fader, A.N. Uterine Papillary Serous Carcinoma. *Clin. Obstet. Gynecol.* **2011**, *54*, 278–291. [CrossRef]
32. Wilson, T.O.; Podratz, K.C.; Gaffey, T.A.; Malkasian, G.D., Jr.; O'Brien, P.C.; Naessens, J.M. Evaluation of unfavorable histologic subtypes in endometrial adenocarcinoma. *Am. J. Obstet. Gynecol.* **1990**, *162*, 418–426. [CrossRef]
33. Wright, J.D.; Fiorelli, J.; Schiff, P.B.; Burke, W.M.; Kansler, A.L.; Cohen, C.J.; Herzog, T.J. Racial disparities for uterine corpus tumors: Changes in clinical characteristics and treatment over time. *Cancer* **2009**, *115*, 1276–1285. [CrossRef]
34. Bokhman, J.V. Two pathogenetic types of endometrial carcinoma. *Gynecol. Oncol.* **1983**, *15*, 10–17. [CrossRef]
35. Massarotti, C.; Buonomo, B.; Dellino, M.; Campanella, M.; De Stefano, C.; Ferrari, A.; Anserini, P.; Lambertini, M.; Peccatori, F.A. Contraception and Hormone Replacement Therapy in Healthy Carriers of Germline BRCA1/2 Genes Pathogenic Variants: Results from an Italian Survey. *Cancers* **2022**, *14*, 3457. [CrossRef]
36. Braun, M.M.; Overbeek-Wager, E.A.; Grumbo, R.J. Diagnosis and Management of Endometrial Cancer. *Am. Fam. Physician* **2016**, *93*, 468–474.
37. Levine, D.A.; The Cancer Genome Atlas Research Network. Integrated genomic characterization of endometrial carcinoma. *Nature* **2013**, *497*, 67–73. [CrossRef]
38. Bergamini, A.; Luisa, F.M.; Dellino, M.; Erica, S.; Loizzi, V.; Bocciolone, L.; Rabaiotti, E.; Cioffi, R.; Sabetta, G.; Cormio, G.; et al. Fertility sparing surgery in sex-cord stromal tumors: Oncological and reproductive outcomes. *Int. J. Gynecol. Cancer* **2022**, *32*, 1063–1070. [CrossRef]
39. Stelloo, E.; Nout, R.A.; Osse, E.M.; Jürgenliemk-Schulz, I.J.; Jobsen, J.J.; Lutgens, L.C.; van der Steen-Banasik, E.M.; Nijman, H.W.; Putter, H.; Bosse, T.; et al. Improved Risk Assessment by Integrating Molecular and Clinicopathological Factors in Early-stage Endometrial Cancer—Combined Analysis of the PORTEC Cohorts. *Clin. Cancer Res.* **2016**, *22*, 4215–4224. [CrossRef]
40. Daniele, A.; Divella, R.; Pilato, B.; Tommasi, S.; Pasanisi, P.; Patruno, M.; Digennaro, M.; Minoia, C.; Dellino, M.; Pisconti, S.; et al. Can harmful lifestyle, obesity and weight changes increase the risk of breast cancer in BRCA 1 and BRCA 2 mutation carriers? A Mini review. *Hered. Cancer Clin. Pract.* **2021**, *19*, 45. [CrossRef]
41. Wortman, B.G.; Bosse, T.; Nout, R.A.; Lutgens, L.C.H.W.; van der Steen-Banasik, E.M.; Westerveld, H.; van den Berg, H.; Slot, A.; De Winter, K.A.J.; Verhoeven-Adema, K.W.; et al. Molecular-integrated risk profile to determine adjuvant radiotherapy in endometrial cancer: Evaluation of the pilot phase of the PORTEC-4a trial. *Gynecol. Oncol.* **2018**, *151*, 69–75. [CrossRef]

42. Setiawan, V.W.; Yang, H.P.; Pike, M.C.; McCann, S.E.; Yu, H.; Xiang, Y.-B.; Wolk, A.; Wentzensen, N.; Weiss, N.S.; Webb, P.M.; et al. Type I and II endometrial cancers: Have they different risk factors? *J. Clin. Oncol.* **2013**, *31*, 2607–2618. [CrossRef]
43. Fleming, C.A.; Heneghan, H.M.; O'Brien, D.; McCartan, D.P.; McDermott, E.W.; Prichard, R.S. Meta-analysis of the cumulative risk of endometrial malignancy and systematic review of endometrial surveillance in extended tamoxifen therapy. *Br. J. Surg.* **2018**, *105*, 1098–1106. [CrossRef]
44. Banno, K.; Kisu, I.; Yanokura, M.; Umene, K.; Nogami, Y.; Tsuji, K.; Masuda, K.; Ueki, A.; Kobayashi, Y.; Yamagami, W.; et al. Hereditary endometrial cancer: Lynch syndrome. *Curr. Obstet. Gynecol. Rep.* **2013**, *2*, 11–18. [CrossRef]
45. Gayther, S.A.; Pharoah, P.D. The inherited genetics of ovarian and endometrial cancer. *Curr. Opin. Genet. Dev.* **2010**, *20*, 231–238. [CrossRef]
46. Alexa, M.; Hasenburg, A.; Battista, M.J. The TCGA Molecular Classification of Endometrial Cancer and Its Possible Impact on Adjuvant Treatment Decisions. *Cancers* **2021**, *13*, 1478. [CrossRef]
47. McAlpine, J.; Leon-Castillo, A.; Bosse, T. The rise of a novel classification system for endometrial carcinoma; Integration of molecular subclasses. *J. Pathol.* **2018**, *244*, 538–549. [CrossRef]
48. Kandoth, C.; McLellan, M.D.; Vandin, F.; Ye, K.; Niu, B.; Lu, C.; Xie, M.; Zhang, Q.; McMichael, J.F.; Wyczalkowski, M.A.; et al. Mutational landscape and significance across 12 major cancer types. *Nature* **2013**, *502*, 333–339. [CrossRef]
49. Edmondson, R.; Crosbie, E.; Nickkho-Amiry, M.; Kaufmann, A.; Stelloo, E.; Nijman, H.; Leary, A.; Auguste, A.; Mileshkin, L.; Pollock, P.; et al. Markers of the p53 pathway further refine molecular profiling in high-risk endometrial cancer: A Trans PORTEC initiative. *Gynecol. Oncol.* **2017**, *146*, 327–333. [CrossRef]
50. Karnezis, A.N.; Leung, S.; Magrill, J.; McConechy, M.K.; Yang, W.; Chow, C.; Kobel, M.; Lee, C.-H.; Huntsman, D.G.; Talhouk, A.; et al. Evaluation of endometrial carcinoma prognostic immunohistochemistry markers in the context of molecular classification. *J. Pathol. Clin. Res.* **2017**, *3*, 279–293. [CrossRef]
51. Mustea, A.; Ralser, D.J.; Egger, E.; Ziehm, U.; Vivas, S.; Brock, S.; Jackson, D.; Condic, M.; Meisel, C.; Otten, L.; et al. Determination of the Cancer Genome Atlas (TCGA) Endometrial Cancer Molecular Subtypes Using the Variant Interpretation and Clinical Decision Support Software MH Guide. *Cancers* **2023**, *15*, 2053. [CrossRef]
52. Mitric, C.; Bernardini, M.Q. Endometrial Cancer: Transitioning from Histology to Genomics. *Curr. Oncol.* **2022**, *29*, 741–757. [CrossRef]
53. Dellino, M.; Gargano, G.; Tinelli, R.; Carriero, C.; Minoia, C.; Tetania, S.; Silvestris, E.; Loizzi, V.; Paradiso, A.; Casamassima, P.; et al. A strengthening the reporting of observational studies in epidemiology (STROBE): Are HE4 and CA 125 suitable to detect a Paget disease of the vulva? *Medicine* **2021**, *100*, e24485. [CrossRef]
54. Santoro, A.; Angelico, G.; Travaglino, A.; Inzani, F.; Arciuolo, D.; Valente, M.; D'alessandris, N.; Scaglione, G.; Fiorentino, V.; Raffone, A.; et al. New Pathological and Clinical Insights in Endometrial Cancer in View of the Updated ESGO/ESTRO/ESP Guidelines. *Cancers* **2021**, *13*, 2623. [CrossRef]
55. Crosbie, E.J.; Kitson, S.J.; McAlpine, J.N.; Mukhopadhyay, A.; Powell, M.E.; Singh, N. Endometrial cancer. *Lancet* **2022**, *399*, 1412–1428. [CrossRef]
56. Talhouk, A.; McConechy, M.K.; Leung, S.; Yang, W.; Lum, A.; Senz, J.; Boyd, N.; Pike, J.; Anglesio, M.; Kwon, J.S.; et al. Confirmation of ProMisE: A simple, genomics-based clinical classifier for endometrial cancer. *Cancer* **2017**, *123*, 802–813. [CrossRef]
57. Raffone, A.; Travaglino, A.; Cerbone, M.; Gencarelli, A.; Mollo, A.; Insabato, L.; Zullo, F. Diagnostic Accuracy of Immunohistochemistry for Mismatch Repair Proteins as Surrogate of Microsatellite Instability Molecular Testing in Endometrial Cancer. *Pathol. Oncol. Res.* **2020**, *26*, 1417–1427. [CrossRef]
58. Sahin, I.H.; Akce, M.; Alese, O.; Shaib, W.; Lesinski, G.B.; El-Rayes, B.; Wu, C. Immune checkpoint inhibitors for the treatment of MSI-H/MMR-D colorectal cancer and a perspective on resistance mechanisms. *Br. J. Cancer* **2019**, *121*, 809–818. [CrossRef]
59. Dellino, M.; Carriero, C.; Silvestris, E.; Capursi, T.; Paradiso, A.; Cormio, G. Primary Vaginal Carcinoma Arising on Cystocele Mimicking Vulvar Cancer. *J. Obstet. Gynecol. Can.* **2020**, *42*, 1543–1545. [CrossRef]
60. Raffone, A.; Catena, U.; Travaglino, A.; Masciullo, V.; Spadola, S.; Della Corte, L.; Piermattei, A.; Insabato, L.; Zannoni, G.F.; Scambia, G.; et al. Mismatch repair-deficiency specifically predicts recurrence of atypical endometrial hyperplasia and early endometrial carcinoma after conservative treatment: A multi-center study. *Gynecol. Oncol.* **2021**, *161*, 795–801. [CrossRef]
61. Bolivar, A.M.; Luthra, R.; Mehrotra, M.; Chen, W.; Barkoh, B.A.; Hu, P.; Zhang, W.; Broaddus, R.R. Targeted next-generation sequencing of endometrial cancer and matched circulating tumor DNA: Identification of plasma based, tumor-associated mutations in early stage patients. *Mod. Pathol.* **2019**, *32*, 405–414. [CrossRef]
62. Raffone, A.; Travaglino, A.; Raimondo, D.; Neola, D.; Maletta, M.; Santoro, A.; Insabato, L.; Casadio, P.; Fanfani, F.; Zannoni, G.F.; et al. Lymphovascular space invasion in endometrial carcinoma: A prognostic factor independent from molecular signature. *Gynecol. Oncol.* **2022**, *165*, 192–197. [CrossRef] [PubMed]
63. Kommoss, F.K.; Karnezis, A.N.; Kommoss, F.; Talhouk, A.; Taran, F.-A.; Staebler, A.; Gilks, C.B.; Huntsman, D.G.; Krämer, B.; Brucker, S.Y.; et al. L1CAM further stratifies endometrial carcinoma patients with no specific molecular risk profile. *Br. J. Cancer* **2018**, *119*, 480–486. [CrossRef]
64. Puechl, A.M.; Spinosa, D.; Berchuck, A.; Secord, A.A.; Drury, K.E.; Broadwater, G.; Wong, J.; Whitaker, R.; Devos, N.; Corcoran, D.L.; et al. Molecular Classification to Prognosticate Response in Medically Managed Endometrial Cancers and Endometrial Intraepithelial Neoplasia. *Cancers* **2021**, *13*, 2847. [CrossRef]

65. Ran, X.; Hu, T.; Li, Z. Molecular Classification in Patients with Endometrial Cancer After Fertility-Preserving Treatment: Application of ProMisE Classifier and Combination of Prognostic Evidence. *Front. Oncol.* **2022**, *12*, 810631. [CrossRef] [PubMed]
66. Shapley, M.; Jordan, K.; Croft, P.R. An epidemiological survey of symptoms of menstrual loss in the community. *Br. J. Gen. Pract.* **2004**, *54*, 359–363. [PubMed]
67. Fraser, I.S.; Critchley, H.O.; Broder, M.; Munro, M.G. The FIGO Recommendations on Terminologies and Definitions for Normal and Abnormal Uterine Bleeding. *Semin. Reprod. Med.* **2011**, *29*, 383–390. [CrossRef]
68. Whitaker, L.; Critchley, H.O. Abnormal uterine bleeding. *Best Pract. Res. Clin. Obstet. Gynaecol.* **2016**, *34*, 54–65. [CrossRef]
69. Kimura, T.; Kamiura, S.; Yamamoto, T.; Seino-Noda, H.; Ohira, H.; Saji, F. Abnormal uterine bleeding and prognosis of endometrial cancer. *Int. J. Gynecol. Obstet.* **2004**, *85*, 145–150. [CrossRef]
70. Timmermans, A.; Opmeer, B.C.; Khan, K.S.; Bachmann, L.M.; Epstein, E.; Clark, T.J.; Gupta, J.K.; Bakour, S.H.; van den Bosch, T.; van Doorn, H.C.; et al. Endometrial Thickness Measurement for Detecting Endometrial Cancer in Women with Postmenopausal Bleeding: A systematic review and meta-analysis. *Obstet. Gynecol.* **2010**, *116*, 160–167. [CrossRef]
71. American College of Obstetricians and Gynecologists. ACOG Committee Opinion No. 734: The Role of Transvaginal Ultrasonography in Evaluating the Endometrium of Women with Postmenopausal Bleeding. *Obstet. Gynecol.* **2018**, *131*, e124–e129. [CrossRef]
72. Endometrio e Cervice. 2021. Available online: https://www.aiom.it/wp-content/uploads/2021/10/2021_LGAIOM_Utero_endometrio_cervice.pdf (accessed on 12 May 2023).
73. Saccardi, C.; Vitagliano, A.; Marchetti, M.; Turco, A.L.; Tosatto, S.; Palumbo, M.; De Lorenzo, L.S.; Vitale, S.G.; Scioscia, M.; Noventa, M. Endometrial Cancer Risk Prediction According to Indication of Diagnostic Hysteroscopy in Post-Menopausal Women. *Diagnostics* **2020**, *10*, 257. [CrossRef] [PubMed]
74. Prat, J. Prognostic parameters of endometrial carcinoma. *Hum. Pathol.* **2004**, *35*, 649–662. [CrossRef] [PubMed]
75. Zaino, R.J.; Kurman, R.; Herbold, D.; Gliedman, J.; Bundy, B.N.; Voet, R.; Advani, H. The significance of squamous differentiation in endometrial carcinoma. Data from a gynecologic oncology group study. *Cancer* **1991**, *68*, 2293–2302. [CrossRef] [PubMed]
76. DiSaia, P.; Creasman, W.; Boronow, R.; Blessing, J. Risk factors and recurrent patterns in Stage I endometrial cancer. *Am. J. Obstet. Gynecol.* **1985**, *151*, 1009–1015. [CrossRef]
77. Lin, G.; Ng, K.-K.; Chang, C.J.; Wang, J.-J.; Ho, K.-C.; Yen, T.-C.; Wu, T.-I.; Wang, C.-C.; Chen, Y.-R.; Huang, Y.-T.; et al. Myometrial Invasion in Endometrial Cancer: Diagnostic Accuracy of Diffusion-weighted 3.0-T MR Imaging—Initial Experience. *Radiology* **2009**, *250*, 784–792. [CrossRef]
78. Kinkel, K.; Kaji, Y.; Yu, K.K.; Segal, M.R.; Lu, Y.; Powell, C.B.; Hricak, H. Radiologic staging in patients with endometrial cancer: A meta-analysis. In *Database of Abstracts of Reviews of Effects (DARE): Quality-Assessed Reviews*; Centre for Reviews and Dissemination: York, UK, 1999.
79. Takeuchi, M.; Matsuzaki, K.; Harada, M. Evaluating Myometrial Invasion in Endometrial Cancer: Comparison of Reduced Field-of-view Diffusion-weighted Imaging and Dynamic Contrast-enhanced MR Imaging. *Magn. Reson. Med. Sci.* **2018**, *17*, 28–34. [CrossRef]
80. Rockall, A.G.; Meroni, R.; Sohaib, S.A.; Reynolds, K.; Alexander-Sefre, F.; Shepherd, J.H.; Jacobs, I.; Reznek, R.H. Evaluation of endometrial carcinoma on magnetic resonance imaging. *Int. J. Gynecol. Cancer* **2007**, *17*, 188–196. [CrossRef]
81. Alcázar, J.L.; Gastón, B.; Navarro, B.; Salas, R.; Aranda, J.; Guerriero, S. Transvaginal ultrasound versus magnetic resonance imaging for preoperative assessment of myometrial infiltration in patients with endometrial cancer: A systematic review and meta-analysis. *J. Gynecol. Oncol.* **2017**, *28*, e86. [CrossRef]
82. van Hanegem, N.; Prins, M.M.; Bongers, M.Y.; Opmeer, B.C.; Sahota, D.S.; Mol, B.W.J.; Timmermans, A. The accuracy of endometrial sampling in women with postmenopausal bleeding: A systematic review and meta-analysis. *Eur. J. Obstet. Gynecol. Reprod. Biol.* **2016**, *197*, 147–155. [CrossRef]
83. DeJong, S.R.; Bakkum-Gamez, J.N.; Clayton, A.C.; Henry, M.R.; Keeney, G.L.; Zhang, J.; Kroneman, T.N.; Laughlin-Tommaso, S.K.; Ahlberg, L.J.; VanOosten, A.L.; et al. Tao brush endometrial cytology is a sensitive diagnostic tool for cancer and hyperplasia among women presenting to clinic with abnormal uterine bleeding. *Cancer Med.* **2021**, *10*, 7040–7047. [CrossRef]
84. NCCN. Endometrial Carcinoma. 2022. Available online: https://www.nccn.org/professionals/physician_gls/pdf/uterine.pdf (accessed on 12 May 2023).
85. Choi, H.J.; Roh, J.-W.; Seo, S.-S.; Lee, S.; Kim, J.-Y.; Kim, S.-K.; Kang, K.W.; Lee, J.S.; Jeong, J.Y.; Park, S.-Y. Comparison of the accuracy of magnetic resonance imaging and positron emission tomography/computed tomography in the presurgical detection of lymph node metastases in patients with uterine cervical carcinoma: A prospective study. *Cancer* **2006**, *106*, 914–922. [CrossRef]
86. Concin, N.; Matias-Guiu, X.; Vergote, I.; Cibula, D.; Mirza, M.R.; Marnitz, S.; Ledermann, J.; Bosse, T.; Chargari, C.; Fagotti, A.; et al. ESGO/ESTRO/ESP guidelines for the management of patients with endometrial carcinoma. *Int. J. Gynecol. Cancer* **2021**, *31*, 12–39. [CrossRef] [PubMed]
87. AlHilli, M.; Dowdy, S.; Weaver, A.; Sauver, J.S.; Keeney, G.; Mariani, A.; Podratz, K.; Bakkum-Gamez, J. Incidence and factors associated with synchronous ovarian and endometrial cancer: A population-based case-control study. *Gynecol. Oncol.* **2012**, *125*, 109–113. [CrossRef] [PubMed]
88. Koskas, M.; Amant, F.; Mirza, M.R.; Creutzberg, C.L. Cancer of the corpus uteri: 2021 update. *Int. J. Gynecol. Obstet.* **2021**, *155* (Suppl. S1), 45–60. [CrossRef] [PubMed]

89. American Joint Committee on Cancer. Corpus uteri-carcinoma and carcinosarcoma. In *AJCC Cancer Staging Manual*, 8th ed.; Springer: New York, NY, USA, 2017; pp. 661–669.
90. Shields, L.E.; Goffman, D.; Caughey, A.B. ACOG practice bulletin: Clinical management guidelines for obstetrician-gynecologist. *Obstet. Gynecol.* **2017**, *130*, e168–e186.
91. Koskas, M.; Bendifallah, S.; Luton, D.; Daraï, E.; Rouzier, R. Safety of uterine and/or ovarian preservation in young women with grade 1 intramucous endometrial adenocarcinoma: A comparison of survival according to the extent of surgery. *Fertil. Steril.* **2012**, *98*, 1229–1235. [CrossRef]
92. Walker, J.L.; Piedmonte, M.R.; Spirtos, N.M.; Eisenkop, S.M.; Schlaerth, J.B.; Mannel, R.S.; Spiegel, G.; Barakat, R.; Pearl, M.L.; Sharma, S.K. Laparoscopy Compared with Laparotomy for Comprehensive Surgical Staging of Uterine Cancer: Gynecologic Oncology Group Study LAP2. *J. Clin. Oncol.* **2009**, *27*, 5331–5336. [CrossRef]
93. Mourits, M.J.; Bijen, C.B.; Arts, H.J.; ter Brugge, H.G.; van der Sijde, R.; Paulsen, L.; Wijma, J.; Bongers, M.Y.; Post, W.J.; van der Zee, A.G.; et al. Safety of laparoscopy versus laparotomy in early-stage endometrial cancer: A randomised trial. *Lancet Oncol.* **2010**, *11*, 763–771. [CrossRef]
94. He, H.; Zeng, D.; Ou, H.; Tang, Y.; Li, J.; Zhong, H. Laparoscopic Treatment of Endometrial Cancer: Systematic Review. *J. Minim. Invasive Gynecol.* **2013**, *20*, 413–423. [CrossRef]
95. Coronado, P.J.; Herraiz, M.A.; Magrina, J.F.; Fasero, M.; Vidart, J.A. Comparison of perioperative outcomes and cost of robotic-assisted laparoscopy, laparoscopy and laparotomy for endometrial cancer. *Eur. J. Obstet. Gynecol. Reprod. Biol.* **2012**, *165*, 289–294. [CrossRef]
96. ASTEC Study Group; Kitchener, H.; Swart, A.; Qian, Q.; Amos, C.; Parmar, M. Efficacy of systematic pelvic lymphadenectomy in endometrial cancer (MRC ASTEC trial): A randomised study. *Lancet* **2008**, *373*, 125–136.
97. Panici, P.B.; Basile, S.; Maneschi, F.; Lissoni, A.A.; Signorelli, M.; Scambia, G.; Angioli, R.; Tateo, S.; Mangili, G.; Katsaros, D.; et al. Systematic Pelvic Lymphadenectomy vs No Lymphadenectomy in Early-Stage Endometrial Carcinoma: Randomized Clinical Trial. *JNCI J. Natl. Cancer Inst.* **2008**, *100*, 1707–1716. [CrossRef] [PubMed]
98. Frederick, P.J.; Straughn, J.M., Jr. The role of comprehensive surgical staging in patients with endometrial cancer. *Cancer Control* **2009**, *16*, 23–29. [CrossRef] [PubMed]
99. Barlin, J.N.; Puri, I.; Bristow, R.E. Cytoreductive surgery for advanced or recurrent endometrial cancer: A meta-analysis. *Gynecol. Oncol.* **2010**, *118*, 14–18. [CrossRef]
100. Musacchio, L.; Boccia, S.M.; Caruso, G.; Santangelo, G.; Fischetti, M.; Tomao, F.; Perniola, G.; Palaia, I.; Muzii, L.; Pignata, S.; et al. Immune checkpoint inhibitors: A promising choice for endometrial cancer patients? *J. Clin. Med.* **2020**, *9*, 1721. [CrossRef]
101. Annaratone, L.; Cascardi, E.; Vissio, E.; Sarotto, I.; Chmielik, E.; Sapino, A.; Berrino, E.; Marchiò, C. The Multifaceted Nature of Tumor Microenvironment in Breast Carcinomas. *Pathobiology* **2020**, *87*, 125–142. [CrossRef]
102. Nixon, N.A.; Blais, N.; Ernst, S.; Kollmannsberger, C.; Bebb, M.; Butler, M.; Smylie, M.; Verma, S. Current Landscape of Immunotherapy in the Treatment of Solid Tumours, with Future Opportunities and Challenges. *Curr. Oncol.* **2018**, *25*, e373–e384. [CrossRef]
103. Patel, S.P.; Kurzrock, R. PD-L1 Expression as a Predictive Biomarker in Cancer Immunotherapy. *Mol. Cancer Ther.* **2015**, *14*, 847–856. [CrossRef]
104. Nduom, E.K.; Wei, J.; Yaghi, N.K.; Huang, N.; Kong, L.-Y.; Gabrusiewicz, K.; Ling, X.; Zhou, S.; Ivan, C.; Chen, J.Q.; et al. PD-L1 expression and prognostic impact in glioblastoma. *Neuro-Oncology* **2016**, *18*, 195–205. [CrossRef]
105. Gatalica, Z.; Snyder, C.; Maney, T.; Ghazalpour, A.; Holterman, D.A.; Xiao, N.; Overberg, P.; Rose, I.; Basu, G.D.; Vranic, S.; et al. Programmed Cell Death 1 (PD-1) and Its Ligand (PD-L1) in Common Cancers and Their Correlation with Molecular Cancer Type. *Cancer Epidemiol. Biomark. Prev.* **2014**, *23*, 2965–2970. [CrossRef]
106. Howitt, B.E.; Shukla, S.A.; Sholl, L.M.; Ritterhouse, L.L.; Watkins, J.C.; Rodig, S.; Stover, E.; Strickland, K.C.; D'andrea, A.D.; Wu, C.J.; et al. Association of Polymerase e–Mutated and Microsatellite-Instable Endometrial Cancers with Neoantigen Load, Number of Tumor-Infiltrating Lymphocytes, and Expression of PD-1 and PD-L1. *JAMA Oncol.* **2015**, *1*, 1319–1323. [CrossRef] [PubMed]
107. Le, D.T.; Durham, J.N.; Smith, K.N.; Wang, H.; Bartlett, B.R.; Aulakh, L.K.; Lu, S.; Kemberling, H.; Wilt, C.; Luber, B.S.; et al. Mismatch repair deficiency predicts response of solid tumors to PD-1 blockade. *Science* **2017**, *357*, 409–413. [CrossRef] [PubMed]
108. Murali, R.; Grisham, R.N.; Soslow, R.A. The roles of pathology in targeted therapy of women with gynecologic cancers. *Gynecol. Oncol.* **2018**, *148*, 213–221. [CrossRef] [PubMed]
109. Oza, A.M.; Elit, L.; Tsao, M.-S.; Kamel-Reid, S.; Biagi, J.; Provencher, D.M.; Gotlieb, W.H.; Hoskins, P.J.; Ghatage, P.; Tonkin, K.S.; et al. Phase II Study of Temsirolimus in Women with Recurrent or Metastatic Endometrial Cancer: A Trial of the NCIC Clinical Trials Group. *J. Clin. Oncol.* **2011**, *29*, 3278–3285. [CrossRef]
110. Emons, G.; Kurzeder, C.; Schmalfeldt, B.; Neuser, P.; de Gregorio, N.; Pfisterer, J.; Park-Simon, T.-W.; Mahner, S.; Schröder, W.; Lück, H.-J.; et al. Temsirolimus in women with platinum-refractory/resistant ovarian cancer or advanced/recurrent endometrial carcinoma. A phase II study of the AGO-study group (AGO-GYN8). *Gynecol. Oncol.* **2016**, *140*, 450–456. [CrossRef]
111. Chang, L.; Chang, M.; Chang, H.M.; Chang, F. Microsatellite Instability: A Predictive Biomarker for Cancer Immunotherapy. *Appl. Immunohistochem. Mol. Morphol.* **2018**, *26*, e15–e21. [CrossRef] [PubMed]
112. Reck, M.; Rodríguez-Abreu, D.; Robinson, A.G.; Hui, R.; Csőszi, T.; Fülöp, A.; Gottfried, M.; Peled, N.; Tafreshi, A.; Cuffe, S.; et al. Updated Analysis of KEYNOTE-024: Pembrolizumab Versus Platinum-Based Chemotherapy for Advanced Non–Small-Cell Lung Cancer with PD-L1 Tumor Proportion Score of 50% or Greater. *J. Clin. Oncol.* **2019**, *37*, 537–546. [CrossRef]

113. Lee, M.; Samstein, R.M.; Valero, C.; Chan, T.A.; Morris, L.G. Tumor mutational burden as a predictive biomarker for checkpoint inhibitor immunotherapy. *Hum. Vaccines Immunother.* **2020**, *16*, 112–115. [CrossRef]
114. Gandhi, L.; Rodríguez-Abreu, D.; Gadgeel, S.; Esteban, E.; Felip, E.; De Angelis, F.; Domine, M.; Clingan, P.; Hochmair, M.J.; Powell, S.F.; et al. Pembrolizumab plus Chemotherapy in Metastatic Non–Small-Cell Lung Cancer. *N. Engl. J. Med.* **2018**, *378*, 2078–2092. [CrossRef]
115. Hamid, O.; Robert, C.; Daud, A.; Hodi, F.S.; Hwu, W.J.; Kefford, R.; Wolchok, J.D.; Hersey, P.; Joseph, R.; Weber, J.S.; et al. Five-year survival outcomes for patients with advanced melanoma treated with pembrolizumab in KEYNOTE-001. *Ann. Oncol.* **2019**, *30*, 582–588. [CrossRef]
116. Di Tucci, C.; Capone, C.; Galati, G.; Iacobelli, V.; Schiavi, M.C.; Di Donato, V.; Muzii, L.; Panici, P.B. Immunotherapy in endometrial cancer: New scenarios on the horizon. *J. Gynecol. Oncol.* **2019**, *30*, e46. [CrossRef] [PubMed]
117. Ott, P.A.; Bang, Y.-J.; Berton-Rigaud, D.; Elez, E.; Pishvaian, M.J.; Rugo, H.S.; Puzanov, I.; Mehnert, J.M.; Aung, K.L.; Lopez, J.; et al. Safety and Antitumor Activity of Pembrolizumab in Advanced Programmed Death Ligand 1–Positive Endometrial Cancer: Results from the KEYNOTE-028 Study. *J. Clin. Oncol.* **2017**, *35*, 2535–2541. [CrossRef] [PubMed]
118. Makker, V.; Rasco, D.W.; Dutcus, C.E.; Stepan, D.E.; Li, D.; Schmidt, E.V.; Shumaker, R.C.; Taylor, M.H. A phase Ib/II trial of lenvatinib (LEN) plus pembrolizumab (Pembro) in patients (Pts) with endometrial carcinoma. *J. Clin. Oncol.* **2017**, *35*, 5598. [CrossRef]
119. Gunderson, C.C.; Fader, A.N.; Carson, K.A.; Bristow, R.E. Oncologic and Reproductive outcomes with progestin therapy in women with endometrial hyperplasia and grade 1 Adenocarcinoma: A systematic review. *Gynecol. Oncol.* **2012**, *125*, 477–482. [CrossRef] [PubMed]
120. Gracia, C.R.; Jeruss, J.S. Lives in the Balance: Women with Cancer and the Right to Fertility Care. *J. Clin. Oncol.* **2013**, *31*, 668–669. [CrossRef] [PubMed]
121. Baker, J.; Obermair, A.; Gebski, V.; Janda, M. Efficacy of oral or intrauterine device-delivered progestin in patients with complex endometrial hyperplasia with atypia or early endometrial adenocarcinoma: A meta-analysis and systematic review of the literature. *Gynecol. Oncol.* **2012**, *125*, 263–270. [CrossRef]
122. Lee, A.J.; Yang, E.J.; Kim, N.K.; Kim, Y.; Suh, D.H.; Kim, J.; Son, J.-H.; Kong, T.-W.; Chang, S.-J.; Hwang, D.W.; et al. Fertility-sparing hormonal treatment in patients with stage I endometrial cancer of grade 2 without myometrial invasion and grade 1–2 with superficial myometrial invasion: Gynecologic Oncology Research Investigators coLLaborAtion study (GORILLA-2001). *Gynecol. Oncol.* **2023**, *174*, 106–113. [CrossRef]
123. Yang, Z.; Yang, X.; Liu, X.; Ma, K.; Meng, Y.; Yin, H. Clinical characteristics and prognostic charac-terization of endometrial carcinoma: A comparative analysis of molecular typing protocols. *BMC Cancer* **2023**, *2023*, 23.
124. Church, D.N.; Stelloo, E.; Nout, R.A.; Valtcheva, N.; Depreeuw, J.; ter Haar, N.; Noske, A.; Amant, F.; Tomlinson, I.P.M.; Wild, P.J.; et al. Prognostic Significance of POLE Proofreading Mutations in Endometrial Cancer. *Gynecol. Oncol.* **2014**, *107*, 402. [CrossRef]
125. Cavaliere, A.F.; Perelli, F.; Zaami, S.; D'indinosante, M.; Turrini, I.; Giusti, M.; Gullo, G.; Vizzielli, G.; Mattei, A.; Scambia, G.; et al. Fertility Sparing Treatments in Endometrial Cancer Patients: The Potential Role of the New Molecular Classification. *Int. J. Mol. Sci.* **2021**, *22*, 12248. [CrossRef]
126. Park, J.Y.; Nam, J.H. Progestins in the Fertility-Sparing Treatment and Retreatment of Patients with Primary and Recurrent Endometrial Cancer. *Oncologist* **2015**, *20*, 270–278. [CrossRef]
127. Floyd, J.L.; Campbell, S.; Rauh-Hain, J.A.; Woodard, T. Fertility preservation in women with early-stage gynecologic cancer: Optimizing oncologic and reproductive outcomes. *Int. J. Gynecol. Cancer* **2021**, *31*, 345–351. [CrossRef] [PubMed]
128. Trojano, G.; Olivieri, C.; Tinelli, R.; Damiani, G.R.; Pellegrino, A.; Cicinelli, E. Conservative treatment in early stage endometrial cancer: A review. *Acta Biomed.* **2019**, *90*, 405–410. [CrossRef] [PubMed]
129. Piatek, S.; Michalski, W.; Sobiczewski, P.; Bidzinski, M.; Szewczyk, G. The results of different fertility-sparing treatment modalities and obstetric outcomes in patients with early endometrial cancer and atypical endometrial hyperplasia: Case series of 30 patients and systematic review. *Eur. J. Obstet. Gynecol. Reprod. Biol.* **2021**, *263*, 139–147. [CrossRef]
130. Shah, J.S.; Jooya, N.D.; Woodard, T.L.; Ramirez, P.T.; Fleming, N.D.; Frumovitz, M. Reproductive counseling and pregnancy outcomes after radical trachelectomy for early stage cervical cancer. *J. Gynecol. Oncol.* **2019**, *30*, e45. [CrossRef]
131. Lee, N.K.; Cheung, M.K.; Shin, J.Y.; Husain, A.; Teng, N.N.; Berek, J.S.; Kapp, D.S.; Osann, K.; Chan, J.K. Prognostic Factors for Uterine Cancer in Reproductive-Aged Women. *Obstet. Gynecol.* **2007**, *109*, 655–662. [CrossRef]
132. Garzon, S.; Uccella, S.; Zorzato, P.C.; Bosco, M.; Franchi, M.; Student, V.; Mariani, A. Fertility-sparing management for endometrial cancer: Review of the literature. *Minerva Med.* **2020**, *112*, 55–69. [CrossRef]
133. Zhang, Z.; Huang, H.; Feng, F.; Wang, J.; Cheng, N. A pilot study of gonadotropin-releasing hormone agonist combined with aromatase inhibitor as fertility-sparing treatment in obese patients with endometrial cancer. *J. Gynecol. Oncol.* **2019**, *30*, e61. [CrossRef]
134. Fujimoto, A.; Ichinose, M.; Harada, M.; Hirata, T.; Osuga, Y.; Fujii, T. The outcome of infertility treatment in patients undergoing assisted reproductive technology after conservative therapy for endometrial cancer. *J. Assist. Reprod. Genet.* **2014**, *31*, 1189–1194. [CrossRef]
135. Yang, B.; Xie, L.; Zhang, H.; Zhu, Q.; Du, Y.; Luo, X.; Chen, X. Insulin resistance and overweight prolonged fertility-sparing treatment duration in endometrial atypical hyperplasia patients. *J. Gynecol. Oncol.* **2018**, *29*, e35. [CrossRef]

136. Cappelletti, E.H.; Humann, J.; Torrejón, R.; Gambadauro, P. Chances of pregnancy and live birth among women undergoing conservative management of early-stage endometrial cancer: A systematic review and meta-analysis. *Hum. Reprod. Updat.* **2022**, *28*, 282–295. [CrossRef] [PubMed]
137. Andress, J.; Pasternak, J.; Walter, C.; Kommoss, S.; Krämer, B.; Hartkopf, A.; Steinmacher, S. Fertility preserving management of early endometrial cancer in a patient cohort at the department of women's health at the university of Tuebingen. *Arch. Gynecol. Obstet.* **2021**, *304*, 215–221. [CrossRef] [PubMed]
138. Ayhan, A.; Tohma, Y.A.; Tunc, M. Fertility preservation in early-stage endometrial cancer and endometrial intraepithelial neoplasia: A single-center experience. *Taiwan. J. Obstet. Gynecol.* **2020**, *59*, 415–419. [CrossRef] [PubMed]
139. Cade, T.J.; Quinn, M.A.; Rome, R.M.; Neesham, D. Long-term outcomes after progestogen treatment for early endometrial cancer. *Aust. N. Z. J. Obstet. Gynaecol.* **2013**, *53*, 566–570. [CrossRef]
140. Casadio, P.; Guasina, F.; Paradisi, R.; Leggieri, C.; Caprara, G.; Seracchioli, R. Fertility-sparing treatment of endometrial cancer with initial infiltration of myometrium by resectoscopic surgery: A pilot study. *Oncologist* **2018**, *23*, 478–480. [CrossRef]
141. Chen, M.; Jin, Y.; Li, Y.; Bi, Y.; Shan, Y.; Pan, L. Oncologic and reproductive outcomes after fertility-sparing management with oral progestin for women with complex endometrial hyperplasia and endometrial cancer. *Int. J. Gynaecol. Obstet.* **2016**, *132*, 34–38. [CrossRef]
142. Choi, M.C.; Jung, S.G.; Park, H.; Cho, Y.H.; Lee, C.; Kim, S.J. Fertility preservation via photodynamic therapy in young patients with early-stage uterine endometrial cancer: A long-term follow-up study. *Int. J. Gynecol. Cancer* **2013**, *23*, 698–704. [CrossRef]
143. Duska, L.R.; Garrett, A.; Rueda, B.R.; Haas, J.; Chang, Y.; Fuller, A.F. Endometrial cancer in women 40 years old or younger. *Gynecol. Oncol.* **2001**, *83*, 388–393. [CrossRef]
144. Falcone, F.; Laurelli, G.; Losito, S.; Di Napoli, M.; Granata, V.; Greggi, S. Fertility preserving treatment with hysteroscopic resection followed by progestin therapy in young women with early endometrial cancer. *J. Gynecol. Oncol.* **2017**, *28*, e2. [CrossRef]
145. Giampaolino, P.; Di Spiezio Sardo, A.; Mollo, A.; Raffone, A.; Travaglino, A.; Boccellino, A.; Bifulco, G. Hysteroscopic endometrial focal resection followed by levonorgestrel intrauterine device insertion as a fertility-sparing treatment of atypical endometrial hyperplasia and early endometrial cancer: A retrospective study. *J. Minim. Invasive Gynecol.* **2019**, *26*, 648–656. [CrossRef]
146. Kaku, T.; Yoshikawa, H.; Tsuda, H.; Sakamoto, A.; Fukunaga, M.; Kuwabara, Y.; Kamura, T. Conservative therapy for adenocarcinoma and atypical endometrial hyperplasia of the endometrium in young women: Central pathologic review and treatment outcome. *Cancer Lett.* **2001**, *167*, 39–48. [CrossRef] [PubMed]
147. Kim, M.K.; Seong, S.J.; Kim, Y.S.; Song, T.; Kim, M.-L.; Yoon, B.S.; Lee, Y.H. Combined medroxyprogesterone acetate/levonorgestrel-intrauterine system treatment in young women with early-stage endometrial cancer. *Am. J. Obstet. Gynecol.* **2013**, *209*, e1–e4. [CrossRef] [PubMed]
148. Kudesia, R.; Singer, T.; Caputo, T.A.; Holcomb, K.M.; Kligman, I.; Rosenwaks, Z.; Gupta, D. Reproductive and oncologic outcomes after progestin therapy for endometrial complex atypical hyperplasia or carcinoma. *Am. J. Obstet. Gynecol.* **2014**, *210*, 255.e1–255.e4. [CrossRef]
149. Leone Roberti Maggiore, U.; Martinelli, F.; Dondi, G.; Bogani, G.; Chiappa, V.; Evangelista, M.T.; Raspagliesi, F. Efficacy and fertility outcomes of levonorgestrel-releasing intra-uterine system treatment for patients with atypical complex hyperplasia or endometrial cancer: A retrospective study. *J. Gynecol. Oncol.* **2019**, *30*, e57. [CrossRef] [PubMed]
150. Minaguchi, T.; Nakagawa, S.; Takazawa, Y.; Nei, T.; Horie, K.; Fujiwara, T.; Taketani, Y. Combined phospho-Akt and PTEN expressions associated with post-treatment hysterectomy after conservative progestin therapy in complex atypical hyperplasia and stage Ia, G1 adenocarcinoma of the endometrium. *Cancer Lett.* **2007**, *248*, 112–122. [CrossRef]
151. Minig, L.; Franchi, D.; Boveri, S.; Casadio, C.; Bocciolone, L.; Sideri, M. Progestin intrauterine device and GnRH analogue for uterus-sparing treatment of endometrial precancers and well-differentiated early endometrial carcinoma in young women. *Ann. Oncol.* **2011**, *22*, 643–649. [CrossRef]
152. Niwa, K.; Tagami, K.; Lian, Z.; Onogi, K.; Mori, H.; Tamaya, T. Outcome of fertility-preserving treatment in young women with endometrial carcinomas. *BJOG Int. J. Obstet. Gynaecol.* **2005**, *112*, 317–320. [CrossRef]
153. Ohyagi-Hara, C.; Sawada, K.; Aki, I.; Mabuchi, S.; Kobayashi, E.; Ueda, Y.; Kimura, T. Efficacies and pregnant outcomes of fertility-sparing treatment with medroxyprogesterone acetate for endometrioid adenocarcinoma and complex atypical hyperplasia: Our experience and a review of the literature. *Arch. Gynecol. Obstet.* **2015**, *291*, 151–157. [CrossRef]
154. Ota, T.; Yoshida, M.; Kimura, M.; Kinoshita, K. Clinicopathologic study of uterine endometrial carcinoma in young women aged 40 years and younger. *Int. J. Gynecol. Cancer Off. J. Int. Gynecol. Cancer Soc.* **2005**, *15*, 657–662. [CrossRef]
155. Park, J.Y.; Seong, S.J.; Kim, T.J.; Kim, J.W.; Kim, S.M.; Bae, D.S.; Nam, J.H. Pregnancy outcomes after fertility-sparing management in young women with early endometrial cancer. *Obstet. Gynecol.* **2013**, *121*, 136–142. [CrossRef]
156. Pashov, A.I.; Tskhay, V.B.; Ionouchene, S.V. The combined GnRH-agonist and intrauterine levonorgestrel-releasing system treatment of complicated atypical hyperplasia and endometrial cancer: A pilot study. *Gynecol. Endocrinol. Off. J. Int. Soc. Gynecol. Endocrinol.* **2012**, *28*, 559–561. [CrossRef] [PubMed]
157. Perri, T.; Korach, J.; Gotlieb, W.H.; Beiner, M.; Meirow, D.; Friedman, E.; Ben-Baruch, G. Prolonged conservative treatment of endometrial cancer patients: More than 1 pregnancy can be achieved. *Int. J. Gynecol. Cancer* **2011**, *21*, 72–78. [CrossRef] [PubMed]
158. Raffone, A.; Travaglino, A.; Flacco, M.E.; Iasevoli, M.; Mollo, A.; Guida, M.; Zullo, F. Clinical predictive factors of response to treatment in patients undergoing conservative management of atypical endometrial hyperplasia and early endometrial cancer. *J. Adolesc. Young Adult Oncol.* **2021**, *10*, 193–201. [CrossRef] [PubMed]

159. Shan, B.-E.; Ren, Y.-L.; Sun, J.-M.; Tu, X.-Y.; Jiang, Z.-X.; Ju, X.-Z.; Wang, H.-Y. A prospective study of fertility-sparing treatment with megestrol acetate following hysteroscopic curettage for well-differentiated endometrioid carcinoma and atypical hyperplasia in young women. *Arch. Gynecol. Obstet.* **2013**, *288*, 1115–1123. [CrossRef]
160. Shirali, E.; Yarandi, F.; Eftekhar, Z.; Shojaei, H.; Khazaeipour, Z. Pregnancy outcome in patients with stage 1a endometrial adenocarcinoma, who conservatively treated with megestrol acetate. *Arch. Gynecol. Obstet.* **2012**, *285*, 791–795. [CrossRef]
161. Ushijima, K.; Yahata, H.; Yoshikawa, H.; Konishi, I.; Yasugi, T.; Saito, T.; Kamura, T. Multicenter phase II study of fertility-sparing treatment with medroxyprogesterone acetate for endometrial carcinoma and atypical hyperplasia in young women. *J. Clin. Oncol.* **2007**, *25*, 2798–2803. [CrossRef]
162. Wang, C.-J.; Chao, A.; Yang, L.-Y.; Hsueh, S.; Huang, Y.-T.; Chou, H.-H.; Lai, C.-H. Fertility-preserving treatment in young women with endometrial adenocarcinoma: A long-term cohort study. *Int. J. Gynecol. Cancer* **2014**, *24*, 718–728. [CrossRef]
163. Wang, F.; Yu, A.; Xu, H.; Zhang, X.; Li, L.; Lou, H.; Lin, J. Fertility preserved hysteroscopic approach for the treatment of stage Ia endometrioid carcinoma. *Int. J. Gynecol. Cancer* **2017**, *27*, 1919–1925. [CrossRef]
164. Yamagami, W.; Susumu, N.; Makabe, T.; Sakai, K.; Nomura, H.; Kataoka, F.; Aoki, D. Is repeated high-dose medroxyprogesterone acetate (MPA) therapy permissible for patients with early stage endometrial cancer or atypical endometrial hyperplasia who desire preserving fertility? *J. Gynecol. Oncol.* **2018**, *29*, e21. [CrossRef]
165. Yamazawa, K.; Hirai, M.; Fujito, A.; Nishi, H.; Terauchi, F.; Ishikura, H.; Isaka, K. Fertility-preserving treatment with progestin, and pathological criteria to predict responses, in young women with endometrial cancer. *Hum. Reprod.* **2007**, *22*, 1953–1958. [CrossRef]
166. Yu, M.; Yang, J.-X.; Wu, M.; Lang, J.-H.; Huo, Z.; Shen, K. Fertility-preserving treatment in young women with well-differentiated endometrial carcinoma and severe atypical hyperplasia of endometrium. *Fertil. Steril.* **2009**, *92*, 2122–2124. [CrossRef] [PubMed]
167. Zhou, R.; Yang, Y.; Lu, Q.; Wang, J.; Miao, Y.; Wang, S.; Wei, L. Prognostic factors of oncological and reproductive outcomes in fertility-sparing treatment of complex atypical hyperplasia and low-grade endometrial cancer using oral progestin in Chinese patients. *Gynecol. Oncol.* **2015**, *139*, 424–428. [CrossRef] [PubMed]
168. Njoku, K.; Chiasserini, D.; Whetton, A.D.; Crosbie, E.J. Proteomic Biomarkers for the Detection of Endometrial Cancer. *Cancers* **2019**, *11*, 1572. [CrossRef] [PubMed]
169. Banach, P.; Suchy, W.; Dereziński, P.; Matysiak, J.; Kokot, Z.J.; Nowak-Markwitz, E. Mass spectrometry as a tool for biomarkers searching in gynecological oncology. *Biomed. Pharmacother.* **2017**, *92*, 836–842. [CrossRef]
170. Shruthi, B.S.; Vinodhkumar, P.; Selvamani. Proteomics: A new perspective for cancer. *Adv. Biomed. Res.* **2016**, *5*, 67. [CrossRef]
171. Arend, R.C.; Jones, B.A.; Martinez, A.; Goodfellow, P. Endometrial cancer: Molecular markers and management of advanced stage disease. *Gynecol. Oncol.* **2018**, *150*, 569–580. [CrossRef]
172. Finotello, F.; Rieder, D.; Hackl, H.; Trajanoski, Z. Next-generation computational tools for interrogating cancer immunity. *Nat. Rev. Genet.* **2019**, *20*, 724–746. [CrossRef]
173. Mittal, P.; Klingler-Hoffmann, M.; Arentz, G.; Zhang, C.; Kaur, G.; Oehler, M.K.; Hoffmann, P. Proteomics of endometrial cancer diagnosis, treatment, and prognosis. *Proteom. Clin. Appl.* **2015**, *10*, 217–229. [CrossRef]
174. Yurkovetsky, Z.; Ta'Asan, S.; Skates, S.; Rand, A.; Lomakin, A.; Linkov, F.; Marrangoni, A.; Velikokhatnaya, L.; Winans, M.; Gorelik, E.; et al. Development of multimarker panel for early detection of endometrial cancer. High diagnostic power of pro-lactin. *Gynecol. Oncol.* **2007**, *107*, 58–65. [CrossRef]
175. Li, L.-M.; Zhu, Y.; Zhong, Y.; Su, T.; Fan, X.-M.; Xi, Q.; Li, M.-Y.; Fu, J.; Tan, H.; Liu, S. Human epididymis protein 4 in en-dometrial cancer: A meta-analysis. *Clin. Chim. Acta* **2018**, *482*, 215–223. [CrossRef]
176. Zanotti, L.; Bignotti, E.; Calza, S.; Bandiera, E.; Ruggeri, G.; Galli, C.; Tognon, G.; Ragnoli, M.; Romani, C.; Tassi, R.A.; et al. Human epididymis protein 4 as a serum marker for diagnosis of endometrial carcinoma and prediction of clinical outcome. *Clin. Chem. Lab. Med.* **2012**, *50*, 2189–2198. [CrossRef] [PubMed]
177. Omer, B.; Genc, S.; Takmaz, O.; Dirican, A.; Kusku-Kiraz, Z.; Berkman, S.; Gurdol, F. The diagnostic role of human epididymis protein 4 and serum amyloid-A in early-stage endometrial cancer patients. *Tumor Biol.* **2013**, *34*, 2645–2650. [CrossRef] [PubMed]
178. Angioli, R.; Plotti, F.; Capriglione, S.; Montera, R.; Damiani, P.; Ricciardi, R.; Aloisi, A.; Luvero, D.; Cafà, E.V.; Dugo, N.; et al. The role of novel biomarker HE4 in endometrial cancer: A case control prospective study. *Tumor Biol.* **2013**, *34*, 571–576. [CrossRef] [PubMed]
179. Moore, R.G.; Brown, A.K.; Miller, M.C.; Badgwell, D.; Lu, Z.; Allard, W.J.; Granai, C.; Bast, R.C., Jr.; Lu, K. Utility of a novel serum tumor biomarker HE4 in patients with endometrioid adenocarcinoma of the uterus. *Gynecol. Oncol.* **2008**, *110*, 196–201. [CrossRef]
180. Martinez-Garcia, E.; Lopez-Gil, C.; Campoy, I.; Vallve, J.; Coll, E.; Cabrera, S.; Cajal, S.R.Y.; Matias-Guiu, X.; Van Oostrum, J.; Reventos, J.; et al. Advances in endometrial cancer protein biomarkers for use in the clinic. *Expert Rev. Proteom.* **2018**, *15*, 81–99. [CrossRef]
181. Diefenbach, C.S.; Shah, Z.; Iasonos, A.; Barakat, R.R.; Levine, D.A.; Aghajanian, C.; Sabbatini, P.; Hensley, M.L.; Konner, J.; Tew, W.; et al. Preoperative serum YKL-40 is a marker for detection and prognosis of endometrial cancer. *Gynecol. Oncol.* **2007**, *104*, 435–442. [CrossRef]
182. Jamaluddin, M.F.B.; Ko, Y.-A.; Ghosh, A.; Syed, S.M.; Ius, Y.; O'sullivan, R.; Netherton, J.K.; Baker, M.A.; Nahar, P.; Jaaback, K.; et al. Proteomic and functional characterization of intra-tumor heterogeneity in human endometrial cancer. *Cell Rep. Med.* **2022**, *3*, 100738. [CrossRef]

183. Berstein, L.M.; Berlev, I.V.; Baltrukova, A.N. Endometrial cancer evolution: New molecular-biologic types and hormo-nal-metabolic shifts. *Future Oncol.* **2017**, *13*, 2593–2605. [CrossRef]
184. Marusyk, A.; Almendro, V.; Polyak, K. Intra-tumour heterogeneity: A looking glass for cancer? *Nat. Rev. Cancer* **2012**, *12*, 323–334. [CrossRef]
185. Marusyk, A.; Polyak, K. Tumor heterogeneity: Causes and consequences. *Biochim. Biophys. Acta* **2010**, *1805*, 105–117. [CrossRef]
186. Michor, F.; Polyak, K. The Origins and Implications of Intratumor Heterogeneity. *Cancer Prev. Res.* **2010**, *3*, 1361–1364. [CrossRef] [PubMed]
187. Liu, F.; Liu, L.; Zheng, J. Expression of annexin A2 in adenomyosis and dysmenorrhea. *Arch. Gynecol. Obstet.* **2019**, *300*, 711–716. [CrossRef] [PubMed]
188. Romano, A.; Rižner, T.L.; Werner, H.M.J.; Semczuk, A.; Lowy, C.; Schröder, C.; Griesbeck, A.; Adamski, J.; Fishman, D.; Tokarz, J. Endometrial cancer diagnostic and prognostic algorithms based on proteomics, metabolomics, and clinical data: A systematic review. *Front. Oncol.* **2023**, *13*, 1120178. [CrossRef] [PubMed]

Disclaimer/Publisher's Note: The statements, opinions and data contained in all publications are solely those of the individual author(s) and contributor(s) and not of MDPI and/or the editor(s). MDPI and/or the editor(s) disclaim responsibility for any injury to people or property resulting from any ideas, methods, instructions or products referred to in the content.

 International Journal of *Molecular Sciences*

Review

KRAS-Dependency in Pancreatic Ductal Adenocarcinoma: Mechanisms of Escaping in Resistance to KRAS Inhibitors and Perspectives of Therapy

Enrico Gurreri [1,2], Giannicola Genovese [2,3,4,5], Luigi Perelli [2], Antonio Agostini [1], Geny Piro [1], Carmine Carbone [1,*] and Giampaolo Tortora [1,6]

[1] Medical Oncology, Fondazione Policlinico Universitario Agostino Gemelli, IRCCS, 00168 Rome, Italy; enrico.gurreri01@icatt.it (E.G.); antonio.agostini@unicatt.it (A.A.); geny.piro@policlinicogemelli.it (G.P.); giampaolo.tortora@policlinicogemelli.it (G.T.)
[2] Department of Genitourinary Medical Oncology, The University of Texas MD Anderson Cancer Center, Houston, TX 77025, USA; ggenovese@mdanderson.org (G.G.); lperelli@mdanderson.org (L.P.)
[3] Department of Genomic Medicine, The University of Texas MD Anderson Cancer Center, Houston, TX 77025, USA
[4] David H. Koch Center for Applied Research of Genitourinary Cancers, The University of Texas MD Anderson Cancer Center, Houston, TX 77025, USA
[5] Translational Research to Advance Therapeutics and Innovation in Oncology, The University of Texas MD Anderson Cancer Center, Houston, TX 77025, USA
[6] Medical Oncology, Università Cattolica del Sacro Cuore, 00168 Rome, Italy
* Correspondence: carmine.carbone@policlinicogemelli.it

Citation: Gurreri, E.; Genovese, G.; Perelli, L.; Agostini, A.; Piro, G.; Carbone, C.; Tortora, G. KRAS-Dependency in Pancreatic Ductal Adenocarcinoma: Mechanisms of Escaping in Resistance to KRAS Inhibitors and Perspectives of Therapy. *Int. J. Mol. Sci.* **2023**, *24*, 9313. https://doi.org/10.3390/ijms24119313

Academic Editor: Laura Paleari

Received: 5 May 2023
Revised: 18 May 2023
Accepted: 24 May 2023
Published: 26 May 2023

Copyright: © 2023 by the authors. Licensee MDPI, Basel, Switzerland. This article is an open access article distributed under the terms and conditions of the Creative Commons Attribution (CC BY) license (https:// creativecommons.org/licenses/by/ 4.0/).

Abstract: Pancreatic ductal adenocarcinoma (PDAC) is still one of the deadliest cancers in oncology because of its increasing incidence and poor survival rate. More than 90% of PDAC patients are KRAS mutated (KRASmu), with KRASG12D and KRASG12V being the most common mutations. Despite this critical role, its characteristics have made direct targeting of the RAS protein extremely difficult. KRAS regulates development, cell growth, epigenetically dysregulated differentiation, and survival in PDAC through activation of key downstream pathways, such as MAPK-ERK and PI3K-AKT-mammalian target of rapamycin (mTOR) signaling, in a KRAS-dependent manner. KRASmu induces the occurrence of acinar-to-ductal metaplasia (ADM) and pancreatic intraepithelial neoplasia (PanIN) and leads to an immunosuppressive tumor microenvironment (TME). In this context, the oncogenic mutation of KRAS induces an epigenetic program that leads to the initiation of PDAC. Several studies have identified multiple direct and indirect inhibitors of KRAS signaling. Therefore, KRAS dependency is so essential in KRASmu PDAC that cancer cells have secured several compensatory escape mechanisms to counteract the efficacy of KRAS inhibitors, such as activation of MEK/ERK signaling or YAP1 upregulation. This review will provide insights into KRAS dependency in PDAC and analyze recent data on inhibitors of KRAS signaling, focusing on how cancer cells establish compensatory escape mechanisms.

Keywords: pancreatic cancer; KRAS; KRAS inhibitors; KRAS-dependency; resistance; escaping

1. Introduction

Pancreatic ductal adenocarcinoma (PDAC) is still one of the deadliest cancers in oncology and, due to its increasing incidence and overall five-year survival rate of less than 5% [1], it is expected to be the second leading cause of cancer-related death in the US by 2030 [2,3]. In addition, the annual incidence of PDAC is increasing in people younger than 30 years of age [4].

Among modifiable risk factors, current cigarette smoking, alcohol use, chronic pancreatitis, and obesity have strong associations with PDAC [3].

In most cases, PDAC is diagnosed at an advanced stage, locally advanced (30–35%) or metastatic (50–55%) [2], and treated with polychemotherapy regimens, including FOLFIRINOX, Gemcitabine/Nab-Paclitaxel, and nanoliposomal Irinotecan/Fluorouracil, with a survival benefit of 2–6 months compared with a single-agent Gemcitabine [3,5–7].

The tumor microenvironment (TME) plays a central role in PDAC biology and represents a desmoplastic scaffold characterized by intricate cellular and acellular crosstalk between activated cancer-associated fibroblasts (CAFs), tumor-associated macrophages (TAMs), myeloid-derived suppressive cells (MDSCs), regulatory T cells (Tregs), and bioactive specialized extracellular matrix (ECM), with low numbers of tumor-infiltrating lymphocytes (TILs). In this scenario, cell–cell interactions are obstructed by the ECM, which explains not only chemoresistance, but also poor response to immunotherapy.

In the heterogenous mutational landscape of pancreatic cancer, *KRAS*, *TP53*, *SMAD4*, and *CDKN2A* represent major oncogenic events involved in key molecular pathways such as DNA damage repair, cell cycle regulation, TGF-β signaling, chromatin regulation, and axonal guidance [8].

About 30% of all human cancers bear activating rat sarcoma (*RAS*) mutations; in particular, *Kirsten rat sarcoma* (*KRAS*) mutations are considerably more frequent than *Harvey rat sarcoma virus oncogene* (*HRAS*) and *Neuroblastoma RAS Viral Oncogene Homolog* (*NRAS*) mutations. As for PDAC, more than 90% of patients are *KRAS* mutated (KRASmu), and KRASG12D and KRASG12V are specifically the most common mutations [9].

The KRAS protein is a molecular switch that cycles between an active, Guanosine-5'-triphosphate (GTP)–bound state and an inactive, Guanosine-5'-diphosphate (GDP)–bound form. In cancer tumorigenesis, *KRAS* mutations typically increase the steady-state levels of the active form, driving protumorigenic pathways, such as the mitogen-activated protein kinase (MAPK) and Phosphatidylinositol 3-kinase (PI3K) pathways.

The first genetic event that leads the earliest precancerous lesions to invasive pancreatic cancer is the mutational activation of KRAS [10,11]. Despite this critical role, its high affinity for nucleotide and the lack of viable binding pockets for small-molecule inhibitors have made direct targeting of the RAS protein extremely difficult over the past four decades [9,12]. In this context, understanding PDAC tumorigenesis is crucial for both the identification of early diagnostic markers and the development of multiple alternative modes of intervention.

In this review, we explore KRAS-dependency in PDAC and analyze recent data on KRAS signaling inhibitors, focusing on how cancer cells establish compensatory escape mechanisms.

2. KRAS-Dependent Tumorigenesis in PDAC

Several studies have demonstrated the strong association between PDAC and inflammation. In the context of chronic pancreatitis, the inflammatory microenvironment can activate survival and proliferation programs and induce chromatin changes in cancer cells, promoting tumor growth. Oncogenic *KRAS* accelerates this process in pancreatic tissue (Figure 1) [13–15], inducing, along with inflammatory damage (e.g., cerulean-induced pancreatitis) and other tumor suppressor deficiencies (e.g., protein (p)16INK4a/p14ARF, Tumor Protein p53 (TP53) and/or Suppressor of Mothers against Decapentaplegic 4 (SMAD4)) loss, the appearance of neoplastic precursor lesions, such as acinar-to-ductal metaplasia (ADM) and pancreatic intraepithelial neoplasia (PanIN) [15].

2.1. KRASmu, Inflammation and Precursor Lesions

KRASmu expression is not sufficient to initiate tumorigenesis in the pancreas, and for ADM induction and development of pancreatic acinar cells, further subsequent events are necessary, such as additional genetic lesions, chronic inflammation, or upregulation of growth factor signaling [11,16,17].

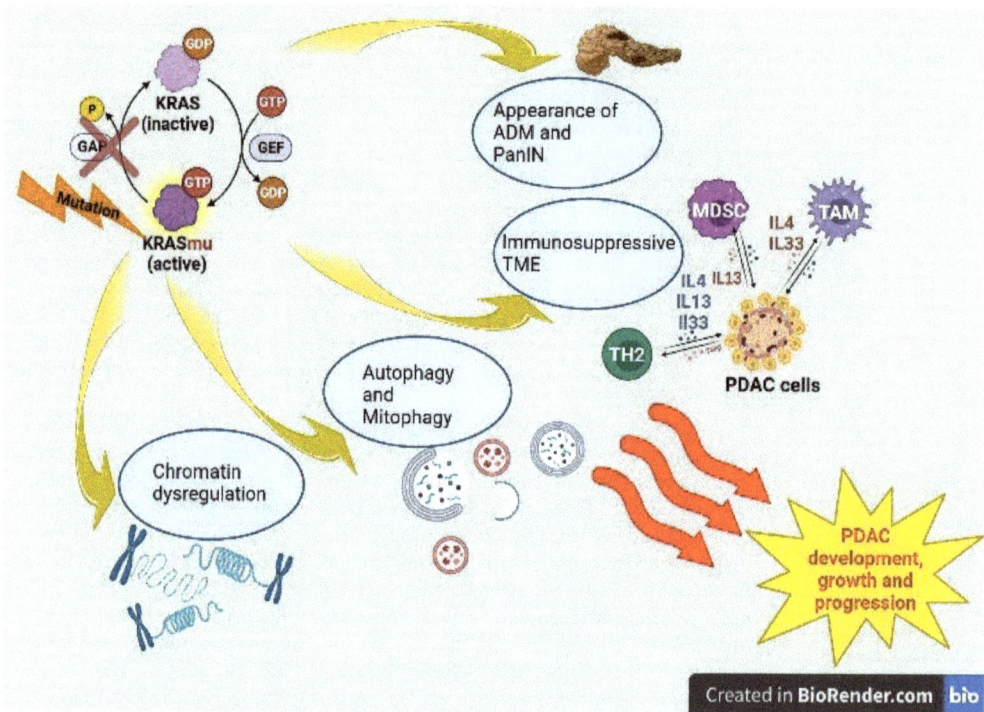

Figure 1. KRAS-dependent tumorigenesis in PDAC. In PDAC tumorigenesis, KRAS mutations typically increase the steady-state levels of the active form driving protumorigenic pathways. KRASmu negatively influences the regeneration program by inducing the appearance of neoplastic precursor lesions, such as acinar-to-ductal metaplasia (ADM) and pancreatic intraepithelial neoplasia (PanIN). Its influence on metabolism leads to fluxes of autophagy and mitophagy. Higher levels of cytokines IL4, IL13 and IL-33, secreted by GATA-3+ TH2-polarized CD4+ T cells, are found in KRASmu PDACs, resulting in an immunosuppressive tumor microenvironment (TME). Oncogenic KRAS mutation induces an epigenetic program, an alternative to physiological regeneration, which leads to PDAC initiation.

PanIN is considered the major pathological basis of PDAC development with properties of ductal cells as well as tumor cells in PDAC. However, recent data in engineered mouse models (GEMM) of pancreatic cancer have evidenced that acinar cells are the main cellular origin of PDAC. Acinar cells, through ADM, show increased expression of ductal cell markers such as cytokeratin-19 (CK-19) or sex-determining region Y box 9 (SOX9) and reduced expression of acinar cell markers, such as amylase or MIST-1 [18–20]. In the presence of KRASmu, the ADM process become irreversible and leads to a change in cellular identity (transdifferentiation) and progression to PanIN. Transcription factors controlling pancreatic duct development, such as SOX9 and hepatocyte nuclear factor 6 (HNF6), or other ones critical for somatic stem cell reprogramming, such as Kruppel-like factor 4 (KLF4), have been demonstrated to regulate ADM process [18]. Ge W and colleagues demonstrated that SOX9 and Phos-SOX9 (S181) levels in acinar cells are both regulated by miR-802, a pancreatic microRNA (miRNA), which controls ADM formation in the presence of oncogenic KRAS [21]. Costamagna A and colleagues demonstrated that integrin and growth factor receptor signaling converge on p130Cas, an adaptor protein encoded by *BRCA1* and a downstream effector of the KRAS pathway, to induce tumorigenesis and boost acinar to ductal metaplasia and subsequent tumorigenesis through PI3K activation [11].

The nuclear factor kappa-light-chain-enhancer of activated B cells (NFκB) family of transcription factors is composed of NFκB2 (p100/p52), NFκB1 (p105/p50), RelA/p65, RelB, and Rel. NFκB2 is necessary for KRASG12D-dependent ADM development, PanIN progression, and tumor proliferation. In the same way, RelB promotes PanIN progression in the KRASG12D PDAC cells [22].

Recent data have shown that *ring finger protein 43* (*RNF43*) is the most frequently mutated gene in IPMNs together with *KRAS*. RNF43 is a member of the RING finger protein family, and an E3 ubiquitin ligase. It mediates the ubiquitination and degradation of the Wnt receptor complex component frizzled, and chains transcription factor 4 (TCF4) to the nuclear membrane by silencing its transcriptional activity. Thus, the loss of RNF43 activates the Wnt/β-catenin signaling pathway [23].

RNA sequencing analysis of PDAC patients in recent works showed high expression of angiopoietin-like 4 (ANGPTL4), which is involved in epidermal differentiation and development of PDAC. Yan HH et al. revealed a tumor-promoting role of ANGPTL4 through regulation of periostin/integrin signaling during PDAC initiation and maintenance. These data suggested the ANGPTL4/periostin axis as a potential molecular target for the prevention of PDAC [24].

IL-33 has a central role in this program, bridging tissue damage with KRAS-dependent epithelial plasticity and accelerating the appearance of early precursor lesions (PanIN) after injury. Additionally, in early neoplasia this cytokine induces an immunosuppressive TME [25].

2.2. KRASmu and Metabolism

KRASmu PDACs depend on glucose and glutamine for energy production and maintenance of the redox balance and show decreased levels of intracellular amino acid. On the other hand, autophagy is required for the maintenance of KRASmu cells [26]. Moreover, recent studies showed how autophagy increases MHC degradation as an escape mechanism from T lymphocytes surveillance and makes tumor cells resistant to immunomodulatory drugs [27].

KRASmu pancreatic cancer cells can channel their glucose metabolism away from the mitochondria through programmed mitophagy via the mediator BCL2/adenovirus E1B 19-kDa-interacting protein 3-like (BNIP3L/NIX). NIX ablation in vivo has been shown to delay the progression of PanIN to PDAC. In this context, Viale A et al. demonstrated that quiescent KRASmu pancreatic cancer cells that survive oncogene ablation, which are responsible for tumor recurrence, have cancer stem cells characteristics and depend on oxidative phosphorylation (OXPHOS) for survival [28]. Glutamine metabolism is involved in redox homeostasis and plays a key role in tumor growth [29].

Previous work has highlighted the crucial role of branched-chain amino acid (BCAA) metabolism in metabolic adaptation in cancer. In particular, BCAA transaminase 1 (BCAT1) or BCAT2 has been shown to be upregulated and important for proliferation in PDAC [30].

2.3. KRASmu and TME

Mutant KRASG12D upregulates peroxisome proliferator activated receptor-delta (PPARδ) in human and murine PanIN lesions, promoting inflammation-related signaling pathways and pancreatic tumorigenesis. PPARδ hyperactivation recruits TAMs and MDSCs via the CCL2/CCR2 axis, remodeling the immune TME. This process has been demonstrated in mouse models fed with high-fat diets enriched with fatty acids that are natural ligands of PPARδ [31].

KRASmu upregulates IL2Rγ and IL4R, two members of the type I cytokine receptor family, the former of which contributes to PDAC tumorigenesis. In fact, the cytokines IL4 and IL13 signaling via IL4R drive JAK-STAT-cMYC activation, resulting in increased glycolysis and tumor growth [30]. Previous works have shown that cMYC expression enhances tumor cell proliferation [32]. In KRASmu PDACs are found higher levels of cytokines IL4, IL13, and, as mentioned in this work, IL-33, secreted by GATA-3+ TH2 polarized CD4+ T cells, resulting in immunosuppressive TME and a self-perpetuating cascade of pro-tumorigenic effects [31]. In addition, KRASmu is crucial for the crosstalk between PDAC cells and activated cancer-associated fibroblasts (CAFs). CAFs are activated through transforming growth factor beta (TGF-β) and sonic hedgehog pathways and regulate the tumor stroma, including extracellular matrix, collagen fibers, and hyaluronic acid, promoting PDAC cell growth and an immunosuppressive TME [10].

Moreover, oncogenic KRASG12D suppresses programmed death ligand 1 (PD-L1) expression and has the strongest suppressive power compared with other KRAS mutations [33].

2.4. KRASmu and Chromatin

In the injury-prone pancreas, oncogenic KRAS mutation induces an epigenetic program alternative to physiological regeneration, which leads to PDAC initiation. Alonso-Curbelo D et al. showed in vivo that pancreatic metaplasia is characterized by epigenetic silencing of acinar identity loci that is enhanced by bromodomain containing 4 (BRD4) suppression [25]. The bromodomain and extraterminal (BET) family member BRD4 is a chromatin reader which binds acetylated and active chromatin enhancing transcription of cell-identity genes.

In KRASmu cells, within 48 h of tissue damage, progression to neoplasia is facilitated by interactions between genetic and environmental insults leading to an 'acinar-to-neoplasia' chromatin switch that alters DNA accessibility. This chromatin remodeling program is present from the early stage of disease to metastasis in advanced PDAC [25].

Sangrador I et al. demonstrated that the transcriptional repressor zinc finger E-box binding homebox 1 (ZEB1) is a key mediator of KRAS-dependent oncogenesis in vivo; indeed, in the presence of a KRAS mutation, ZEB1 haploinsufficiency delays PDAC development. Notably, ZEB1 is predominantly expressed in stromal myofibroblasts associated with PanIN and PDAC [34]. Genovese G et al. highlight the crucial tumor-suppressive role of SWI/SNF related, matrix associated, actin dependent regulator of chromatin, subfamily B, member 1 (SMARCB1) as a differentiation checkpoint and a gatekeeper of epithelial–mesenchymal transition. This is a novel mechanism of KRAS-dependent tumorigenesis of PDAC cells that fails to activate downstream KRAS signaling (e.g., through MAPK) [35]. SMARCB1 is a switch/sucrose non-fermentable (SWI/SNF) chromatin remodeling factor whose activity restrains growth and metabolic programs via MYC activation [35].

3. KRAS Signaling Inhibitors in PDAC
3.1. KRASG12C Inhibition

The frequency of KRASG12C mutations in PDAC patients is abnormally high in some regions such as Japan, while its frequency in PDAC patients worldwide is quite low. However, none of the KRASG12C inhibitors, such as sotorasib or adagrasib, have been approved as a treatment for PDAC [36]. A previous work has demonstrated the growth inhibition power of Adagrasib (MRTX849) in a pancreatic cancer cell line [37,38] and it is undergoing clinical trials for patients with KRASG12C mutant pancreatic cancer (NCT03785249) (Table 1). Additionally, a confirmed partial response has been reported in the phase I/Ib cohort in a patient with PDAC (NCT03785249). Ostrem et al. pinpointed the druggable switch-II pocket in KRASG12C through X-ray crystallography and mass spectrometry [38]. Only a small percentage of PDAC patients harbor the G12C mutation; on the other hand, as previously mentioned in this review, G12D is the most prevalent KRAS mutation in PDAC.

Table 1. Registered trials of KRASmu signaling inhibitor combination therapy on clinicaltrials.gov, accessed on 6 Marzh 2023.

ClinicalTrials.Gov Identifier	Title	Phase	Drugs	Targets
NCT03785249	Phase 1/2 Study of MRTX849 in Patients with Cancer Having a KRAS G12C Mutation KRYSTAL-1	I/II	MRTX849 (Adagrasib)	KRAS^{G12C}
NCT03948763	A Study of mRNA-5671/V941 as Monotherapy and in Combination with Pembrolizumab (V941-001)	I	mRNA-5671/V941 Pembrolizumab	KRAS^{mu} PD-1
NCT03592888	DC Vaccine in Pancreatic Cancer	I	mDC3/8	KRAS^{mu}
NCT04117087	Pooled Mutant KRAS-Targeted Long Peptide Vaccine Combined with Nivolumab and Ipilimumab for Patients with Resected MMR-p Colorectal and Pancreatic Cancer	I	KRAS peptide vaccine Nivolumab Ipilimumab	KRAS^{mu} PD-1 CTLA-4
NCT03745326	Administering Peripheral Blood Lymphocytes Transduced with a Murine T-Cell Receptor Recognizing the G12D Variant of Mutated RAS in HLA-A*11:01 Patients	I/II	Cyclophosphamide Fludarabine Aldesleukin Anti-KRAS G12D mTCR PBL	KRAS^{G12D}
NCT03190941	Administering Peripheral Blood Lymphocytes Transduced with a Murine T-Cell Receptor Recognizing the G12V Variant of Mutated RAS in HLA-A*11:01 Patients	I/II	Cyclophosphamide Fludarabine Aldesleukin Anti-KRAS G12V mTCR PBL	KRAS^{G12V}
NCT04330664	Adagrasib in Combination with TNO155 in Patients with Cancer (KRYSTAL 2)	I/II	MRTX849 (Adagrasib) TNO155	KRAS^{G12C} SHP2
NCT04185883	Sotorasib Activity in Subjects with Advanced Solid Tumors with KRAS p.G12C Mutation (CodeBreak 101)	I/II	Sotorasib AMG 404 Trametinib RMC-4630 Afatinib Pembrolizumab Panitumumab Carboplatin, pemetrexed, docetaxel, paclitaxel Atezolizumab Everolimus Palbociclib MVASI® (bevacizumab-awwb) TNO155 FOLFIRI FOLFOX BI 1701963	KRAS^{G12C} PD-1 MAP2K1 SHP2 EGFR PD-L1 mTOR CDK4/6 VEGF SOS1

PD-1—programmed cell death protein 1; CTLA-4—cytotoxic T-lymphocyte antigen 4; SHP2—Src homology region 2-containing protein tyrosine phosphatase 2; MAP2K1—mitogen-activated protein kinase kinase; EGFR—epithelial growth factor receptor; PD-L1—programmed death ligand 1; mTOR—mammalian target of rapamycin; CDK4/6—cyclin-dependent kinase 4/6; VEGF—vascular endothelial growth factor; SOS1—son of sevenless.

3.2. KRASG12D Inhibition

Various groups have shown that it is possible to pharmacologize GTP-bound KRASG12D: KD2 is a cyclic peptide that can selectively target the switch-II groove in mutant GTP-bound KRASG12D; both in vitro and in vivo, KS-58, a bicyclic peptide, has shown activity against KRASG12D mutated pancreatic cancer [39]. MRTX1133 is a small molecule that selectively targets KRASG12D by blocking downstream pathways through inhibition of nucleotide exchange and binding of effector rapidly accelerated fibrosarcoma 1 (RAF1); in vivo, MRTX1133 reduced phosphorylation of extracellular signal-regulated kinase (ERK), resulting in tumor regression [33]. Using a medicinal chemistry approach, other compounds were discovered: TH-Z827 and TH-Z835 are two inhibitors that bind with Asp12 inside the switch-II pocket, specifically inhibiting KRAS signaling, and not KRASG12C or WT, in G12D mutant PDAC in vitro and in vivo; in vitro, KD-8 is another inhibitor of KRASG12D PDAC tumor growth [33]. These works provide proof of concept evidence that the KRASG12D mutation could potentially be targeted, benefiting a larger number of PDAC patients [38].

3.3. Other Inhibitors

RNA interference (RNAi) has been demonstrated to suppress KRASmu expression in pancreatic cancer cells inhibiting anchorage-independent growth and tumorigenic proliferation. These effects suggest RNAi as a potential drug for KRASmu PDAC [10]. Messenger ribonucleic acid (mRNA)-5671/V941 is a novel mRNA vaccine encoding mutant KRAS which is being studied in patients with KRASmu cancers in phase I clinical trial (NCT03948763), with or without Pembrolizumab. Similarly, dendritic cell vaccines (NCT03592888) and peptide vaccines (NCT04117087) for KRASmu patients are in clinical trials (Table 1) [38].

Engineering patient lymphocytes to express receptors that specifically target tumor neoantigens is known as adoptive cell therapy [40]. A previous work has generated murine T cells that can recognize KRASG12D PDAC in an HLA-A*11:01 restricted manner and inhibit tumor growth in vivo [41]. Two ongoing phase I/II clinical trials are investigating the transfer of such cells engineered to express the murine T-cell receptor (TCR) specific for KRASG12D (NCT03745326) or KRASG12V (NCT03190941) (Table 1) in an HLA-restricted manner, in patients with solid tumors, including KRASG12D PDAC [33,38].

Tricomplex inhibitors bind cyclophilin A, a chaperone protein ubiquitously present inside the cell [42], which in turn binds the target protein, creating a target-inhibitor-cyclophilin-A complex. RMC-9805 is a novel specific tricomplex inhibitor of KRASG12D which suppresses tumor growth in xenograft PDAC [33]. RMC-6236, on the other hand, is a multi-RAS inhibitor tricomplex, capable of targeting multiple different mutations of KRAS, such as KRASG12V [38].

For KRAS to localize on the membrane and become active, giving rise to signal transduction, its post-translational prenylation catalyzed by farnesyl transferases (FTases) is required [43]. For this reason, a number of FTase inhibitors (FTIs), e.g., lonafarnib and tipifarnib, have been designed and clinically studied [44], reaching phase III clinical trials for various cancer types, with disappointing results in PDAC [38].

4. Mechanisms of Escaping in Resistance to KRAS Inhibitors

Despite promising results in preclinical and clinical studies of several inhibitors outlined in this paper, there are still few studies on mechanisms of escaping in resistance to KRASG12D inhibitors. So far, even MRTX1133, a KRASG12D inhibitor, has not entered clinical trials yet. Thus, most of the data about mechanisms of escaping in resistance to KRASG12D need to be translated from studies upon clinical application of KRASG12C inhibitors.

Pre- and post-treatment comparison of samples from patients treated with the KRASG12C inhibitor sotorasib showed multiple genetic escape mutations after treatment in 63% of patients, including *KRAS*, *NRAS*, *BRAF*, *EGFR*, *FGFR2*, and *MYC* [36]. Further exploration revealed that the inhibitory effect of sotorasib was reduced after the occurrence of KRASG12V, NRASQ61K or MRASQ71R (a small GTPase regulating the dimerization and activation of CRAF, a RAF family protein in KRAS downstream of the MAPK-ERK pathway [36]). The binding site for a series of KRASG12C inhibitors, including sotorasib, is the cysteine 12 residue of the KRASG12C protein, which is located near the switch-II pocket [45]. Mutations at this site, such as KRASY96D found in a patient after treatment with the KRASG12C inhibitor MRTX849, disrupt the hydrogen bond between the binding site and KRASG12C inhibitor [36]. In the development of resistance, other acquired genetic mutations in *KRAS* are G12D/R/W, G13D, Q61H, R68S, H95D/Q/R, Y96C, and KRASG12C allele amplification, as well as genetic changes such as *MET* (also known as the N-methyl-N0-nitroso-guanidine human osteosarcoma transforming gene) amplification, mutational activation of mitogen-activated protein kinase kinase 1 (MAP2K1) also known as *MEK* and rearranged during transfection (*RET*), oncogene fusion of anaplastic lymphoma kinase (*ALK*), *RET*, *B-RAF*, *RAF1*, and fibroblast growth factor receptor 3 (*FGFR3*), loss-of-function mutations in neurofibromatosis 1 (*NF1*) and phosphatase and tensin homolog (*PTEN*) [12,38]. These mutations increase the levels of active GTP-bound KRAS protein, preventing drug binding [46]. It has been demonstrated that KRASmu inhibitors, such as sotorasib, or MEK inhibitors, such as trametinib, to induce escape mechanisms in PDAC, cause activation of the mechanistic target of rapamycin 2 (mTORC2) molecule rapamycin-insensitive companion of mammalian target of rapamycin (RICTOR) and phosphorylation of AKT at Ser-473 by integrin-linked kinase (ILK) in several PDAC mouse models and human tumors [38]. After these results, it was shown that inhibition of mTORC2 alone stimulates ERK activation [47], promoting cell survival. On the other hand, mTORC2 signaling has an important role in the development of resistance in PDAC; in fact, its inhibitors have anti-tumor activity in PDAC cells when combined with an inhibitor of KRASmu or MEK [38]. Because PDAC is characterized by high cell heterogeneity, it is highly likely that future use of KRASG12D inhibitors in PDAC may lead to the same result by increasing mutational burden with the previously mentioned mutations and genetic changes [36].

Cheng DK et al. demonstrated that oncogenic KRASmu inhibits wild type (WT) RAS signaling through NF1/Ribosomal S6 kinase 1 (RSK1) [48]. In fact, the inhibition of this negative feedback pathway with KRAS inhibitors activates WT RAS signaling and promotes adaptive resistance, as evidenced by PDAC cells that survived after treatment with KRASG12V inhibitor. In addition, inhibition of both WT RAS, through son of sevenless 1 (SOS1) inhibition, and KRASG12C mutation, through AMG 510, showed the best response in vitro, demonstrating a synergy between KRASG12C inhibitors and upstream effectors (SOS1) inhibitors. These data strengthen the idea of WT RAS as a central actor in the acquired resistance to KRASmu inhibitors [49].

Alterations of multiple receptor tyrosine kinase (RTK)-RAS-MAPK pathways contribute to the resistance to KRASG12C inhibitor adagrasib, and the combination of inhibitors of RTK, Src homology region 2-containing protein tyrosine phosphatase 2 (SHP2), and KRASmu is the subject of ongoing clinical trials (NCT04330664, NCT04185883) (Table 1) [12,38].

Moreover, KRASmu tumors have another intrinsic mechanism of resistance; in fact, in the presence of FTIs described above, KRAS is prenylated by geranylgeranyl transferase 1 (GGTase1), the so-called alternative prenylation [38].

As previously stated in this paper, the TME is composed of various types of cells and plays a central role in chemoresistance and immunoresistance in PDAC. CAFs have been shown to contribute to resistance to therapies such as KRASmu inhibitors, and to express vitamin D receptor, which plays an important role in the development of resistance; confirming this, a work has shown how calcipotriol, a vitamin D receptor agonist, can override the influence of CAFs in a murine model [38]. As stated above, autophagy is necessary for the maintenance of KRASmu cells; in fact, cancer cells utilize several metabolites derived from autophagic degradation of CAFs, resulting in resistance to different therapies including KRAS inhibitors [50]. As well as CAFs, neurons present in the TME release amino acid serine, promoting PDAC growth [38].

5. Alternative Targets in KRAS Signaling Pathways and Future Perspectives

5.1. Alternative Targets in KRAS Signaling Pathway

An effective strategy to overcome resistance to KRAS inhibitors is combining different types of therapies, e.g., the combination of KRASG12C inhibitors with SHP2 (an upstream effector of RAS) inhibitors has been shown to surpass resistance to KRASG12C inhibitors alone and remodel TME in PDAC models, reducing the number of activated CAFs [49]. On the other hand, the combination of KRAS inhibitors with downstream target therapies, such as MAPK-ERK and PI3K-AKT-mammalian target of rapamycin (mTOR) pathways inhibitors, showed disappointing results in pancreatic cancer models. Nevertheless, the inhibitory effect of histone deacetylase (HDAC) synergizes with the combined targeting of the MAPK-ERK and PI3K-AKT-mTOR pathways [36]. Promising new combinations include KRASG12C inhibitors with cell cycle checkpoints or immune checkpoints inhibitors (ICIs) [37,45], or the triple inhibition of KRASG12C/SHP2/PD-L1, tested in PDAC murine models [51]. Recent data suggest nuclear export protein exportin 1 (XPO1), which transports protein cargo from the nucleus to the cytoplasm, as an important factor in relieving tumor cells from resistance to KRASG12C inhibitors [36].

Another promising factor in resistance to KRAS inhibition is the deubiquitinase ubiquitin specific peptidase 21 (USP21), which at is amplified and overexpressed in about 20% of PDAC patient samples. Its nuclear activity promotes pancreatic tumor growth and tumor stem cell properties. USP21 deubiquitinates the transcription factor 7 (TCF7) and amplifies canonical Wnt signaling [52]. Further explorations are needed to test USP21 inhibitors with or without KRAS inhibitors as possible therapies in PDAC [53].

Different molecules that prevent the interaction between son of sevenless (SOS1) and KRAS have been identified [38]. SOS1 is a guanine nucleotide exchange factor (GEF) that binds GDP-bound KRAS and catalyzes the switch of GDP for GTP, activating KRAS [54]. Its low levels result in inhibition of tumor growth. BI-3406 is an SOS1 inhibitor active only in KRASmu cells with anti-tumoral activity synergistic with MEK inhibitors. BI-1701963 is another SOS1 inhibitor which is being studied in an ongoing phase I clinical trial (NCT04111458) with or without MEK inhibitor (trametinib) in patients with KRASmu cancers (Table 2) [38].

SHP2 is a tyrosine phosphatase protein encoded by the gene *PTPN11* with an intrinsic regulatory mechanism [38]. Its role in KRAS signaling seems to be linked to other proteins such as SOS1 and growth factor receptor-bound protein 2 (GRB2) and it is involved in various pathways such as KRAS-MAPK signaling [55]. SHP2 plays a central role in cancer development in KRASmu PDAC and NSCLC models. Its inhibition has been seen as synergistic with MEK inhibition, stopping tumor growth in PDAC and NSCLC models in vivo [38]. SHP099 is a compound that locks SHP2 in its inactive, autoinhibited state and is able to inhibit tumor growth via the MAPK pathway in vivo [56]. Several potential inhibitors of SHP2 are in different ongoing clinical trials [57]: RMC-4630 is currently in phase I/Ib clinical trials with or without an ERK inhibitor (NCT03634982, NCT04916236); RMC-4550 has shown to inhibit KRASmu cells proliferation in preclinical models [58]; TNO155, an allosteric inhibitor of SHP2 with demonstrated anti-tumoral activity through MAPK, is in several ongoing phase I and II clinical trials with or without several synergistic targeted therapies (NCT04000529, NCT04330664, NCT03114319) (Tables 1 and 2) [59]; other SHP2 inhibitors are in clinical trials, such as ERAS-601 (NCT04670679), JAB-3312 (NCT04121286), BBP-398 (NCT04528836), and RLY-1971 (NCT04252339) (Table 2) [38].

5.2. Future Perspectives

Proteolysis targeting chimeras (PROTACs) consist of two peptides, one that binds the target protein linked to another peptide that recruits an E3 ubiquitin ligase for proteasomal degradation of the target protein [60]. LC-2 is a PROTAC that targets KRASG12C composed of MRTX849 bound to a von Hippel–Lindau (VHL) recruiter peptide; the former binds covalently to KRASmu, and the latter induces sustained proteasomal degradation of

KRASmu and subsequent MAPK inhibition. The compound 17f is a PROTAC that targets phosphodiesterase 6 (PDEδ), an important prenylation factor [38].

Table 2. Registered trials of alternative targets inhibitor combination therapy on https://clinicaltrials.gov, accessed on 6 March 2023.

ClinicalTrials.Gov Identifier	Title	Phase	Drugs	Targets
NCT04111458	A Study to Test Different Doses of BI 1701963 Alone and Combined with Trametinib in Patients with Different Types of Advanced Cancer (Solid Tumors with KRAS Mutation)	I	BI 1701963 Trametinib	SOS1 MAP2K1
NCT03634982	Dose Escalation of RMC-4630 Monotherapy in Relapsed/Refractory Solid Tumors	I	RMC-4630	SHP2
NCT04916236	Combination Therapy of RMC-4630 and LY3214996 in Metastatic KRAS Mutant Cancers (SHERPA)	I	RMC-4630 LY3214996	SHP2 ERK
NCT04000529	Phase Ib Study of TNO155 in Combination with Spartalizumab or Ribociclib in Selected Malignancies	I	TNO155 Spartalizumab Ribociclib	SHP2 CDK4/6 PD-1
NCT03114319	Dose Finding Study of TNO155 in Adult Patients with Advanced Solid Tumors	I	TNO155 EGF816 (nazartinib)	SHP2 EGFR
NCT04670679	A Dose Escalation/Expansion Study of ERAS-601 in Patients with Advanced or Metastatic Solid Tumors (FLAGSHP-1)	I	ERAS-601 Cetuximab Pembrolizumab	SHP2 EGFR PD-1
NCT04121286	A Study of JAB-3312 in Adult Patients with Advanced Solid Tumors in China	I	JAB-3312	SHP2
NCT04528836	First-in-Human Study of the SHP2 Inhibitor BBP-398 in Patients with Advanced Solid Tumors	I	BBP-398 (Formerly known as IACS-15509)	SHP2
NCT04252339	RLY-1971 in Subjects with Advanced or Metastatic Solid Tumors	I	RLY-1971	SHP2

SOS1—son of sevenless; MAP2K1—mitogen-activated protein kinase kinase; EGFR—epithelial growth factor receptor; SHP2—Src homology region 2-containing protein tyrosine phosphatase 2; ERK—extracellular signal-regulated kinase; CDK4/6—cyclin-dependent kinase 4/6; PD-L1—programmed death ligand 1.

NS-1 is a monobody inhibitor of RAS dimerization that targets the α4–α5 interface of KRAS, able to prevent the development and progression of pancreatic cancer in murine model. The unique problem with this potential new strategy is the size of the molecule NS-1, which makes its intracellular localization difficult [38].

Peptide nucleic acids (PNAs) are synthetic nucleotide analogs capable of binding to specific complementary DNA and RNA sequences [36] or to the mRNA of the target gene, inhibiting its translation process [61]. In a previous work, PNAs significantly inhibited tumor cell activity and reduced KRASG12D gene expression in the human metastatic pancreatic adenocarcinoma cell line AsPC-1 [36].

As previously mentioned in this paper, WT KRAS can induce resistance to KRAS inhibitors with its compensatory effect, and in this setting pan-RAS inhibitors could overcome this mechanism. Jin Wang et al. have developed several small-molecule pan-RAS inhibitors that stabilize the "open non-signaling intermediate conformation" of RAS [62]: NSC290956 (also called Spiclomazine or APY606) inhibited the proliferation of KRAS-dependent pancreatic cancer cell lines CFPAC-1 (KRASG12V), MIA PaCa-2 (KRASG12C), Capan-1 (KRASG12V), SW1990 (KRASG12T), and BxPC-3 (WT KRAS) [36]; NSC48160 showed similar effect on CPFAC-1 (KRASG12V) and BxPC-3 (WT KRAS) [63] and induced apoptosis in MIA PaCa-2 (KRASG12C) [64]; inhibitory effects of NSC48693 on KRAS-dependent cancer cells were superior to those of NSC48160 on CFPAC-1(KRASG12V), MIA PaCa-2 (KRASG12C), and BxPC-3 (WT KRAS) cells [36].

We have already mentioned a class of tricomplex inhibitors that has shown promising results; another similar class of compounds is the tricomplex RAS-ON (RASON) inhibitors, comprising KRASG12C, G12D, and G13C inhibitors, and a G12X inhibitor which targets multiple different G12 mutations, which are currently being studied after initial preclinical data [65].

As mentioned earlier in this review, IL4/IL13 cytokines play a central role in TME remodeling via the IL4-IL4R-Janus kinase (JAK)-signal transducer and activator of transcription proteins (STAT) signaling cascade. However, trials with JAK1/2 inhibitors have yielded disheartening results [66]. Further explorations are necessary to study interactions between cancer cells and TME cells.

We have already discussed RNAi, which still poses a challenge because of enzymatic breakdown, renal clearance, and precise targeting of the tissue of interest. Loaded siRNAs targeting KRAS into local drug eluter (LODER), a biodegradable polymetric matrix that protects siRNA allowing constant local release of siRNA inside tumor tissues, have shown anti-tumoral activity towards human pancreatic tumor cells by increasing the survival of murine models. In an open-label phase I–IIa study with 15 patients enrolled, LODER siRNAs in combination with FOLFIRINOX showed a median overall survival rate of 15 months and an 18-month survival rate of 38.5%. A phase II study of patients with locally advanced PDAC is ongoing to test siRNAG12D LODER in combination with standard therapy [10]. Another approach for siRNA delivery against KRASmu is represented by exosomes, also termed inhibitory exosomes (iExosomes), which have shown efficacy in several preclinical models of pancreatic cancer [10]. In addition, CD47, a 'do not eat me' signal present on iExosomes, enables them to enter cells via micropinocytosis that is enhanced by KRASmu-dependent tumorigenesis. A clinical trial is pending to address the feasibility, safety, and efficacy of iExosomes in patients with metastatic pancreatic cancer [10,67]. One, 2-dioleoyl-sn-glycero-3-phosphatidylcholine (DOPC) is the major component of a promising nanoliposomal platform used in vivo for KRAS-targeting siRNA delivery [10].

Recent data have already challenged conventional knowledge that *KRAS* mutations (e.g., G12C, G12D, G12V or G12S) cannot hydrolyze GTP and return to the GDP-bound state and have demonstrated that these mutations are still able to hydrolyze GTP, suggesting a further fine-tuning regulation of RAS. The aforementioned RASON, a novel protein encoded by the long non-protein intergenic coding RNA 00673 (LINC00673), is the first identified positive regulator of KRAS that binds directly to it, stabilizing its hyperactive state in a way that differs from guanine nucleotide exchange factors (GEFs) or GTPase-activating proteins (GAPs) [9].

As we stated above, the BET family member BRD4 is a chromatin reader which binds acetylated and active chromatin, enhancing transcription of cell-identity genes with a tumor-permissive role. Principe et al. have developed XP-524, a promising BET inhibitor that has shown encouraging results in combination with gemcitabine or PARP inhibitors by restraining the effects of *KRAS* activating mutations. In addition, XP-524 increases CD4+ and CD8+ T cell populations [68], suggesting a possible strategy to sensitize KRASmu PDAC to immune checkpoint inhibition.

Jonghwa Jin et al. explored glutamine metabolism in PDAC, targeting glutamine transporters as a promising strategy for advanced or drug-resistant cancers [29].

Dixon defined ferroptosis as a type of iron-dependent non-apoptotic cell death in KRASmu cancer cells in 2012 [69], however oncogenic KRAS makes cells more resistant to ferroptosis through upregulation of ferroptosis suppressor protein 1 (FSP1). Based on these data, Müller F and colleagues suggested ferroptosis induction combined with FSP1 inhibition as a new therapeutic strategy against KRASmu cancers [70].

6. KRAS Dependency in PDAC

KRAS regulates development, cell growth, epigenetically dysregulated differentiation, and survival in PDAC through activation of key downstream pathways, such as MAPK-ERK and PI3K-AKT-mTOR signaling, in a KRAS-dependent manner [71]. Previous works showed that some pancreatic cancer cell lines survive to KRASmu silencing and then depend on PI3K. This mechanism induces overexpression of yes-associated protein 1 (YAP1), an important transcriptional co-activator of the hippo pathway, thus escaping KRAS inhibition [36]. Furthermore, higher dosage of the mutant allele of *KRAS* than its WT counterpart has been associated with poor prognosis in cancer patients. This phenomenon is called "mutant allele-specific imbalance" (MASI) [71]. In addition, recent data have shown that restoration of WT KRAS in pancreatic cancer cells induces inhibition of nuclear translocation of YAP1 [71,72], which is associated with poor prognosis in PDAC patients [73]. In mouse models, after KRAS inactivation, one-third of spontaneous tumor recurrences that escape KRASG12D dependency shows deficit of KRAS expression and YAP1 amplification, resulting in an aggressive quasi-mesenchymal phenotype with the activation of cell cycle and DNA repair pathways through cooperation of the transcription factor E2F [10].

Chen et al. observed that murine PDAC cells, when KRAS is sustainedly silenced, undergo a reversible cell state change without mutational or transcriptional alterations, characterized by morphological changes and tumor-promoting activity with activation of the focal adhesion pathway, suggesting that the latter is a possible manifestation of acquired KRAS independency [38].

USP21's ability to bypass KRAS dependency occurs in the cytoplasm with a novel mechanism different from the previously mentioned KRAS escape mechanisms, such as activation of MEK/ERK signaling or YAP1 upregulation. Hou et al. evidenced that USP21 reduces autophagy and increases amino acid levels, resulting in upregulation of MTOR-associated signaling [74]. Further analysis demonstrated that microtubule affinity regulating kinase 3 (MARK3), a microtubule-binding kinase and regulator of microtubule dynamics, is directly deubiquitinated by USP21, leading to KRAS-independent growth, cancer development, and macropinocytosis, a key metabolism mechanism for KRASmu PDAC cell survival.

Similarly, Hou et al. showed that an upregulation of HDAC5 enhances the recruitment of TAMs into the TME, promoting resistance to KRAS inhibitors through the activation of the C-C motif chemokine ligand 2 (CCL2)/C-C motif chemokine receptor 2 (CCR2) axis and of transforming growth factor-β (TGF-β) in a SMAD-4 dependent manner, bypassing KRAS-dependency [38,75].

Several clinical studies have shown that KRAS inhibition blocks both PI3K-AKT-MTOR and MAPK signaling in KRAS-dependent tumors [76]. On the other hand, inhibition of the MAPK pathway alone leads to hyperactivation of the PI3K-AKT-mTOR pathway via different RTKs such as AXL and Platelet derived growth factor receptor alpha (PDGFRa), and activation of several escape circuits, such as recruitment of insulin receptor kinase by MTORC1 and MEK inhibition-mediated hyperactivation of the ERBB receptors epithelial growth factor receptor (EGFR), human epidermal growth factor receptor 2 (HER2), and ERBB3. However, RTK activation after MEK inhibition has been demonstrated in both RASmu and WT RAS tumor models and recent data have evidenced that combinations of inhibitors are highly toxic [76].

Thus, KRAS dependency is so strong and essential in KRASmu PDAC that cancer cells have secured several compensatory escape mechanisms to counteract the effectiveness of KRAS inhibitors [77].

7. Conclusions

Pancreatic ductal adenocarcinoma remains a real challenge in oncology. A lot of new possibilities are now actual therapies, but more studies are needed to refine these new strategies.

In the past 30 years, around 70 reviews on KRAS's role in PDAC have been published. We conducted this review because we felt it was necessary to collect the most up-to-date data covering the past 5 years and create a new comprehensive information base to rely on for future research. We sought to summarize and make the topic clear. We have focused on KRAS-dependency, enlightening this key role in PDAC tumorigenesis. We found that KRASmu influences cells at early stages of tumorigenesis in different ways. There are several oncogenic mutations of *KRAS*, and we focused on the more frequent ones in PDAC.

We reviewed all types of direct and undirect inhibitors of KRAS signaling, its upstream and downstream effectors, critically analyzing what is now affirmed as reality and current therapies. We showed how these new strategies are limited by several mechanisms of escape, highlighting the necessity of much more studies to understand how to overcome these limitations.

ADM and to a greater extent PanIN represent the critical elements in the establishment of oncogenic *KRAS* dependency and its further effects. In this setting, PPARδ is a potential target to prevent PanIN cancerization [31].

Chromatin remodeling plays a central role in identity changes of ADM in the inflamed and injured pancreas, suggesting a tool for early detection of epigenetically dysregulated programs in PDAC development [25]. These changes often lead to cancer stem cells characteristics and oxidative phosphorylation (OXPHOS) for cancer survival, showing other new viable targets [28]. Often, the PDAC phenotype changes to aggressive mesenchymal type, showing how SWI/SNF-controlled proteostasis, with its chromatin remodeler SMARCB1, need to be further explored to better understand the epithelial–mesenchymal transition in PDAC [35].

Several emerging therapies use inhibitors of different players in KRAS signaling, such as p130Cas [11], dual inhibition of FTase and GGTase activity [38], or RASON [9] in KRASmu PDAC.

However, the endless combination possibilities of inhibitors lead to infinite possibilities of mechanisms of escaping and resistance that we have yet to fully understand and overcome in order to definitively win against PDAC.

Author Contributions: Conceptualization, writing—original draft preparation, writing—review and editing, supervision, E.G., G.G., L.P., A.A., G.P., C.C. and G.T. All authors have read and agreed to the published version of the manuscript.

Funding: This work was supported by the AIRC IG grant number 26330 to G.T., My First AIRC Grant "Luigi Bonatti e Anna Maria Bonatti Rocca" grant number 23681 to C.C., Convenzione Gemelli-FIMP Progetto CUP J38D19000690001 to G.T. and Ministry of Health CO 2019-12369662 to G.T.

Institutional Review Board Statement: Not applicable.

Informed Consent Statement: Not applicable.

Data Availability Statement: Not applicable.

Conflicts of Interest: The authors declare no conflict of interest.

References

1. Ha, C.S.R.; Müller-Nurasyid, M.; Petrera, A.; Hauck, S.M.; Marini, F.; Bartsch, D.K.; Slater, E.P.; Strauch, K. Proteomics Biomarker Discovery for Individualized Prevention of Familial Pancreatic Cancer Using Statistical Learning. *PLoS ONE* **2023**, *18*, e0280399. [CrossRef]
2. Siegel, R.L.; Miller, K.D.; Fuchs, H.E.; Jemal, A. Cancer Statistics, 2021. *CA. Cancer J. Clin.* **2021**, *71*, 7–33. [CrossRef]
3. Park, W.; Chawla, A.; O'Reilly, E.M. Pancreatic Cancer: A Review. *Jama* **2021**, *326*, 851–862. [CrossRef]
4. Sung, H.; Siegel, R.L.; Rosenberg, P.S.; Jemal, A. Emerging Cancer Trends among Young Adults in the USA: Analysis of a Population-Based Cancer Registry. *Lancet Public Health* **2019**, *4*, e137–e147. [CrossRef]
5. Conroy, T.; Desseigne, F.; Ychou, M.; Bouché, O.; Guimbaud, R.; Bécouarn, Y.; Adenis, A.; Raoul, J.-L.; Gourgou-Bourgade, S.; de la Fouchardière, C.; et al. FOLFIRINOX versus Gemcitabine for Metastatic Pancreatic Cancer. *N. Engl. J. Med.* **2011**, *364*, 1817–1825. [CrossRef] [PubMed]

6. Von Hoff, D.D.; Ervin, T.; Arena, F.P.; Chiorean, E.G.; Infante, J.; Moore, M.; Seay, T.; Tjulandin, S.A.; Ma, W.W.; Saleh, M.N.; et al. Increased Survival in Pancreatic Cancer with Nab-Paclitaxel plus Gemcitabine. *N. Engl. J. Med.* **2013**, *369*, 1691–1703. [CrossRef] [PubMed]
7. Wang-Gillam, A.; Li, C.-P.; Bodoky, G.; Dean, A.; Shan, Y.-S.; Jameson, G.; Macarulla, T.; Lee, K.-H.; Cunningham, D.; Blanc, J.F.; et al. Nanoliposomal Irinotecan with Fluorouracil and Folinic Acid in Metastatic Pancreatic Cancer after Previous Gemcitabine-Based Therapy (NAPOLI-1): A Global, Randomised, Open-Label, Phase 3 Trial. *Lancet Lond. Engl.* **2016**, *387*, 545–557. [CrossRef] [PubMed]
8. Bailey, P.; Chang, D.K.; Nones, K.; Johns, A.L.; Patch, A.-M.; Gingras, M.-C.; Miller, D.K.; Christ, A.N.; Bruxner, T.J.C.; Quinn, M.C.; et al. Genomic Analyses Identify Molecular Subtypes of Pancreatic Cancer. *Nature* **2016**, *531*, 47–52. [CrossRef] [PubMed]
9. Cheng, R.; Li, F.; Zhang, M.; Xia, X.; Wu, J.; Gao, X.; Zhou, H.; Zhang, Z.; Huang, N.; Yang, X.; et al. A Novel Protein RASON Encoded by a LncRNA Controls Oncogenic RAS Signaling in KRAS Mutant Cancers. *Cell Res.* **2023**, *33*, 30–45. [CrossRef] [PubMed]
10. Buscail, L.; Bournet, B.; Cordelier, P. Role of Oncogenic KRAS in the Diagnosis, Prognosis and Treatment of Pancreatic Cancer. *Nat. Rev. Gastroenterol. Hepatol.* **2020**, *17*, 153–168. [CrossRef]
11. Costamagna, A.; Natalini, D.; Camacho Leal, M.D.P.; Simoni, M.; Gozzelino, L.; Cappello, P.; Novelli, F.; Ambrogio, C.; Defilippi, P.; Turco, E.; et al. Docking Protein P130Cas Regulates Acinar to Ductal Metaplasia During Pancreatic Adenocarcinoma Development and Pancreatitis. *Gastroenterology* **2022**, *162*, 1242–1255.e11. [CrossRef]
12. Awad, M.M.; Liu, S.; Rybkin, I.I.; Arbour, K.C.; Dilly, J.; Zhu, V.W.; Johnson, M.L.; Heist, R.S.; Patil, T.; Riely, G.J.; et al. Acquired Resistance to KRASG12C Inhibition in Cancer. *N. Engl. J. Med.* **2021**, *384*, 2382–2393. [CrossRef] [PubMed]
13. Gidekel Friedlander, S.Y.; Chu, G.C.; Snyder, E.L.; Girnius, N.; Dibelius, G.; Crowley, D.; Vasile, E.; DePinho, R.A.; Jacks, T. Context-Dependent Transformation of Adult Pancreatic Cells by Oncogenic K-Ras. *Cancer Cell* **2009**, *16*, 379–389. [CrossRef] [PubMed]
14. Guerra, C.; Collado, M.; Navas, C.; Schuhmacher, A.J.; Hernández-Porras, I.; Cañamero, M.; Rodriguez-Justo, M.; Serrano, M.; Barbacid, M. Pancreatitis-Induced Inflammation Contributes to Pancreatic Cancer by Inhibiting Oncogene-Induced Senescence. *Cancer Cell* **2011**, *19*, 728–739. [CrossRef] [PubMed]
15. Del Poggetto, E.; Ho, I.-L.; Balestrieri, C.; Yen, E.-Y.; Zhang, S.; Citron, F.; Shah, R.; Corti, D.; Diaferia, G.R.; Li, C.-Y.; et al. Epithelial Memory of Inflammation Limits Tissue Damage While Promoting Pancreatic Tumorigenesis. *Science* **2021**, *373*, eabj0486. [CrossRef]
16. Storz, P.; Crawford, H.C. Carcinogenesis of Pancreatic Ductal Adenocarcinoma. *Gastroenterology* **2020**, *158*, 2072–2081. [CrossRef]
17. Guerra, C.; Schuhmacher, A.J.; Cañamero, M.; Grippo, P.J.; Verdaguer, L.; Pérez-Gallego, L.; Dubus, P.; Sandgren, E.P.; Barbacid, M. Chronic Pancreatitis Is Essential for Induction of Pancreatic Ductal Adenocarcinoma by K-Ras Oncogenes in Adult Mice. *Cancer Cell* **2007**, *11*, 291–302. [CrossRef]
18. Wang, L.; Xie, D.; Wei, D. Pancreatic Acinar-to-Ductal Metaplasia and Pancreatic Cancer. *Methods Mol. Biol. Clifton NJ* **2019**, *1882*, 299–308. [CrossRef]
19. Strobel, O.; Dor, Y.; Alsina, J.; Stirman, A.; Lauwers, G.; Trainor, A.; Castillo, C.F.-D.; Warshaw, A.L.; Thayer, S.P. In Vivo Lineage Tracing Defines the Role of Acinar-to-Ductal Transdifferentiation in Inflammatory Ductal Metaplasia. *Gastroenterology* **2007**, *133*, 1999–2009. [CrossRef]
20. Shi, G.; DiRenzo, D.; Qu, C.; Barney, D.; Miley, D.; Konieczny, S.F. Maintenance of Acinar Cell Organization Is Critical to Preventing Kras-Induced Acinar-Ductal Metaplasia. *Oncogene* **2013**, *32*, 1950–1958. [CrossRef]
21. Ge, W.; Goga, A.; He, Y.; Silva, P.N.; Hirt, C.K.; Herrmanns, K.; Guccini, I.; Godbersen, S.; Schwank, G.; Stoffel, M. MiR-802 Suppresses Acinar-to-Ductal Reprogramming During Early Pancreatitis and Pancreatic Carcinogenesis. *Gastroenterology* **2022**, *162*, 269–284. [CrossRef] [PubMed]
22. Hassan, Z.; Schneeweis, C.; Wirth, M.; Müller, S.; Geismann, C.; Neuß, T.; Steiger, K.; Krämer, O.H.; Schmid, R.M.; Rad, R.; et al. Important Role of Nfkb2 in the KrasG12D-Driven Carcinogenesis in the Pancreas. *Pancreatology* **2021**, *21*, 912–919. [CrossRef] [PubMed]
23. Zhou, X.; Sun, Z.; Zhang, M.; Qu, X.; Yang, S.; Wang, L.; Jing, Y.; Li, L.; Deng, W.; Liu, F.; et al. Deficient Rnf43 Potentiates Hyperactive Kras-Mediated Pancreatic Preneoplasia Initiation and Malignant Transformation. *Anim. Models Exp. Med.* **2022**, *5*, 61–71. [CrossRef]
24. Yan, H.H.; Jung, K.H.; Lee, J.E.; Son, M.K.; Fang, Z.; Park, J.H.; Kim, S.J.; Kim, J.Y.; Lim, J.H.; Hong, S. S. ANGPTL4 Accelerates KRASG12D-Induced Acinar to Ductal Metaplasia and Pancreatic Carcinogenesis. *Cancer Lett.* **2021**, *519*, 185–198. [CrossRef] [PubMed]
25. Alonso-Curbelo, D.; Ho, Y.-J.; Burdziak, C.; Maag, J.L.V.; Morris, J.P.; Chandwani, R.; Chen, H.-A.; Tsanov, K.M.; Barriga, F.M.; Luan, W.; et al. A Gene-Environment-Induced Epigenetic Program Initiates Tumorigenesis. *Nature* **2021**, *590*, 642–648. [CrossRef] [PubMed]
26. Rosenfeldt, M.T.; O'Prey, J.; Morton, J.P.; Nixon, C.; MacKay, G.; Mrowinska, A.; Au, A.; Rai, T.S.; Zheng, L.; Ridgway, R.; et al. P53 Status Determines the Role of Autophagy in Pancreatic Tumour Development. *Nature* **2013**, *504*, 296–300. [CrossRef]
27. Carbone, C.; Piro, G.; Agostini, A.; Delfino, P.; De Sanctis, F.; Nasca, V.; Spallotta, F.; Sette, C.; Martini, M.; Ugel, S.; et al. Intratumoral Injection of TLR9 Agonist Promotes an Immunopermissive Microenvironment Transition and Causes Cooperative Antitumor Activity in Combination with Anti-PD1 in Pancreatic Cancer. *J. Immunother. Cancer* **2021**, *9*, e002876. [CrossRef]

28. Viale, A.; Pettazzoni, P.; Lyssiotis, C.A.; Ying, H.; Sánchez, N.; Marchesini, M.; Carugo, A.; Green, T.; Seth, S.; Giuliani, V.; et al. Oncogene Ablation-Resistant Pancreatic Cancer Cells Depend on Mitochondrial Function. *Nature* **2014**, *514*, 628–632. [CrossRef]
29. Jin, J.; Byun, J.-K.; Choi, Y.-K.; Park, K.-G. Targeting Glutamine Metabolism as a Therapeutic Strategy for Cancer. *Exp. Mol. Med.* **2023**, *55*, 706–715. [CrossRef]
30. Dey, P.; Baddour, J.; Muller, F.; Wu, C.C.; Wang, H.; Liao, W.-T.; Lan, Z.; Chen, A.; Gutschner, T.; Kang, Y.; et al. Genomic Deletion of Malic Enzyme 2 Confers Collateral Lethality in Pancreatic Cancer. *Nature* **2017**, *542*, 119–123. [CrossRef]
31. Liu, Y.; Deguchi, Y.; Wei, D.; Liu, F.; Moussalli, M.J.; Deguchi, E.; Li, D.; Wang, H.; Valentin, L.A.; Colby, J.K.; et al. Rapid Acceleration of KRAS-Mutant Pancreatic Carcinogenesis via Remodeling of Tumor Immune Microenvironment by PPARδ. *Nat. Commun.* **2022**, *13*, 2665. [CrossRef]
32. Berta, M.A.; Baker, C.M.; Cottle, D.L.; Watt, F.M. Dose and Context Dependent Effects of Myc on Epidermal Stem Cell Proliferation and Differentiation. *EMBO Mol. Med.* **2010**, *2*, 16–25. [CrossRef] [PubMed]
33. Bannoura, S.F.; Khan, H.Y.; Azmi, A.S. KRAS G12D Targeted Therapies for Pancreatic Cancer: Has the Fortress Been Conquered? *Front. Oncol.* **2022**, *12*, 1013902. [CrossRef] [PubMed]
34. Sangrador, I.; Molero, X.; Campbell, F.; Franch-Expósito, S.; Rovira-Rigau, M.; Samper, E.; Domínguez-Fraile, M.; Fillat, C.; Castells, A.; Vaquero, E.C. Zeb1 in Stromal Myofibroblasts Promotes Kras-Driven Development of Pancreatic Cancer. *Cancer Res.* **2018**, *78*, 2624–2637. [CrossRef] [PubMed]
35. Genovese, G.; Carugo, A.; Tepper, J.; Robinson, F.S.; Li, L.; Svelto, M.; Nezi, L.; Corti, D.; Minelli, R.; Pettazzoni, P.; et al. Synthetic Vulnerabilities of Mesenchymal Subpopulations in Pancreatic Cancer. *Nature* **2017**, *542*, 362–366. [CrossRef] [PubMed]
36. He, Q.; Liu, Z.; Wang, J. Targeting KRAS in PDAC: A New Way to Cure It? *Cancers* **2022**, *14*, 4982. [CrossRef]
37. Hallin, J.; Engstrom, L.D.; Hargis, L.; Calinisan, A.; Aranda, R.; Briere, D.M.; Sudhakar, N.; Bowcut, V.; Baer, B.R.; Ballard, J.A.; et al. The KRASG12C Inhibitor MRTX849 Provides Insight toward Therapeutic Susceptibility of KRAS-Mutant Cancers in Mouse Models and Patients. *Cancer Discov.* **2020**, *10*, 54–71. [CrossRef]
38. Bannoura, S.F.; Uddin, M.H.; Nagasaka, M.; Fazili, F.; Al-Hallak, M.N.; Philip, P.A.; El-Rayes, B.; Azmi, A.S. Targeting KRAS in Pancreatic Cancer: New Drugs on the Horizon. *Cancer Metastasis Rev.* **2021**, *40*, 819–835. [CrossRef]
39. Sakamoto, K.; Masutani, T.; Hirokawa, T. Generation of KS-58 as the First K-Ras(G12D)-Inhibitory Peptide Presenting Anti-Cancer Activity in Vivo. *Sci. Rep.* **2020**, *10*, 21671. [CrossRef]
40. Wang, Z.; Cao, Y.J. Adoptive Cell Therapy Targeting Neoantigens: A Frontier for Cancer Research. *Front. Immunol.* **2020**, *11*, 176. [CrossRef]
41. Wang, Q.J.; Yu, Z.; Griffith, K.; Hanada, K.; Restifo, N.P.; Yang, J.C. Identification of T-Cell Receptors Targeting KRAS-Mutated Human Tumors. *Cancer Immunol. Res.* **2016**, *4*, 204–214. [CrossRef] [PubMed]
42. Wang, P.; Heitman, J. The Cyclophilins. *Genome Biol.* **2005**, *6*, 226. [CrossRef]
43. Ahearn, I.M.; Haigis, K.; Bar-Sagi, D.; Philips, M.R. Regulating the Regulator: Post-Translational Modification of RAS. *Nat. Rev. Mol. Cell Biol.* **2011**, *13*, 39–51. [CrossRef]
44. Wang, W.-H.; Yuan, T.; Qian, M.-J.; Yan, F.-J.; Yang, L.; He, Q.-J.; Yang, B.; Lu, J.-J.; Zhu, H. Post-Translational Modification of KRAS: Potential Targets for Cancer Therapy. *Acta Pharmacol. Sin.* **2021**, *42*, 1201–1211. [CrossRef] [PubMed]
45. Canon, J.; Rex, K.; Saiki, A.Y.; Mohr, C.; Cooke, K.; Bagal, D.; Gaida, K.; Holt, T.; Knutson, C.G.; Koppada, N.; et al. The Clinical KRAS(G12C) Inhibitor AMG 510 Drives Anti-Tumour Immunity. *Nature* **2019**, *575*, 217–223. [CrossRef]
46. Lito, P.; Solomon, M.; Li, L.-S.; Hansen, R.; Rosen, N. Allele-Specific Inhibitors Inactivate Mutant KRAS G12C by a Trapping Mechanism. *Science* **2016**, *351*, 604–608. [CrossRef] [PubMed]
47. Soares, H.P.; Ming, M.; Mellon, M.; Young, S.H.; Han, L.; Sinnet-Smith, J.; Rozengurt, E. Dual PI3K/MTOR Inhibitors Induce Rapid Overactivation of the MEK/ERK Pathway in Human Pancreatic Cancer Cells through Suppression of MTORC2. *Mol. Cancer Ther.* **2015**, *14*, 1014–1023. [CrossRef]
48. Oncogenic KRAS Engages an RSK1/NF1 Pathway to Inhibit Wild-Type RAS Signaling in Pancreatic Cancer—PubMed. Available online: https://pubmed.ncbi.nlm.nih.gov/34021083/ (accessed on 4 March 2023).
49. Fedele, C.; Li, S.; Teng, K.W.; Foster, C.J.R.; Peng, D.; Ran, H.; Mita, P.; Geer, M.J.; Hattori, T.; Koide, A.; et al. SHP2 Inhibition Diminishes KRASG12C Cycling and Promotes Tumor Microenvironment Remodeling. *J. Exp. Med.* **2021**, *218*, e20201414. [CrossRef]
50. Zhao, H.; Yang, L.; Baddour, J.; Achreja, A.; Bernard, V.; Moss, T.; Marini, J.C.; Tudawe, T.; Seviour, E.G.; San Lucas, F.A.; et al. Tumor Microenvironment Derived Exosomes Pleiotropically Modulate Cancer Cell Metabolism. *eLife* **2016**, *5*, e10250. [CrossRef]
51. Fedele, C.; Ran, H.; Diskin, B.; Wei, W.; Jen, J.; Geer, M.J.; Araki, K.; Ozerdem, U.; Simeone, D.M.; Miller, G.; et al. SHP2 Inhibition Prevents Adaptive Resistance to MEK Inhibitors in Multiple Cancer Models. *Cancer Discov.* **2018**, *8*, 1237–1249. [CrossRef]
52. Hou, P.; Ma, X.; Zhang, Q.; Wu, C.-J.; Liao, W.; Li, J.; Wang, H.; Zhao, J.; Zhou, X.; Guan, C.; et al. USP21 Deubiquitinase Promotes Pancreas Cancer Cell Stemness via Wnt Pathway Activation. *Genes Dev.* **2019**, *33*, 1361–1366. [CrossRef]
53. Crawford, H.C. Anticipating Resistance to KRAS Inhibition: A Novel Role for USP21 in Macropinocytosis Regulation. *Genes Dev.* **2021**, *35*, 1325–1326. [CrossRef] [PubMed]
54. Hennig, A.; Markwart, R.; Esparza-Franco, M.A.; Ladds, G.; Rubio, I. Ras Activation Revisited: Role of GEF and GAP Systems. *Biol. Chem.* **2015**, *396*, 831–848. [CrossRef] [PubMed]
55. Li, W.; Nishimura, R.; Kashishian, A.; Batzer, A.G.; Kim, W.J.; Cooper, J.A.; Schlessinger, J. A New Function for a Phosphotyrosine Phosphatase: Linking GRB2-Sos to a Receptor Tyrosine Kinase. *Mol. Cell. Biol.* **1994**, *14*, 509–517. [CrossRef] [PubMed]

56. Chen, Y.-N.P.; LaMarche, M.J.; Chan, H.M.; Fekkes, P.; Garcia-Fortanet, J.; Acker, M.G.; Antonakos, B.; Chen, C.H.-T.; Chen, Z.; Cooke, V.G.; et al. Allosteric Inhibition of SHP2 Phosphatase Inhibits Cancers Driven by Receptor Tyrosine Kinases. *Nature* **2016**, *535*, 148–152. [CrossRef] [PubMed]
57. Kerr, D.L.; Haderk, F.; Bivona, T.G. Allosteric SHP2 Inhibitors in Cancer: Targeting the Intersection of RAS, Resistance, and the Immune Microenvironment. *Curr. Opin. Chem. Biol.* **2021**, *62*, 1–12. [CrossRef]
58. Nichols, R.J.; Haderk, F.; Stahlhut, C.; Schulze, C.J.; Hemmati, G.; Wildes, D.; Tzitzilonis, C.; Mordec, K.; Marquez, A.; Romero, J.; et al. RAS Nucleotide Cycling Underlies the SHP2 Phosphatase Dependence of Mutant BRAF-, NF1- and RAS-Driven Cancers. *Nat. Cell Biol.* **2018**, *20*, 1064–1073. [CrossRef]
59. Liu, C.; Lu, H.; Wang, H.; Loo, A.; Zhang, X.; Yang, G.; Kowal, C.; Delach, S.; Wang, Y.; Goldoni, S.; et al. Combinations with Allosteric SHP2 Inhibitor TNO155 to Block Receptor Tyrosine Kinase Signaling. *Clin. Cancer Res.* **2021**, *27*, 342–354. [CrossRef]
60. Chamberlain, P.P.; Hamann, L.G. Development of Targeted Protein Degradation Therapeutics. *Nat. Chem. Biol.* **2019**, *15*, 937–944. [CrossRef]
61. Chiarantini, L.; Cerasi, A.; Fraternale, A.; Millo, E.; Benatti, U.; Sparnacci, K.; Laus, M.; Ballestri, M.; Tondelli, L. Comparison of Novel Delivery Systems for Antisense Peptide Nucleic Acids. *J. Control. Release* **2005**, *109*, 24–36. [CrossRef]
62. Zheng, X.; Liu, Z.; Li, D.; Wang, E.; Wang, J. Rational Drug Design: The Search for Ras Protein Hydrolysis Intermediate Conformation Inhibitors with Both Affinity and Specificity. *Curr. Pharm. Des.* **2013**, *19*, 2246–2258. [CrossRef]
63. Li, D.; Liu, Z.; Zhao, W.; Zheng, X.; Wang, J.; Wang, E. A Small-Molecule Induces Apoptosis and Suppresses Metastasis in Pancreatic Cancer Cells. *Eur. J. Pharm. Sci.* **2013**, *48*, 658–667. [CrossRef] [PubMed]
64. Liu, Z.; Li, D.; Zheng, X.; Wang, E.; Wang, J. Selective Induction of Apoptosis: Promising Therapy in Pancreatic Cancer. *Curr. Pharm. Des.* **2013**, *19*, 2259–2268. [CrossRef]
65. Khan, H.Y.; Nagasaka, M.; Li, Y.; Aboukameel, A.; Uddin, M.H.; Sexton, R.; Bannoura, S.; Mzannar, Y.; Al-Hallak, M.N.; Kim, S.; et al. Inhibitor of the Nuclear Transport Protein XPO1 Enhances the Anticancer Efficacy of KRAS G12C Inhibitors in Preclinical Models of KRAS G12C-Mutant Cancers. *Cancer Res. Commun.* **2022**, *2*, 342–352. [CrossRef] [PubMed]
66. Dey, P.; Li, J.; Zhang, J.; Chaurasiya, S.; Strom, A.; Wang, H.; Liao, W.-T.; Cavallaro, F.; Denz, P.; Bernard, V.; et al. Oncogenic KRAS-Driven Metabolic Reprogramming in Pancreatic Cancer Cells Utilizes Cytokines from the Tumor Microenvironment. *Cancer Discov.* **2020**, *10*, 608–625. [CrossRef] [PubMed]
67. M.D. Anderson Cancer Center. Phase I Study of Mesenchymal Stromal Cells-Derived Exosomes with KrasG12D SiRNA for Metastatic Pancreas Cancer Patients Harboring KrasG12D Mutation. 2023. Available online: https://clinicaltrials.gov (accessed on 5 March 2023).
68. XP-524 Is a Dual-BET/EP300 Inhibitor That Represses Oncogenic KRAS and Potentiates Immune Checkpoint Inhibition in Pancreatic Cancer—PubMed. Available online: https://pubmed.ncbi.nlm.nih.gov/35064087/ (accessed on 4 March 2023).
69. Dixon, S.J.; Lemberg, K.M.; Lamprecht, M.R.; Skouta, R.; Zaitsev, E.M.; Gleason, C.E.; Patel, D.N.; Bauer, A.J.; Cantley, A.M.; Yang, W.S.; et al. Ferroptosis: An Iron-Dependent Form of Nonapoptotic Cell Death. *Cell* **2012**, *149*, 1060–1072. [CrossRef]
70. Müller, F.; Lim, J.K.M.; Bebber, C.M.; Seidel, E.; Tishina, S.; Dahlhaus, A.; Stroh, J.; Beck, J.; Yapici, F.I.; Nakayama, K.; et al. Elevated FSP1 Protects KRAS-Mutated Cells from Ferroptosis during Tumor Initiation. *Cell Death Differ.* **2023**, *30*, 442–456. [CrossRef]
71. Yan, H.; Yu, C.-C.; Fine, S.A.; Youssof, A.L.; Yang, Y.-R.; Yan, J.; Karg, D.C.; Cheung, E.C.; Friedman, R.A.; Ying, H.; et al. Loss of the Wild-Type KRAS Allele Promotes Pancreatic Cancer Progression through Functional Activation of YAP1. *Oncogene* **2021**, *40*, 6759–6771. [CrossRef]
72. Cordenonsi, M.; Zanconato, F.; Azzolin, L.; Forcato, M.; Rosato, A.; Frasson, C.; Inui, M.; Montagner, M.; Parenti, A.R.; Poletti, A.; et al. The Hippo Transducer TAZ Confers Cancer Stem Cell-Related Traits on Breast Cancer Cells. *Cell* **2011**, *147*, 759–772. [CrossRef]
73. Salcedo Allende, M.T.; Zeron Medina, J.; Hernandez, J.; Macarulla, T.; Balsells, J.; Merino, X.; Allende, H.; Tabernero, J.; Ramon, Y.; Cajal, S. Overexpression of Yes Associated Protein 1, an Independent Prognostic Marker in Patients with Pancreatic Ductal Adenocarcinoma, Correlated with Liver Metastasis and Poor Prognosis. *Pancreas* **2017**, *46*, 913–920. [CrossRef]
74. Hou, P.; Ma, X.; Yang, Z.; Zhang, Q.; Wu, C.-J.; Li, J.; Tan, L.; Yao, W.; Yan, L.; Zhou, X.; et al. USP21 Deubiquitinase Elevates Macropinocytosis to Enable Oncogenic KRAS Bypass in Pancreatic Cancer. *Genes Dev.* **2021**, *35*, 1327–1332. [CrossRef] [PubMed]
75. Hou, P.; Kapoor, A.; Zhang, Q.; Li, J.; Wu, C.-J.; Li, J.; Lan, Z.; Tang, M.; Ma, X.; Ackroyd, J.J.; et al. Tumor Microenvironment Remodeling Enables Bypass of Oncogenic KRAS Dependency in Pancreatic Cancer. *Cancer Discov.* **2020**, *10*, 1058–1077. [CrossRef] [PubMed]
76. Pettazzoni, P.; Viale, A.; Shah, P.; Carugo, A.; Ying, H.; Wang, H.; Genovese, G.; Seth, S.; Minelli, R.; Green, T.; et al. Genetic Events That Limit the Efficacy of MEK and RTK Inhibitor Therapies in a Mouse Model of KRAS-Driven Pancreatic Cancer. *Cancer Res.* **2015**, *75*, 1091–1101. [CrossRef] [PubMed]
77. Waters, A.M.; Der, C.J. KRAS: The Critical Driver and Therapeutic Target for Pancreatic Cancer. *Cold Spring Harb. Perspect. Med.* **2018**, *8*, a031435. [CrossRef] [PubMed]

Disclaimer/Publisher's Note: The statements, opinions and data contained in all publications are solely those of the individual author(s) and contributor(s) and not of MDPI and/or the editor(s). MDPI and/or the editor(s) disclaim responsibility for any injury to people or property resulting from any ideas, methods, instructions or products referred to in the content.

Review

The Comprehensive Analysis of Specific Proteins as Novel Biomarkers Involved in the Diagnosis and Progression of Gastric Cancer

Elżbieta Pawluczuk [1], Marta Łukaszewicz-Zając [2,*] and Barbara Mroczko [1,2]

[1] Department of Neurodegeneration Diagnostics, Medical University of Bialystok, 15-269 Bialystok, Poland; elzbieta.pawluczuk16@wp.pl (E.P.); mroczko@umb.edu.pl (B.M.)
[2] Department of Biochemical Diagnostics, Medical University of Bialystok, Waszyngtona 15a, 15-269 Bialystok, Poland
* Correspondence: marta.lukaszewicz-zajac@umb.edu.pl; Tel.: +48-85-7468785; Fax: +48-85-7468585

Abstract: Gastric cancer (GC) cases are predicted to rise by 2040 to approximately 1.8 million cases, while GC-caused deaths to 1.3 million yearly worldwide. To change this prognosis, there is a need to improve the diagnosis of GC patients because this deadly malignancy is usually detected at an advanced stage. Therefore, new biomarkers of early GC are sorely needed. In the present paper, we summarized and referred to a number of original pieces of research concerning the clinical significance of specific proteins as potential biomarkers for GC in comparison to well-established tumor markers for this malignancy. It has been proved that selected chemokines and their specific receptors, vascular endothelial growth factor (VEGF) and epidermal growth factor receptor (EGFR), specific proteins such as interleukin 6 (IL-6) and C-reactive protein (CRP), matrix metalloproteinases (MMPs) and their tissue inhibitors (TIMPs), a disintegrin and metalloproteinase with thrombospondin motifs (ADAMTS), as well as DNA- and RNA-based biomarkers, and c-MET (tyrosine-protein kinase Met) play a role in the pathogenesis of GC. Based on the recent scientific literature, our review indicates that presented specific proteins are potential biomarkers in the diagnosis and progression of GC as well as might be used as prognostic factors of GC patients' survival.

Keywords: biomarkers; gastric cancer; DNA-based markers; chemokines; MMPs; TIMPs; novel biomarker

Citation: Pawluczuk, E.; Łukaszewicz-Zając, M.; Mroczko, B. The Comprehensive Analysis of Specific Proteins as Novel Biomarkers Involved in the Diagnosis and Progression of Gastric Cancer. *Int. J. Mol. Sci.* **2023**, *24*, 8833. https://doi.org/10.3390/ijms24108833

Academic Editor: Laura Paleari

Received: 29 March 2023
Revised: 12 May 2023
Accepted: 14 May 2023
Published: 16 May 2023

Copyright: © 2023 by the authors. Licensee MDPI, Basel, Switzerland. This article is an open access article distributed under the terms and conditions of the Creative Commons Attribution (CC BY) license (https://creativecommons.org/licenses/by/4.0/).

1. Gastric Cancer

According to the newest data from WHO, published in 2020, gastric cancer (GC) remains the fifth most common cancer worldwide. Moreover, it is the third most often cause of cancer-related death in the world [1]. The Lauren classification divides gastric adenocarcinoma into two histopathological types: diffuse and intestinal type [2]. The diffuse type is more often observed among young women and subjects with cancer-positive histories. The intestinal type of GC is mostly connected with chronic atrophic gastritis, which can develop into intestinal metaplasia via dysplasia and then can transform into carcinoma in situ. This type occurs more often in older patients, men and people from high-risk countries [3,4]. The risk factors of GC are *Helicobacter pylori* (*H. pylori*) and Epstein-Barr virus (EBV) infections, as well as chronic inflammation process, alcohol consumption, smoking, a diet high in salt, obesity, and also a lack of fruit and vegetables. Risk factors are connected with working conditions such as chemical exposure and working with metal, wood and rubber [5]. The stomach has several anatomical parts: cardia, fundus, body, antrum and pylorus. There are also risk factors connected with adenocarcinoma arising from the cardia (cardia GC) and from other parts of the stomach (non-cardia GC). Cardia and non-cardia GC are both correlated with older age, male sex, tobacco smoking, race, past cancer family history, low intake of fiber, low physical activity and radiation. Development of cardia GC is associated with obesity and gastroesophageal reflux disease (GERD), while non-cardia GC is correlated with *H. pylori* infection, low socioeconomic status, high intake of salty and smoked food and low consumption of fruits and vegetables [6]. Despite the knowledge about risk factors, the prognosis of GC is still poor, and the five-year survival rate is lower than 30% [7].

The diagnostic methods of GC are mainly invasive, including gastroscopy with a biopsy. In clinical practice, there is also a supportive role of imaging methods like computed tomography (CT) and magnetic resonance imaging (MRI), positron emission tomography (PET) and endoscopic ultrasound scanning (EUS) [8]. In addition, laboratory tests are also important tools in the diagnostic process of patients with this malignancy. Well-established tumor markers useful in the routine diagnosis of GC patients are carbohydrate antigen 72-4 (CA 72-4), carcinoembryonic antigen (CEA) and carbohydrate antigen 19-9 (CA 19-9). The concentrations of these proteins may also be elevated in patients with other nonmalignant diseases; thus, their diagnostic specificity and sensitivity are not significantly high enough to be used in screening tests that can accurately diagnose early-stage GC. Therefore, there is still a need to find new potential biomarkers for the early detection of this malignancy. In this review, we present a number of research papers, concerning the significance of potential biomarkers as novel perspectives in the diagnostic process of GC, in comparison to classical tumor markers for this common malignancy (Figure 1, Table 1) [9–14].

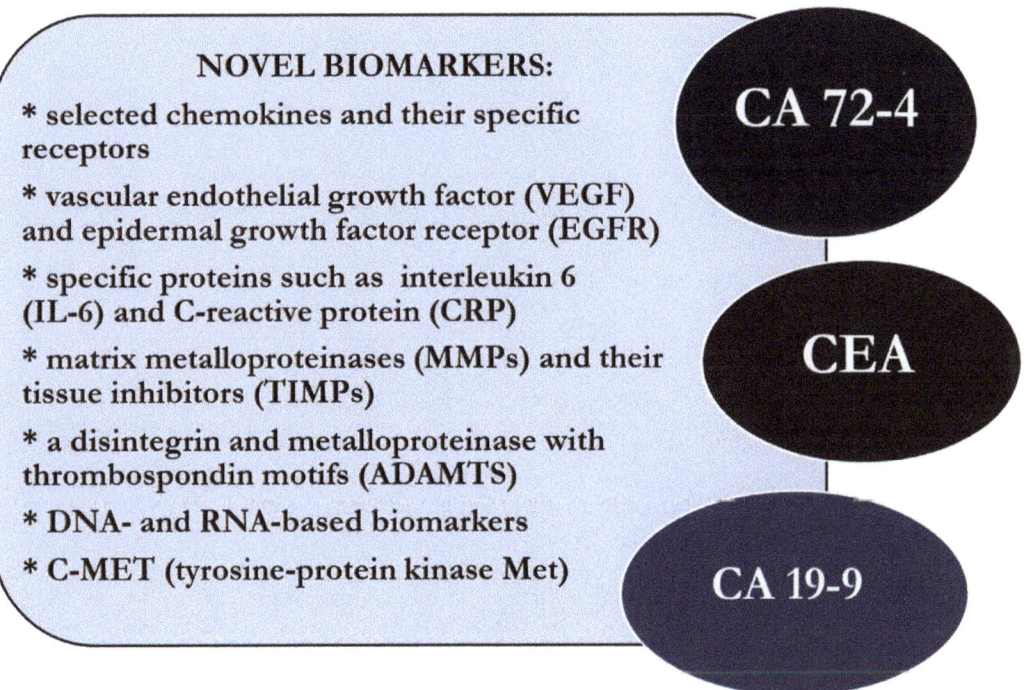

Figure 1. Potential, novel biomarkers and well-established tumor markers for gastric cancer (GC).

Table 1. Significance of novel biomarkers in gastric cancer (GC).

Potential Biomarker	Source		Results	References
CXCL8 and CXCR2	Serum level	✓ ✓	Higher levels of CXCL8 and CXCR2 in GC patients than in the healthy control group. Potential significance in the diagnosis of GC patients.	[15]
CXCL13	Expression	✓ ✓	Elevated expression of CXCL13 is associated with a larger tumor diameter and a shorter overall survival rate. Low expression of CXCL13 is associated with a longer survival rate, especially in the group of patients who received adjuvant chemotherapy.	[16]
CXCR3	Expression	✓ ✓ ✓	Higher expression of CXCR3 in GC tissues than in precancerous tissues. Overexpression of CXCR3 correlated with decreased M2 macrophage infiltration. Lower CXCR3 expression correlated with worse differentiation, more advanced GC stage and higher depth of invasion.	[17]
VEGF and EGFR	Expression	✓	Elevated VEGF and EGFR expressions associated with nodal invasion and tumor progression.	[18]
IL-6 and CRP	Expression	✓	Higher IL-6 expression in GC cell lines than in colorectal cancer cell lines.	[19]
	Expression	✓	Higher IL-6 expression in GC patients than in the control group.	[20]
	Serum level	✓	Potential significance of IL-6 and CRP in the diagnosis of GC patients.	[21]
TIMPs	Serum level	✓	TIMP-2 is useful in predicting GC tumor progression, especially for nodal involvement.	[22]
	Serum and plasma level	✓ ✓	Plasma TIMP-1 is a better GC biomarker than serum TIMP-1. Plasma TIMP-1 may be useful in the prognosis of a GC patient's survival.	[23]
MMPSs	Mucosal mRNA expression	✓	MMP-1 mRNA levels are higher in GC with active *H. pylori* infections than in GC without *H. pylori* infections.	[24]
	Serum level	✓ ✓	GC patients with elevated serum MMP-14 levels had a 5-year disease-specific survival of 22.1%. GC patients with lower concentrations of MMP-14 had a 5-year specific survival of 49.2%.	[25]
		✓	MMP-2 may be used as an independent prognostic factor of a GC patient's survival.	[22]
	Serum and plasma level	✓	Plasma and serum levels of MMP-9 were significantly higher in GC patients than in the control group.	[23]
ADAMTS	Expression	✓	ADAMTS-12 upregulation in GC patients and prediction of worse survival rate of GC patients.	[26]
	Expression	✓	Lower expression of ADAMTS-2 was a good prognostic factor of GC patient survival.	[27]
	Expression		Downregulation of ADAMTS-8 expression in GC cell lines and tissues.	[28]

2. Biomarkers of GC Applied in Clinical Practice

The diagnostic and prognostic significance of biochemical classical tumor markers in routine practice has been demonstrated. Cancer antigen—CA 72-4 was proved to be the first-line tumor marker with the highest diagnostic sensitivity and specificity for GC. This protein was detected in the 1980s as an antigen reactive to murine antibodies produced by mice immunized with human metastatic breast cancer. Its level is elevated among patients with gastric cancer, colorectal cancer, liver cancer, pancreatic cancer, breast cancer and ovarian cancer [11–13,29]. It is considered to be one of the most specific and sensitive GC biomarkers. In the study by Chen et al., the diagnostic sensitivity of CA 72-4 in GC was 32.2%, and the specificity was 97.1% [11–13,29].

The second line tumor marker in the diagnosis of GC patients is CEA. This protein was first isolated by Gold and Freedman in 1965, and it became a first-line biomarker for colorectal cancer [30]. However, it is used mainly as a marker for digestive system tumors, including GC. The diagnostic specificity and sensitivity of only CEA concentration measurements are not high enough to establish a recognition of this cancer. Thus, in routine practice, the assessment of serum CEA levels is used to monitor the treatment process in colorectal cancer patients and to identify recurrences after surgical resection of a tumor. Serum levels of CEA are elevated in the blood of patients with pancreatic cancer, gastric cancer, breast cancer, lung cancer, cervical cancer, bladder cancer, lymphomas, as well as patients with nonmalignant diseases such as gastric ulcer and duodenal ulcer, colitis ulcerosa, cirrhosis and chronic pancreatitis [13,14,31–34]. In the study performed by Hao et al., elevated serum level of CEA was identified in 49 diseases [31]. The highest CEA concentrations in descending order had patients with lung fibrosis, pancreatic cancer, uremia, chronic obstructive pulmonary disease, colon cancer, Alzheimer's disease, rectum cancer, and lung cancer [31]. In the study by Chen et al., the diagnostic sensitivity of CEA in GC diagnosis was 28.7%, and the diagnostic specificity was 96.2% [11].

Cancer antigen—CA 19-9 is also useful in the diagnosis of GC. It is a high-molecular-weight mucin that takes part in the adhesion of endothelial and cancer cells. This protein was detected by mice immunization by a human colorectal cell line. Some researchers indicated the usefulness of CA 19-9 measurement as a marker for gallbladder cancer, bile duct cancer, pancreatic cancer and colorectal cancer [13,35–37]. In the study by Chen et al., it was established that the serum level of CA 19-9 was higher in GC patients than in the control group and in benign lesions [11]. The diagnostic sensitivity of serum CA 19-9 measurement in GC patients was 26.4%, and the specificity was 99%. Moreover, elevated serum CA 19-9 concentration was proved to be an independent prognosis predictor in GC patients with metastasis or recurrent cancer [38–40].

Some clinical investigations have indicated that human epidermal growth factor receptor 2 (HER2) might also be an important biomarker of GC, especially among patients with an advanced stage. The evaluation of HER2 levels is used to select subgroups for treatment with trastuzumab, and its level is assessed by immunohistochemical technique and FISH methods (fluorescence in situ hybridization) [41,42]. The introduction of trastuzumab led to HER2-positive GC diagnostics. In addition, the association between HER2 expression and treatment of colon cancer, bladder cancer and biliary cancer patients has also been evaluated [9,10]. It is expected that more anti-HER2 medications will be introduced into clinical practice in the treatment process of HER2-positive cancers, including GC [9,10].

3. Perspective for Novel Biomarkers in GC

3.1. Chemokines and Their Specific Receptors

Chemokines are small proteins from 8 to 14 kDa secreted by leukocytes, endothelium cells, fibroblasts, monocytes and tumor microenvironment cells. These cytokines are involved in various physiological processes such as cell migration, adhesion and activation of leukocytes as well as inflammation and immunological response. Moreover, chemokines may also regulate pathological processes, including autoimmunological diseases as well as tumor growth via the promotion of the proliferation of malignant cells and neoangiogenesis [43–45].

The biological functions of chemokines are activated via their specific, seven-transmembrane G-coupled receptors [45]. Some investigators suggest that selected chemokines and their specific receptors are connected with GC pathogenesis and may be used as potential GC biomarkers. Our previous findings indicated the significance of various chemokines and their specific receptors as novel biomarkers of gastrointestinal cancers, including GC. Our results suggest the role of serum C-X-C motif chemokine ligand 8 (CXCL8) and its specific receptor (CXCR2) as a promising candidate for a biomarker in the GC diagnosis [15]. We assessed that the serum level of CXCL8 and CXCR2 was significantly higher among GC patients than in the healthy control group. The diagnostic sensitivity of CXCL8 was higher than the well-known tumor marker, CEA, and higher than the specific receptor for CXCL8 (CXCR2). Combined analysis of CXCL8 and CA19-9 increased the diagnostic sensitivity up to 89%. Positive predictive values (PPV) for CXCL8 and CXCR2 were higher than for the classical tumor marker CA19-9. In addition, we proved the role of CXCL8 as a potential GC biomarker, especially in combined measurement with classical tumor markers [15]. In the study of Wei et al., the expression of other chemokines from C-X-C family chemokines—

CXCL13—was analyzed. High CXCL13 expression was associated with a larger tumor diameter and a shorter overall survival rate. In addition, low CXCL13 expression was connected with longer survival, especially in the group of patients who received adjuvant chemotherapy. Authors conclude that CXCL13 expression may be a predictive biomarker for postoperative adjuvant chemotherapy benefit in GC patients [16].

A study by Chen et al. revealed that CXCR3 expression was higher in GC tissues than in precancerous tissues. Moreover, overexpression of CXCR3 correlated with decreased alternatively activated macrophage infiltration (M2 macrophages), and patients with decreased infiltration had also a better overall survival rate [17]. The lower expression of CXCR3 correlated with worse differentiation, more advanced GC stage and higher depth of invasion. They assumed that CXCR3 expression might be used as an independent prognostic parameter for the overall survival of GC patients [17].

3.2. Vascular Endothelial Growth Factor (VEGF) and Epidermal Growth Factor Receptor (EGFR)

Vascular endothelial growth factor (VEGF) or vascular permeability factor (VPF) is a growth factor that stimulates the formation of blood vessels. It also increases vascular permeability, promotes cell migration and regulates the normal and pathological angiogenic processes [46]. VEGF binds to its receptor, VEGFR, and promotes vascularization, vasodilatation and vascular growth [47]. Epidermal growth factor receptor (EGFR) belongs to the tyrosine kinase receptors family. It is a key regulator in cell proliferation, differentiation, survival and cancer development [48]. In the Lieto et al. study, VEGF and EGFR levels were assessed to evaluate whether these molecules might be used as novel biomarkers of GC. VEGF expression was positive in 48% of assessed samples and EGFR in 44% of samples. Curatively treated GC patients had lower VEGF and EGFR expression than noncurative-treated GC cases. In addition, elevated VEGF and EGFR expression was associated with nodal invasion and tumor progression. EGFR expression had a linear relationship with the number of metastatic lymph nodes [18]. Study authors suggested that expression of VEGF and EGFR were independent prognostic indicators of worse outcomes for GC patients. In the study performed by Rao et al., serum VEGF levels were elevated in comparison to healthy controls and significantly increased in patients with advanced tumor TNM stages ($p < 0.05$). They concluded that VEGF might be an effective indicator for the evaluation of the prognosis of GC patients [49]. Vidal et al. assessed VEGF levels in GC patients. Elevated VEGF concentrations were correlated with advanced TNM stage, lymph nodes invasion, a lower probability of recurrence-free status and shorter disease-specific survival of GC patients [50].

3.3. Specific Proteins—Interleukin 6 (IL-6) and C-Reactive Protein (CRP)

Interleukin 6 is a proinflammatory cytokine that stimulates the production of other acute-phase proteins, such as CRP, α1-antichymotrypsin, fibrinogen, SAA and haptoglobin. This cytokine is involved in the immune response, inflammation processes and hematopoiesis, as well as in many pathological conditions, including malignancies such as breast cancer, lung cancer, pancreatic cancer and GC [51]. It was proved that IL-6 stimulates the synthesis of C-reactive protein (CRP). CRP is an acute-phase protein produced by hepatocytes in the liver. The serum concentration of CRP in healthy patients is below 10 mg/L; however, its level may be elevated during many pathological conditions, such as bacterial infections, as well as malignant diseases, including GC [52]. Matsuo et al. assessed IL-6 expression in nine GC cell lines and nine colorectal cancer cell lines. Two GC and one colorectal cancer cell lines expressed IL-6. The level of IL-6 secretion was higher in GC cell lines than in colorectal cancer cell lines, and this difference was statistically significant [19]. The study of Wang et al. has indicated that the expression of IL-6 in GC patients was higher than in the control group, and this difference was statistically significant [20]. In addition, Chang et al. revealed that elevated CRP levels were observed in 38% of GC patients and in 4.9% of the patients from the control group, and these differences were statistically significant [53]. A higher concentration of serum CRP was also associated with larger tumor size, the presence of lymph nodes and distant metastases, as well as more advanced stages of GC and worse survival rates of patients [53]. In our previous research, we assessed serum IL-6, CEA and CA19-9 levels in GC patients and compared them with healthy controls [21]. Moreover, our data revealed that serum CRP concentration correlated with the presence of lymph node and distant metastases, advanced cancer stage and gastric wall invasion, while IL-6, CEA and CA 19-9 levels correlated with nodal involvement [21]. Additionally, the diagnostic sensitivity of IL-6 was higher than for CRP, CA 19-9 and CEA and increased in combined assessment with CRP or CEA [21]. Our study revealed better usefulness of inflammation proteins IL-6 and CRP than classical GC markers such as CEA and CA 19-9 in the diagnosis and progression of this malignancy [21].

3.4. Matrix Metalloproteinases (MMPs) and Their Tissue Inhibitors (TIMPs)

Matrix metalloproteinases (MMPs), also known as matrix metallopeptidases, are calcium-dependent zinc-containing endopeptidases. They are produced by leukocytes, macrophages, endothelial cells, fibroblasts and tumor cells [54]. Tissue inhibitors of metalloproteinases (TIMPs) are specific endogenous protease inhibitors that inhibit MMPs and regulate the process of their activation [55]. There are four types of TIMPs: TIMP-1, TIMP-2, TIMP-3 and TIMP-4. Several studies suggested the usefulness of selected MMPs and their tissue inhibitors as novel biomarkers of malignancies, including GC. Mucosal MMP-1 mRNA levels were higher in GC with active *H. pylori* infection than in GC without active *H. pylori* infection [24]. Laitinen et al. revealed that serum MMP-9 and TIMP-1 might be used as prognostic biomarkers of GC. In addition, a worse prognosis was observed among GC patients with high serum levels of TIMP-1 [56]. Kasurinen et al. assessed that patients with elevated serum MMP-14 levels had a 5-year disease-specific survival of 22.1%, while GC patients with lower concentrations of MMP-14 had a 5-year specific survival of 49.2%, and this difference was statistically significant ($p = 0.01$) [25]. The authors concluded that serum MMP-14 was a marker of poor prognosis as well as an indicator of the presence of distant metastases [25]. In our previous study, we assessed the diagnostic significance of gelatinases such as MMP-9 and MMP-2 as well as their tissue inhibitors (TIMP-1 and TIMP-2) in GC patients. We indicated that the diagnostic sensitivity of MMP-2 and TIMP-2 was higher in GC and inflammatory cells compared to normal tissue, whereas serum levels of these proteins were statistically lower in GC patients than in healthy subjects. In addition, our paper demonstrated that there was a significant positive correlation between TIMP-2 immunoreactivity in inflammatory cells and the presence of lymph node metastasis. In addition, we revealed that the area under the ROC curve (AUC) for TIMP-2 was higher than MMP-2, while serum MMP-2 was proved to be an independent prognostic factor of GC patients' survival. Thus, our findings suggest that TIMP-2 might be a predictor of GC progression, especially for nodal involvement, while serum MMP-2 can be used as an independent prognostic factor of patient survival [22]. Previously we evaluated that plasma and serum MMP-9, its tissue inhibitor TIMP-1 and classical tumor marker (CEA) levels were significantly higher in GC patients compared with healthy controls. Moreover, diagnostic criteria such as AUC, diagnostic sensitivity and accuracy of plasma TIMP-1 were higher than those for MMP-9 and CEA, whereas an increased plasma TIMP-1 level was a significant independent prognostic factor for the survival of GC patients. Based on our findings, we suggest that plasma TIMP-1 is a better biomarker than serum TIMP-1 and might be useful for the diagnosis of GC and the prognosis of patient survival [23].

3.5. A Disintegrin and Metalloproteinase with Thrombospondin Motifs (ADAMTS)

A disintegrin and metalloproteinase with thrombospondin motifs (ADAMTS) is a family of multidomain extracellular protease enzymes. Some clinical investigations suggest that ADAMTS12 is related to oncogenesis. In cancer-related processes, ADAMTS12 shows dual effects, being pro and anti-tumor in a proteolytic or non-proteolytic manner. Hou et al. indicated that ADAMTS12 was upregulated in GC patients and predicted a worse overall survival rate in GC patients. In addition, based on AUC analysis, the authors demonstrated that ADAMTS12 had a certain predictive value for the diagnosis of GC [26]. In the study of Jiang et al., ADAMTS2 expression was elevated in the cytoplasm of gastric tumor cells and fibroblast cells. Upregulated expression of ADAMTS2 was correlated with the type of Lauren classification and TNM stage of GC. Lower expression of ADAMTS2 was a good prognostic factor of GC patients' survival. Additionally, multivariate analysis indicated that ADAMTS2 expression was an independent prognostic factor, and this protein may be used as a potential biomarker for GC prognosis [27]. The study of Chen et al. assessed the downregulation of ADAMTS8 mRNA expression in GC cell lines and tissues. There was a significant correlation between ADAMTS8 expression and higher depth of tumor invasion, and the presence of lymph node metastasis. ADAMTS mRNA was also significantly lower in methylated primary gastric tumors than in nontumor tissues. Methylation of the ADAMTS8 gene was statistically higher in primary gastric tumors than in nontumor tissues. There was a significant association between ADAMTS18 methylation and lymph node metastasis [28]. The authors suggest that selected ADAMTS might have promising usefulness as novel biomarkers of GC.

3.6. DNA-Based Biomarkers

DNA methylation plays an important role in GC development. It is suggested that methylated DNA can be served as a biomarker in the plasma, serum or in gastric washes of a cancer patient, including GC. Altered DNA methylation was also observed in patients with *H. pylori*-infected gastric mucosa. Frequently methylated genes in primary GC are: *p16, RUNX3, CDH1, APC, DAPK, GSTP1, MLH1, LOX, FLNc, HRASL, HAND, THBD, F2R, NT5E, GREM, ZNF177, CLDN3, PAX6, CTSL, ALX4, TMEFF2, CHCHD10, IGFBP3, NPR1, CHFR ADAMTS9, FOXD3* and *PAX5* [57–63]. In the Li et al. study, *PAX5* hypermethylation was detected in 77% of primary GC tissues compared to 10.5% of normal gastric tissues ($p < 0.0001$) [63]. Moreover, GC patients with *PAX5* methylation had worse survival compared with unmethylated cases. It was also concluded that *PAX5* might be a suppressor in GC, and the detection of methylated *PAX5* can be used as an independent GC prognostic factor [63]. In addition, some authors revealed that the presence of *H. pylori* infection, which is a well-established risk factor of GC, causes epigenetic deregulation of *FOXD3* to promote gastric carcinogenesis. *FOXD3* hypermethylation in GC was significantly elevated compared with adjacent preneoplastic tissues. Patients with higher *FOXD3* methylation levels survived for significantly shorter time than the patients with lower *FXDP3* methylation levels. Circulating cell-free DNA (cfDNA) is cell-free extracellular DNA from both normal and cancer cells. Some investigations have shown the presence of circulating DNA containing tumor-specific genetic information in the peripheral blood of patients with cancer. The most studied subject connected with cfDNA is ctDNA originating from primary tumors, metastases or circulating tumor cells [64]. Park et al. assessed circulating DNA levels in GC patients and the control group. These results revealed that the plasma cfDNA concentration was significantly higher in patients with GC than that in healthy patients. The authors suggested that plasma is a better source of cfDNA as a biomarker than serum. The optimal cut-off value of plasma cfDNA concentration for discriminating between GC and healthy patients was 32.3 ng/mL. The diagnostic sensitivity of this biomarker was estimated at 75%, while the diagnostic specificity was 63% in GC patients. In addition, the AUC for cfDNA was 0.784 [65]. The presented results confirm the potential role of DNA-based biomarkers in the diagnosis of GC patients.

3.7. RNA-Based Biomarkers

A growing body of evidence indicates the importance of novel RNA-based biomarkers in GC. MicroRNAs (miRNAs) consist of 20–25 nucleotides, and they are short non-coding RNAs. Their role may be involved in carcinogenesis [66]. The expression of microRNA-34a was lower in GC tissues, metastatic GC tissues and in more advanced stages of the cancer than in the control group. Moreover, microRNA-34a expression was elevated in GC tissues without metastases. Additionally, serum microRNA-34a level in GC patients was also higher than in healthy controls and correlated with better prognosis of GC patients' survival [67]. Recent studies established some promising biomarkers which can diagnose GC using non-invasive methods in urine samples. These biomarkers are urinary microRNAs such as miR-6807-5p and miR-6856-5p. It was demonstrated that urinary micro-RNAs had higher expression in GC patients than in the healthy control group. Serum levels of miR6807-5p and miR-6856-5P were also higher in stage I of the GC than in the control group, and this difference was statistically significant [68]. Moreover, the authors suggest that this panel of urinary microRNA biomarkers may be used in the diagnosis of GC patients in the early stage of the disease. In the presented study, researchers also compared the levels of assessed parameters in patients before and after the tumor resection and revealed that the urinary levels of these miRNAs decreased to the undetectable level in all cases, which may suggest the promising role of RNA-based biomarkers in the diagnosis of patients with this malignancy [68].

3.8. C-MET (Tyrosine-Protein Kinase Met)

Some clinical investigations have suggested the potential role of c-MET (tyrosine-protein kinase Met), also known as hepatocyte growth factor receptor (HGFR), in GC pathogenesis [69–71]. It is a protein encoded by the *MET* gene. Activation of MET phosphorylates transduction cascade may promote tumor growth, angiogenesis, migration of the cells and metastasis [67]. Zhang et al. indicated that the expression of c-MET was higher in GC tissues than in paracancerous tissues [72]. Moreover, there was a statistically significant difference between the c-MET expression and clinicopathological characteristics of GC. Elevated expression of c-MET correlated with higher M-stage from TNM staging system (M—the presence of distant metastasis). Overexpression of c-MET was associated with poor overall survival of GC patients. In addition, multivariate Cox regression analysis showed that c-Met might be used as an independent risk factor for 5 years of survival after surgery [73]. In a study performed by Tsujio, HER2-positive GC patients with the c-MET positive expression had worse overall survival than subjects with c-MET negative expression. c-MET positive expression was also more often observed with patients in N1 lymph node metastasis than without nodal involvement (N0 subgroup) [74]. Moreover, c-MET may also be an important biomarker in the treatment process of GC patients. Yashiro et al. indicated that the inhibition of c-MET increased the chemosensitivity of cancer stem cells to the irinotecan in GC, which suggested the role of this biomarker in the monitoring of treatment [70].

3.9. Other Molecular Biomarkers of GC

The molecular characterization of GC has been under excessive investigation; therefore, the identification of novel biomarkers and therapeutic targets is sorely needed.

Fibroblast growth factor receptors (FGFRs) are a family of tyrosine kinase receptors (RTKs). Their signaling is important in processes such as the proliferation or invasion of tumor cells [75]. There is a potential GC therapeutic target connected with this factor: bemarituzumab. It is an afucosylated, humanized IgG1 anti-fibroblast growth factor receptor 2 isoform IIb (FGFR2b) monoclonal antibody. There was a two-phase study performed on patients with FGFR2b-selected gastric or gastro-esophageal junction adenocarcinoma. The study revealed that bemarituzumab has promising clinical efficiency [76].

Epstein-Barr virus (EBV) is a member of the herpes virus family. It was originally identified in a human Burkitt lymphoma cell line. EBV infects more than 90% of the population, and most of those infections are asymptomatic. However, in individuals, it increases the risk of Burkitt lymphoma, Hodgkin lymphoma, nasopharyngeal carcinoma and gastric adenocarcinoma [77]. Epstein–Barr-virus-associated gastric cancer (EBVaGC) occurs in 2–20% of GC cases. It occurs more often among males than females and among Caucasians than Asians. It is a distinct molecular GC subtype with a better prognosis and fewer lymph node metastases [78,79]. Bai et al. revealed that in GC patients with DNA mismatch repair proficiency (pMMR), EBV status was concluded to be an independent predictive factor for overall survival and survival without progression [80].

Cytokeratin-19 (CK19) is expressed by cancer cells and may be used as a marker of metastases in GC [81]. Kutun et al. assessed the patients with resectable GC, unresectable GC and the control group and concluded that expression of both CEA and CK-19 in the peripheral blood of GC patients are strong major vascular invasion (MVI) predictors and worse survival [82]. In addition, microsatellite instability (MSI) and Epstein-Barr virus (EBV) status in GC were also proved to be important biomarkers regarding overall survival and progression-free survival, as well as response to perioperative chemotherapy [83]. The authors indicated that female patients with MSI-high malignancy had significantly better overall survival than females with microsatellite stable (MSS) tumors when submitted to response to perioperative chemotherapy, whereas opposite findings were revealed in male patients. The presented results support research concerning the personalized treatment of GC, considering both patients' and disease characteristics [83].

PD1 (programmed cell death 1) receptor and its ligand PD-L1 are involved in immunomodulation. PD1/PD-L1-Immunotherapy is approved for GC. A high PD-L1/PD1 expression was associated with a better outcome for the patient and was an independent factor in overall and tumor-specific survival prognosis [84,85].

The role of neutrophic thyrosine receptor kinase (*NTRK*) in GC was assessed in several studies, but it was suggested that the role is marginal as the occurrence of *NTRK* fusion in GC is very rare [86–88]. In spite of this, there is entrectinib, which has anti-cancer activity in GC cells with *NTRK* overexpression. It stops angiogenesis and cancer progression by apoptosis induction [89]. It may be used in the treatment of solid, advanced or metastatic tumors, such as GC. It was approved by the FDA in 2019 for adult patients with ROS1-positive metastatic non-small cell lung cancer [90].

In GC treatment, a positive association between tumor mutational burden (TMB) and clinical outcomes in GC patients with pembrolizumab was observed [91]. Lee et al. assessed the association between TMB status and first-line pembrolizumab with chemotherapy treatment outcomes in addition to chemotherapy in KEYNOTE-062. The study revealed the clinical efficiency of first-line with pembrolizumab-based therapy for patients with advanced gastric/gastroesophageal junction adenocarcinoma [92].

4. Methods for Evaluating GC Biomarkers

In routine practice, serum concentrations of classical tumor markers (for example, CEA) are measured using the standard, low-cost chemiluminescent method in routine analyzers with ready-to-use calibrators and controls. However, the diagnostic sensitivity and specificity of classical tumor markers are insufficient to use these biomarkers in the early detection of malignancy. There is also a need for further analyses, considering the relationship between the technical difficulties of tumor marker detection methods and their cost-effectiveness with the diagnostic accuracy of GC biomarkers. We summed up sources and methods of assessed biomarkers detection in a flowchart to help to understand their clinical applicability (Figure 2). The most common diagnostic methods for novel GC biomarkers that were used in presented studies differ significantly in terms of technical complexity and costs. Collecting the tumor specimens for immunohistochemistry (IHC) is a more invasive method than the assessment of the biomarkers in serum or in plasma. That assessment may be used in immunoenzyme techniques, such as ELISA or multiplex technology. IHC is also a more complicated and time-consuming technique in comparison to faster, cheaper and simpler immunoenzyme methods. Enzyme-linked immunosorbent assay (ELISA) is a highly sensitive method that is widely used in research. It involves many steps of analysis that take even 6 h to complete the measurement of a single biomarker. On the other hand, multiplex technology is able to measure up to 500 protein targets combined from a single complex biological sample and can also be used to assess RNA and protein targets as well as gene expression.

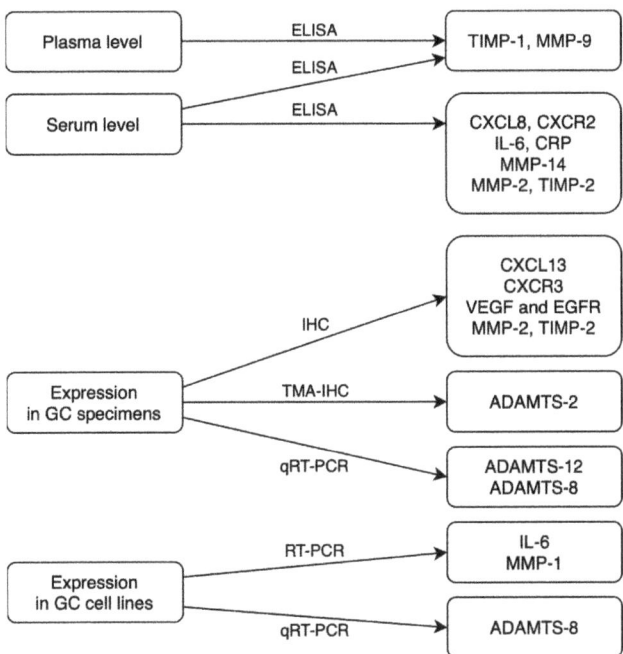

Figure 2. Sources and methods of potential GC biomarkers detection (ELISA—enzyme-linked immunosorbent assay; IHC—immunohistochemistry; TMA-IHC—tissue microarray immunohistochemistry; qRT-PCR—quantitative reverse transcription polymerase chain reaction) [15–19,21–28].

Culture of tissue or fluid remains the current standard of care for cancer diagnosis, including GC. However, molecular techniques seem to be the future of early diagnosis of malignant diseases. The costs of applying diagnostic techniques based on polymerase chain reaction (PCR) are known to be higher than ELISA or multiplex technology, but these methods improve the understanding of the nature and biology of a tumor [93]. PCR is a molecular biology technique used to amplify DNA. It is detected as the reaction progresses in real-time, and reverse transcription-PCR (RT-PCR) allows for quantification of the levels of messenger or ribosomal RNA and evaluates the level of protein synthetic activity. Molecular diagnostic techniques are widely applied, powerful and sensitive. They are used to identify biomarkers in the genome and proteome [94]. An assessment of a biomarker level in body fluids in combination with molecular tests may, in the future, be the best method in the cancer diagnostic process, including GC diagnostics.

5. Conclusions

Biomarkers for early detection of CRC are important to improve the management of patients with GC. The diagnostic specificity and sensitivity of well-established biochemical markers are limited. Therefore, there is a need for new non-invasive, safe, easily measurable and low-cost methods in the cancer diagnostic process. Our review presents the potential usefulness of novel biomarkers and compares their significance with the well-established tumor markers for this malignancy, including CA 72-4, CEA and CA 19-9.

The present paper indicates that selected chemokines and their receptors, VEGF and EGFR, DNA- and RNA-based biomarkers, and MET, IL-6 and CRP, as well as MMPs and their tissue inhibitors, are promising biomarkers in the diagnosis and progression of GC, while selected ADAMs and TIMPs might be used as potential prognostic factors of a GC patient survival.

Presented findings suggest that novel biomarkers may help in establishing an accurate diagnosis and in the treatment of cancer, in choosing medications for a patient or in predicting the drug response. They can also be used in the follow-up of cancer survivors. There is a need for more studies on larger groups connected with this subject to take a new biomarker into routine use; however, there are some promising candidates to improve the GC diagnostic and therapeutic process.

Author Contributions: E.P., M.Ł.-Z. and B.M. put forward the idea of the study. M.Ł.-Z. coordinated project funding. All authors have read and agreed to the published version of the manuscript.

Funding: This research was funded by Medical University of Bialystok. The APC was funded by Medical University of Bialystok, Poland.

Institutional Review Board Statement: Not applicable.

Informed Consent Statement: Not applicable.

Data Availability Statement: Not applicable.

Conflicts of Interest: The authors declare no conflict of interest.

References

1. WHO. Available online: https://gco.iarc.fr/today/data/factsheets/populations/900-world-fact-sheets.pdf (accessed on 21 March 2023).
2. Laurén, P. The two histological main types of gastric carcinoma: Diffuse and so-called intestinal-type carcinoma. *Acta Pathol. Microbiol. Scand.* **1965**, *6*, 31–49. [CrossRef] [PubMed]
3. Lochhead, P.; El-Omar, E.; Bédard, E.; Shore, D.F.; Gatzoulis, M.A. Gastric cancer. *Br. Med. Bull.* **2008**, *85*, 87–100. [CrossRef] [PubMed]
4. Crew, K.D.; Neugut, A.I. Epidemiology of gastric cancer. *World J. Gastroenterol.* **2006**, *12*, 354–362. [CrossRef] [PubMed]
5. Fock, K.M. Review article: The epidemiology and prevention of gastric cancer. *Aliment. Pharmacol. Ther.* **2014**, *40*, 250–260. [CrossRef] [PubMed]
6. Karimi, P.; Islami, F.; Anandasabapathy, S.; Freedman, N.D.; Kamangar, F. Gastric cancer: Descriptive epidemiology, risk factors, screening, and prevention. *Cancer Epidemiol. Biomark. Prev.* **2014**, *23*, 700–713. [CrossRef]
7. Verbeke, H.; Geboes, K.; Van Damme, J.; Struyf, S. The role of CXC chemokines in the transition of chronicn inflammation to esophageal and gastric cancer. *Biochim. Biophys. Acta* **2012**, *1825*, 117–129.
8. Hallinan, J.; Venkatesh, S.K. Gastric carcinoma: Imaging diagnosis, staging and assessment of treatment response. *Cancer Imaging* **2013**, *13*, 212–227. [CrossRef]
9. Boku, N. HER2-positive gastric cancer. *Gastric Cancer* **2014**, *17*, 1–12. [CrossRef]
10. Meric-Bernstam, F.; Johnson, A.M.; Dumbrava, E.E.I.; Raghav, K.; Balaji, K.; Bhatt, M.; Murthy, R.K.; Rodon, J.; Piha-Paul, S.A. Advances in HER2-Targeted Therapy: Novel Agents and Opportunities Beyond Breast and Gastric Cancer. *Clin. Cancer Res.* **2019**, *25*, 2033–2041. [CrossRef]
11. Chen, C.; Chen, Q.; Zhao, Q.; Liu, M.; Guo, J. Value of Combined Detection of Serum CEA, CA72-4, CA19-9, CA15-3 and CA12-5 in the Diagnosis of Gastric Cancer. *Ann. Clin. Lab. Sci.* **2017**, *47*, 260–263.
12. Mattar, R.; de Andrade, C.R.A.; DiFavero, G.M.; Gama-Rodrigues, J.J.; Laudanna, A.A. Preoperative serum levels of CA 72-4, CEA, CA 19-9, and alpha-fetoprotein in patients with gastric cancer. *Revista do Hospital das Clinicas* **2002**, *57*, 89–92. [CrossRef]
13. Rao, H.; Wu, H.; Huang, Q.; Yu, Z.; Zhong, Z. Clinical Value of Serum CEA, CA24-2 and CA19-9 in Patients with Colorectal Cancer. *Clin. Lab.* **2021**, *67*. [CrossRef]
14. Feng, F.; Tian, Y.; Xu, G.; Liu, Z.; Liu, S.; Zheng, G.; Guo, M.; Lian, X.; Fan, D.; Zhang, H. Diagnostic and prognostic value of CEA, CA19–9, AFP and CA125 for early gastric cancer. *BMC Cancer* **2017**, *17*, 737. [CrossRef]
15. Pawluczuk, E.; Łukaszewicz-Zając, M.; Gryko, M.; Kulczyńska-Przybik, A.; Mroczko, B. Serum CXCL8 and Its Specific Receptor (CXCR2) in Gastric Cancer. *Cancers* **2021**, *13*, 5186. [CrossRef]
16. Wei, Y.; Lin, C.; Li, H.; Xu, Z.; Wang, J.; Li, R.; Liu, H.; Zhang, H.; He, H.; Xu, J. CXCL13 expression is prognostic and predictive for postoperative adjuvant chemotherapy benefit in patients with gastric cancer. *Cancer Immunol. Immunother.* **2018**, *67*, 261–269. [CrossRef]
17. Chen, F.; Yuan, J.; Yan, H.; Liu, H.; Yin, S. Chemokine receptor CXCR3 correlates with decreased M2 macrophage infiltration and favorable prognosis in gastric cancer. *BioMed Res. Int.* **2019**, *2019*, 6832867. [CrossRef]
18. Lieto, E.; Ferraraccio, F.; Orditura, M.; Castellano, P.; La Mura, A.; Pinto, M.; Zamboli, A.; De Vita, F.; Galizia, G. Expression of vascular endothelial growth factor (VEGF) and epidermal growth factor receptor (EGFR) is an independent prognostic indicator of worse outcome in gastric cancer patients. *Ann. Surg. Oncol.* **2008**, *15*, 69–79. [CrossRef]
19. Matsuo, K.; Oka, M.; Murase, K.; Soda, H.; Isomoto, H.; Takeshima, F.; Mizuta, Y.; Murata, I.; Kohno, S. Expression of interleukin 6 and its receptor in human gastric and colorectal cancers. *J. Int. Med. Res.* **2003**, *31*, 69–75. [CrossRef]

20. Wang, X.; Li, J.; Liu, W.; Zhang, X.; Xue, L. The diagnostic value of interleukin 6 as a biomarker for gastric cancer: A meta-analysis and systematic review. *Medicine* **2021**, *100*, e27945. [CrossRef]
21. Łukaszewicz-Zając, M.; Mroczko, B.; Gryko, M.; Kędra, B.; Szmitkowski, M. Comparison between clinical significance of serum proinflammatory proteins (IL-6 and CRP) and classic tumor markers (CEA and CA 19-9) in gastric cancer. *Clin. Exp. Med.* **2010**, *11*, 89–96. [CrossRef]
22. Łukaszewicz-Zając, M.; Mroczko, B.; Guzińska-Ustymowicz, K.; Pryczynicz, A.; Gryko, M.; Kemona, A.; Kędra, B.; Szmitkowski, M. Matrix metalloproteinase 2 (MMP-2) and their tissue inhibitor 2 (TIMP-2) in gastric cancer patients. *Adv. Med. Sci.* **2013**, *58*, 235–243. [CrossRef] [PubMed]
23. Mroczko, B.; Groblewska, M.; Łukaszewicz-Zając, M.; Bandurski, R.; Kędra, B.; Szmitkowski, M. Pre-treatment serum and plasma levels of matrix metalloproteinase 9 (MMP-9) and tissue inhibitor of matrix metalloproteinases 1 (TIMP-1) in gastric cancer patients. *Clin. Chem. Lab. Med.* **2009**, *47*, 1133–1139. [CrossRef] [PubMed]
24. Wu, J.Y.; Lu, H.; Sun, Y.; Graham, D.Y.; Cheung, H.S.; Yamaoka, Y. Balance between polyoma enhancing activator 3 and activator protein 1 regulates *Helicobacter pylori*-stimulated matrix metalloproteinase 1 expression. *Cancer Res.* **2006**, *66*, 5111–5120. [CrossRef] [PubMed]
25. Kasurinen, A.; Tervahartiala, T.; Laitinen, A.; Kokkola, A.; Sorsa, T.; Böckelman, C.; Haglund, C. High serum MMP-14 predicts worse survival in gastric cancer. *PLoS ONE* **2018**, *13*, e0208800. [CrossRef]
26. Hou, Y.; Xu, Y.; Wu, D. ADAMTS12 acts as a tumor microenvironment related cancer promoter in gastric cancer. *Sci. Rep.* **2021**, *11*, 10996. [CrossRef]
27. Jiang, C.; Zhou, Y.; Huang, Y.; Wang, Y.; Wang, W.; Kuai, X. Overexpression of ADAMTS-2 in tumor cells and stroma is predictive of poor clinical prognosis in gastric cancer. *Hum. Pathol.* **2019**, *84*, 44–51. [CrossRef]
28. Chen, J.; Zhang, J.; Li, X.; Zhang, C.; Zhang, H.; Jin, J.; Dai, D. Downregulation of ADAMTS8 by DNA Hypermethylation in Gastric Cancer and Its Clinical Significance. *BioMed Res. Int.* **2016**, *2016*, 5083841. [CrossRef]
29. Hu, P.-J.; Chen, M.-Y.; Wu, M.-S.; Lin, Y.-C.; Shih, P.-H.; Lai, C.-H.; Lin, H.-J. Clinical Evaluation of CA72-4 for Screening Gastric Cancer in a Healthy Population: A Multicenter Retrospective Study. *Cancers* **2019**, *11*, 733. [CrossRef]
30. Gold, P.; Freedman, S.O. Demonstration of tumor-specific antigens in human colonic carcinomata by immunological tolerance and absorption techniques. *J. Exp. Med.* **1965**, *121*, 439–462. [CrossRef]
31. Hao, C.; Zhang, G.; Zhang, L. Serum CEA levels in 49 different types of cancer and noncancer diseases. *Prog. Mol. Biol. Transl. Sci.* **2019**, *162*, 213–227. [CrossRef]
32. Wang, W.; Xu, X.; Tian, B.; Wang, Y.; Du, L.; Sun, T.; Shi, Y.; Zhao, X.; Jing, J. The diagnostic value of serum tumor markers CEA, CA19-9, CA125, CA15-3, and TPS in metastatic breast cancer. *Clin. Chim. Acta* **2017**, *470*, 51–55. [CrossRef]
33. Grunnet, M.; Sorensen, J. Carcinoembryonic antigen (CEA) as tumor marker in lung cancer. *Lung Cancer* **2012**, *76*, 138–143. [CrossRef]
34. Van Manen, L.; Groen, J.V.; Putter, H.; Vahrmeijer, A.L.; Swijnenburg, R.-J.; Bonsing, B.A.; Mieog, J.S.D. Elevated CEA and CA19-9 serum levels independently predict advanced pancreatic cancer at diagnosis. *Biomarkers* **2020**, *25*, 186–193. [CrossRef]
35. Chen, Y.; Gao, S.-G.; Chen, J.-M.; Wang, G.-P.; Wang, Z.-F.; Zhou, B.; Jin, C.-H.; Yang, Y.-T.; Feng, X.-S. Serum CA242, CA199, CA125, CEA, and TSGF are Biomarkers for the Efficacy and Prognosis of Cryoablation in Pancreatic Cancer Patients. *Cell Biochem. Biophys.* **2015**, *71*, 1287–1291. [CrossRef]
36. Wang, Y.F.; Feng, F.L.; Zhao, X.H.; Ye, Z.X.; Zeng, H.P.; Li, Z.; Jiang, X.Q.; Peng, Z.H. Combined detection tumor markers for diagnosis and prognosis of gallbladder cancer. *World J. Gastroenterol.* **2014**, *20*, 4085–4092. [CrossRef]
37. Zhou, J.; Zhang, Q.; Li, P.; Shan, Y.; Zhao, D.; Cai, J. Prognostic factors of carcinoma of the ampulla of Vater after surgery. *Tumor Biol.* **2014**, *35*, 1143–1148. [CrossRef]
38. Kato, K.; Taniguchi, M.; Kawakami, T.; Nagase, A.; Matsuda, M.; Onodea, K.; Yamaguchi, H.; Higuchi, M.; Furukawa, H. Gastric Cancer with a Very High Serum CA 19-9 Level. *Case Rep. Gastroenterol.* **2011**, *5*, 258–261. [CrossRef]
39. Koprowski, H.; Steplewski, Z.; Mitchell, K.; Herlyn, M.; Herlyn, D.; Fuhrer, P. Colorectal carcinoma antigens detected by hybridoma antibodies. *Somatic Cell Genet.* **1979**, *5*, 957–971. [CrossRef]
40. Jo, J.-C.; Ryu, M.-H.; Koo, D.-H.; Ryoo, B.-Y.; Kim, H.J.; Kim, T.W.; Choi, K.D.; Lee, G.H.; Jung, H.-Y.; Yook, J.H.; et al. Serum CA 19-9 as a prognostic factor in patients with metastatic gastric cancer. *Asia-Pac. J. Clin. Oncol.* **2013**, *9*, 324–330. [CrossRef]
41. Duffy, M.; Lamerz, R.; Haglund, C.; Nicolini, A.; Kalousova, M.; Holubec, L.; Sturgeon, C. Tumor markers in colorectal cancer, gastric cancer and gastrointestinal stromal cancers: European group on tumor markers 2014 guidelines update. *Int. J. Cancer* **2014**, *134*, 2513–2522. [CrossRef]
42. Rüschoff, J.; Dietel, M.; Baretton, G.; Arbogast, S.; Walch, A.; Monges, G.; Chenard, M.-P.; Penault-Llorca, F.; Nagelmeier, I.; Schlake, W.; et al. HER2 diagnostics in gastric cancer—Guideline validation and development of standardized immunohistochemical testing. *Virchows Arch.* **2010**, *457*, 299–307. [CrossRef] [PubMed]
43. Baj-Krzyworzeka, M.; Węglarczyk, K.; Baran, J.; Szczepanik, A.; Szura, M.; Siedlar, M. Elevated level of some chemokines in plasma of gastric cancer patients. *Cent. Eur. J. Immunol.* **2016**, *41*, 358–362. [CrossRef] [PubMed]
44. Kruizinga, R.C.; Bestebroer, J.; Berghuis, P.; de Haas, C.J.C.; Links, T.P.; de Vries, E.G.E.; Walenkamp, A.M.E. Role of chemokines and their receptors in cancer. *Curr. Pharm. Des.* **2009**, *15*, 3396–3416. [CrossRef] [PubMed]
45. Łukaszewicz-Zając, M.; Mroczko, B.; Szmitkowski, M. Chemokines and their receptors in esophageal cancer—The systematic review and future perspectives. *Tumor Biol.* **2015**, *36*, 5707–5714. [CrossRef]

46. Melincovici, C.S.; Boşca, A.B.; Şuşman, S.; Mărginean, M.; Mihu, C.; Istrate, M.; Moldovan, I.M.; Roman, A.L.; Mihu, C.M. Vascular endothelial growth factor (VEGF)—Key factor in normal and pathological angiogenesis. *Rom. J. Morphol. Embryol.* **2018**, *59*, 455–467.
47. Ylä-Herttuala, S.; Rissanen, T.T.; Vajanto, I.; Hartikainen, J. Vascular endothelial growth factors: Biology and current status of clinical applications in cardiovascular medicine. *J. Am. Coll. Cardiol.* **2007**, *49*, 1015–1026. [CrossRef]
48. Hajjo, R.; Sweidan, K. Review on Epidermal Growth Factor Receptor (EGFR) Structure, Signaling Pathways, Interactions, and Recent Updates of EGFR Inhibitors. *Curr. Top. Med. Chem.* **2020**, *20*, 815–834. [CrossRef]
49. Rao, S.; Zhao, L.; Zheng, X.; Li, S.; Tian, X. Correlation of Serum IL-17, VEGF, and Lactate Dehydrogenase (LDH) Levels with Prognosis of Gastric Cancer. *Evid.-Based Complement. Altern. Med.* **2022**, *2022*, 8126672. [CrossRef]
50. Vidal, Ó.; Metges, J.-P.; Elizalde, I.; Valentíni, M.; Volant, A.; Molina, R.; Castells, A.; Pera, M. High preoperative serum vascular endothelial growth factor levels predict poor clinical outcome after curative resection of gastric cancer. *Br. J. Surg.* **2009**, *96*, 1443–1451. [CrossRef]
51. Tanaka, T.; Narazaki, M.; Kishimoto, T. IL-6 in inflammation, immunity, and disease. *Cold Spring Harb. Perspect. Biol.* **2014**, *6*, a016295. [CrossRef]
52. Clyne, B.; Olshaker, J.S. The C-reactive protein. *J. Emerg. Med.* **1999**, *17*, 1019–1025. [CrossRef]
53. Chang, C.-C.; Sun, C.-F.; Pai, H.-J.; Wang, W.-K.; Hsieh, C.-C.; Kuo, L.-M.; Wang, C.-S. Preoperative serum C-reactive protein and gastric cancer; clinical-pathological correlation and prognostic significance. *Chang Gung Med. J.* **2010**, *33*, 301–312.
54. Nagase, H.; Visse, R.; Murphy, G. Structure and function of matrix metalloproteinases and TIMPs. *Cardiovasc. Res.* **2006**, *69*, 562–573. [CrossRef]
55. Shim, K.-N.; Jung, S.-A.; Joo, Y.-H.; Yoo, K. Clinical significance of tissue levels of matrix metalloproteinases and tissue inhibitors of metalloproteinases in gastric cancer. *J. Gastroenterol.* **2007**, *42*, 120–128. [CrossRef]
56. Laitinen, A.; Hagström, J.; Mustonen, H.; Kokkola, A.; Tervahartiala, T.; Sorsa, T.; Böckelman, C.; Haglund, C. Serum MMP-8 and TIMP-1 as prognostic biomarkers in gastric cancer. *Tumor Biol.* **2018**, *40*. [CrossRef]
57. Calcagno, D.Q.; Gigek, C.O.; Chen, E.S.; Burbano, R.R.; Smith, M.D.A.C. DNA and histone methylation in gastric carcinogenesis. *World J. Gastroenterol.* **2013**, *19*, 1182–1192. [CrossRef]
58. Suzuki, H.; Tokino, T.; Shinomura, Y.; Imai, K.; Toyota, M. DNA methylation and cancer pathways in gastrointestinal tumors. *Pharmacogenomics* **2008**, *9*, 1917–1928. [CrossRef]
59. Enomoto, S.; Maekita, T.; Tsukamoto, T.; Nakajima, T.; Nakazawa, K.; Tatematsu, M.; Ichinose, M.; Ushijima, T. Lack of association between CpG island methylator phenotype in human gastric cancers and methylation in their background non-cancerous gastric mucosae. *Cancer Sci.* **2007**, *98*, 1853–1861. [CrossRef]
60. Chen, H.Y.; Zhu, B.H.; Zhang, C.H.; Yang, D.J.; Peng, J.J.; Chen, J.H.; Liu, F.K.; He, Y.L. High CpG island methylator phenotype is associated with lymph node metastasis and prognosis in gastric cancer. *Cancer Sci.* **2012**, *103*, 73–79. [CrossRef]
61. Du, W.; Wang, S.; Zhou, Q.; Li, X.; Chu, J.; Chang, Z.; Tao, Q.; Ng, E.K.O.; Fang, J.; Sung, J.J.Y.; et al. ADAMTS9 is a functional tumor suppressor through inhibiting AKT/mTOR pathway and associated with poor survival in gastric cancer. *Oncogene* **2013**, *32*, 3319–3328. [CrossRef]
62. Cheng, A.S.; Li, M.S.; Kang, W.; Cheng, V.Y.; Chou, J.L.; Lau, S.S.; Go, M.Y.; Lee, C.C.; Ling, T.K.; Ng, E.K.; et al. Helicobacter pylori causes epigenetic dysregulation of FOXD3 to promote gastric carcinogenesis. *Gastroenterology* **2013**, *144*, 122–133.e9. [CrossRef]
63. Li, X.; Cheung, K.F.; Ma, X.; Tian, L.; Zhao, J.; Go, M.Y.Y.; Shen, B.; Cheng, A.S.; Ying, J.; Tao, Q.; et al. Epigenetic inactivation of paired box gene 5, a novel tumor suppressor gene, through direct upregulation of p53 is associated with prognosis in gastric cancer patients. *Oncogene* **2012**, *31*, 3419–3430. [CrossRef] [PubMed]
64. Qi, Q.; Pan, Y.F.; Shen, J.J.; Gu, X.Q.; Han, S.W.; Liao, H.H.; Jiang, Y.Z.; Zhong, L.P. Circulating DNA for detection of gastric cancer. *Eur. Rev. Med. Pharmacol. Sci.* **2016**, *20*, 2558–2564. [PubMed]
65. Park, J.-L.; Kim, H.J.; Choi, B.Y.; Lee, H.-C.; Jang, H.-R.; Song, K.S.; Noh, S.-M.; Kim, S.-Y.; Han, D.S. Quantitative analysis of cell-free DNA in the plasma of gastric cancer patients. *Oncol. Lett.* **2012**, *3*, 921–926. [CrossRef] [PubMed]
66. Calin, G.A.; Croce, C.M. Chromosomal rearrangements and microRNAs: A new cancer link with clinical implications. *J. Clin. Investig.* **2007**, *117*, 2059–2066. [CrossRef]
67. Li, Z.; Liu, Z.M.; Xu, B.H. A meta-analysis of the effect of microRNA-34a on the progression and prognosis of gastric cancer. *Eur. Rev. Med. Pharmacol. Sci.* **2018**, *22*, 8281–8287. [CrossRef]
68. Iwasaki, H.; Shimura, T.; Yamada, T.; Okuda, Y.; Natsume, M.; Kitagawa, M.; Horike, S.-I.; Kataoka, H. A novel urinary microRNA biomarker panel for detecting gastric cancer. *J. Gastroenterol.* **2019**, *54*, 1061–1069. [CrossRef]
69. Huang, J.-Y.W.T.-J. Overexpression of the c-met protooncogene in human gastric carcinoma–correlation to clinical features. *Acta Oncol.* **2001**, *40*, 638–643. [CrossRef]
70. Yu, S.; Yu, Y.; Zhao, N.; Cui, J.; Li, W.; Liu, T. C-Met as a prognostic marker in gastric cancer: A systematic review and meta-analysis. *PLoS ONE* **2013**, *8*, e79137. [CrossRef]
71. Nakajima, M.; Sawada, H.; Yamada, Y.; Watanabe, A.; Tatsumi, M.; Yamashita, J.; Matsuda, M.; Sakaguchi, T.; Hirao, T.; Nakano, H. The prognostic significance of amplification and overexpression of c-met and c-erb B-2 in human gastric carcinomas. *Cancer* **1999**, *85*, 1894–1902. [CrossRef]
72. Marano, L.; Chiari, R.; Fabozzi, A.; De Vita, F.; Boccardi, V.; Roviello, G.; Petrioli, R.; Marrelli, D.; Roviello, F.; Patriti, A. c-Met targeting in advanced gastric cancer: An open challenge. *Cancer Lett.* **2015**, *365*, 30–36. [CrossRef]

73. Zhang, Z.; Miao, L.; Wang, S.; Zhao, Y.; Xie, Y.; Yun, H.; Ren, Z.; Wang, G.; Teng, M.; Li, Y. Study on the expression of c-Met in gastric cancer and its correlation with preoperative serum tumor markers and prognosis. *World J. Surg. Oncol.* **2022**, *20*, 204. [CrossRef]
74. Tsujio, G.; Maruo, K.; Yamamoto, Y.; Sera, T.; Sugimoto, A.; Kasashima, H.; Miki, Y.; Yoshii, M.; Tamura, T.; Toyokawa, T.; et al. Significance of tumor heterogeneity of p-Smad2 and c-Met in HER2-positive gastric carcinoma with lymph node metastasis. *BMC Cancer* **2022**, *22*, 598. [CrossRef]
75. Ghedini, G.C.; Ronca, R.; Presta, M.; Giacomini, A. Future applications of FGF/FGFR inhibitors in cancer. *Expert Rev. Anticancer Ther.* **2018**, *18*, 861–872. [CrossRef]
76. Wainberg, Z.A.; Enzinger, P.C.; Kang, Y.-K.; Qin, S.; Yamaguchi, K.; Kim, I.-H.; Saeed, A.; Oh, S.C.; Li, J.; Turk, H.M.; et al. Bemarituzumab in patients with FGFR2b-selected gastric or gastro-oesophageal junction adenocarcinoma (FIGHT): A randomised, double-blind, placebo-controlled, phase 2 study. *Lancet Oncol.* **2022**, *23*, 1430–1440. [CrossRef]
77. Lee, J.-H.; Kim, S.-H.; Han, S.-H.; An, J.-S.; Lee, E.-S.; Kim, Y.-S. Clinicopathological and molecular characteristics of Epstein-Barr virus-associated gastric carcinoma: A meta-analysis. *J. Gastroenterol. Hepatol.* **2009**, *24*, 354–365. [CrossRef]
78. Wang, K.; Yuen, S.T.; Xu, J.; Lee, S.P.; Yan, H.H.N.; Shi, S.T.; Siu, H.C.; Deng, S.; Chu, K.M.; Law, S.; et al. Whole-genome sequencing and comprehensive molecular profiling identify new driver mutations in gastric cancer. *Nat. Genet.* **2014**, *46*, 573–582. [CrossRef]
79. Röcken, C. Predictive biomarkers in gastric cancer. *J. Cancer Res. Clin. Oncol.* **2023**, *149*, 467–481. [CrossRef]
80. Bai, Y.; Xie, T.; Wang, Z.; Tong, S.; Zhao, X.; Zhao, F.; Cai, J.; Wei, X.; Peng, Z.; Shen, L. Efficacy and predictive biomarkers of immunotherapy in Epstein-Barr virus-associated gastric cancer. *J. Immunother. Cancer* **2022**, *10*, e004080. [CrossRef]
81. Gęca, K.; Rawicz-Pruszyński, K.; Mielko, J.; Mlak, R.; Sędłak, K.; Polkowski, W.P. Rapid Detection of Free Cancer Cells in Intraoperative Peritoneal Lavage Using One-Step Nucleic Acid Amplification (OSNA) in Gastric Cancer Patients. *Cells* **2020**, *9*, 2168. [CrossRef]
82. Kutun, S.; Celik, A.; Cem Kockar, M.; Erkorkmaz, U.; Eroğlu, A.; Cetin, A.; Erkosar, B.; Yakicier, C. Expression of CK-19 and CEA mRNA in peripheral blood of gastric cancer patients. *Exp. Oncol.* **2010**, *32*, 263–268. [PubMed]
83. do Nascimento, C.N.; Mascarenhas-Lemos, L.; Silva, J.R.; Marques, D.S.; Gouveia, C.F.; Faria, A.; Velho, S.; Garrido, R.; Maio, R.; Costa, A.; et al. EBV and MSI Status in Gastric Cancer: Does It Matter? *Cancers* **2022**, *15*, 74. [CrossRef] [PubMed]
84. Schoemig-Markiefka, B.; Eschbach, J.; Scheel, A.H.; Pamuk, A.; Rueschoff, J.; Zander, T.; Buettner, R.; Schroeder, W.; Bruns, C.J.; Loeser, H.; et al. Optimized PD-L1 scoring of gastric cancer. *Gastric Cancer* **2021**, *24*, 1115–1122. [CrossRef] [PubMed]
85. Böger, C.; Behrens, H.-M.; Mathiak, M.; Krüger, S.; Kalthoff, H.; Röcken, C. PD-L1 is an independent prognostic predictor in gastric cancer of Western patients. *Oncotarget* **2016**, *7*, 24269–24283. [CrossRef]
86. Iannantuono, G.M.; Riondino, S.; Sganga, S.; Rosenfeld, R.; Guerriero, S.; Carlucci, M.; Capotondi, B.; Torino, F.; Roselli, M. NTRK Gene Fusions in Solid Tumors and TRK Inhibitors: A Systematic Review of Case Reports and Case Series. *J. Pers. Med.* **2022**, *12*, 1819. [CrossRef]
87. Shinozaki-Ushiku, A.; Ishikawa, S.; Komura, D.; Seto, Y.; Aburatani, H.; Ushiku, T. The first case of gastric carcinoma with NTRK rearrangement: Identification of a novel ATP1B–NTRK1 fusion. *Gastric Cancer* **2020**, *23*, 944–947. [CrossRef]
88. Arnold, A.; Daum, S.; von Winterfeld, M.; Berg, E.; Hummel, M.; Horst, D.; Rau, B.; Stein, U.; Treese, C. Analysis of NTRK expression in gastric and esophageal adenocarcinoma (AGE) with pan-TRK immunohistochemistry. *Pathol. Res. Pract.* **2019**, *215*, 152662. [CrossRef]
89. Sohn, S.-H.; Sul, H.J.; Kim, B.J.; Kim, H.S.; Zang, D.Y. Entrectinib Induces Apoptosis and Inhibits the Epithelial-Mesenchymal Transition in Gastric Cancer with NTRK Overexpression. *Int. J. Mol. Sci.* **2021**, *23*, 395. [CrossRef]
90. Jiang, Q.; Li, M.; Li, H.; Chen, L. Entrectinib, a new multi-target inhibitor for cancer therapy. *Biomed. Pharmacother.* **2022**, *150*, 112974. [CrossRef]
91. Cheng, Y.; Bu, D.; Zhang, Q.; Sun, R.; Lyle, S.; Zhao, G.; Dong, L.; Li, H.; Zhao, Y.; Yu, J.; et al. Genomic and transcriptomic profiling indicates the prognosis significance of mutational signature for TMB-high subtype in Chinese patients with gastric cancer. *J. Adv. Res.* **2022**. [CrossRef]
92. Lee, K.W.; Van Cutsem, E.; Bang, Y.J.; Fuchs, C.S.; Kudaba, I.; Garrido, M.; Chung, H.C.; Lee, J.; Castro, H.R.; Chao, J.; et al. Association of Tumor Mutational Burden with Efficacy of Pembrolizumab±Chemotherapy as First-Line Therapy for Gastric Cancer in the Phase III KEYNOTE-062 Study. *Clin. Cancer Res.* **2022**, *28*, 3489–3498. [CrossRef]
93. Saab, S.; Ahn, T.; McDaniel, T.; Yanny, B.; Tong, M.J. Economic Comparison of Serologic and Molecular Screening Strategies for Hepatitis C Virus. *Gastroenterol. Hepatol.* **2018**, *14*, 459–462.
94. Dwivedi, S.; Purohit, P.; Misra, R.; Pareek, P.; Goel, A.; Khattri, S.; Pant, K.K.; Misra, S.; Sharma, P. Diseases and Molecular Diagnostics: A Step Closer to Precision Medicine. *Indian J. Clin. Biochem.* **2017**, *32*, 374–398. [CrossRef]

Disclaimer/Publisher's Note: The statements, opinions and data contained in all publications are solely those of the individual author(s) and contributor(s) and not of MDPI and/or the editor(s). MDPI and/or the editor(s) disclaim responsibility for any injury to people or property resulting from any ideas, methods, instructions or products referred to in the content.

Article

Adhesion to the Brain Endothelium Selects Breast Cancer Cells with Brain Metastasis Potential

Bai Zhang [1,2], Xueyi Li [1,2], Kai Tang [1,2], Ying Xin [1,2], Guanshuo Hu [1,2], Yufan Zheng [1,2], Keming Li [1,2], Cunyu Zhang [1,2] and Youhua Tan [1,2,*]

1. The Hong Kong Polytechnic University Shenzhen Research Institute, Shenzhen 518000, China
2. Department of Biomedical Engineering, The Hong Kong Polytechnic University, Hong Kong 999077, China
* Correspondence: youhua.tan@polyu.edu.hk

Abstract: Tumor cells metastasize from a primary lesion to distant organs mainly through hematogenous dissemination, in which tumor cell re-adhesion to the endothelium is essential before extravasating into the target site. We thus hypothesize that tumor cells with the ability to adhere to the endothelium of a specific organ exhibit enhanced metastatic tropism to this target organ. This study tested this hypothesis and developed an in vitro model to mimic the adhesion between tumor cells and brain endothelium under fluid shear stress, which selected a subpopulation of tumor cells with enhanced adhesion strength. The selected cells up-regulated the genes related to brain metastasis and exhibited an enhanced ability to transmigrate through the blood–brain barrier. In the soft microenvironments that mimicked brain tissue, these cells had elevated adhesion and survival ability. Further, tumor cells selected by brain endothelium adhesion expressed higher levels of *MUC1*, *VCAM1*, and *VLA-4*, which were relevant to breast cancer brain metastasis. In summary, this study provides the first piece of evidence to support that the adhesion of circulating tumor cells to the brain endothelium selects the cells with enhanced brain metastasis potential.

Keywords: endothelial adhesion; biomechanics; mechanobiology; fluid shear stress; brain metastasis

1. Introduction

Tumor metastasis accounts for over 90% of cancer-related deaths and refers to the dissemination of tumor cells from a primary lesion to distant organs. This is a sequential process, in which a subpopulation of tumor cells dislodge from the primary tumor, migrate and invade the local stroma, intravasate into and survive in circulation, re-adhere to the endothelium, extravasate into distant organs, and form metastatic colonization [1]. Tumor cells must survive under a variety of rate-limiting factors and less than 0.1% of circulating tumor cells (CTCs) may eventually generate metastatic tumors, implicating the poor metastatic efficiency [2,3]. This also suggests that not all but only a small subpopulation of disseminated tumor cells have the ability to succeed in the entire metastatic process and each metastatic step may enrich a portion of disseminated cells [4]. Therefore, it is important to isolate those rare metastasis-competent tumor cells for the comprehensive characterization of their unique properties and the development of effective targeting strategies.

Tumor cells metastasize to distant organs mainly through the hematogenous route, during which CTCs need to adhere to the endothelium within the target organs before they can extravasate the vasculature. Such arrest often starts from weak adhesion between CTCs and the endothelium under blood flow and is then stabilized through a strong adhesion that is formed during the rolling of cancer cells on the endothelium [5,6]. The weak adhesion that captures cancer cells in circulation needs to be formed quickly to overwhelm the dislodgement caused by blood shear stress. It is known that selectins expressed on endothelial cells mediate such weak adhesions. The adhesion molecules on the surface of CTCs, such as CD44 and MUC1, can bind to selectins so that the weak adhesion between

Citation: Zhang, B.; Li, X.; Tang, K.; Xin, Y.; Hu, G.; Zheng, Y.; Li, K.; Zhang, C.; Tan, Y. Adhesion to the Brain Endothelium Selects Breast Cancer Cells with Brain Metastasis Potential. *Int. J. Mol. Sci.* **2023**, *24*, 7087. https://doi.org/10.3390/ijms24087087

Academic Editor: Laura Paleari

Received: 25 March 2023
Accepted: 4 April 2023
Published: 11 April 2023

Copyright: © 2023 by the authors. Licensee MDPI, Basel, Switzerland. This article is an open access article distributed under the terms and conditions of the Creative Commons Attribution (CC BY) license (https://creativecommons.org/licenses/by/4.0/).

tumor cells and the endothelium is established [7–9]. To further stabilize this interaction, strong adhesion is required for the stable attachment of tumor cells to the endothelium. Integrins universally expressed on cancer cells (such as integrin α5β1 and α4β1), and their ligands on endothelial cells (such as VCAM-1 and ICAM-1), have been reported to mediate this strong adhesion [6,10–12]. Some of these adhesion molecules involved in either weak or strong adhesion between tumor cells and endothelium, such as CD44, VCAM1, and VLA4, play indispensable roles in metastasis [12–15]. This adhesion process may inevitably select CTCs with certain adhesion molecule profiles and thus enrich a subpopulation of these cells with enhanced metastatic potential.

Tumor cells do not randomly choose their metastatic sites; rather, they disseminate to the preferred organs for metastatic colonization, which is referred to as organotropism [16]. In particular, endothelial cells in different organs may exhibit distinct expression profiles of adhesion molecules, which may be critical in the re-attachment of tumor cells to the specific endothelium and in mediating metastatic organotropism in the target site [17,18]. However, it remains unclear whether the tumor cell-endothelium adhesion can select a subpopulation of cells that preferentially metastasize to a specific organ. In this study, we developed a microfluidic system to mimic the interaction between CTCs and endothelial cells in the brain tissue. Tumor cells that stably adhered to the brain endothelium under fluid shear stress were enriched and their brain metastasis ability was characterized, including the gene expression profile, adhesion to the brain endothelium, transmigration ability across the blood–brain barrier (BBB), and adhesion and survival in the brain tissue. In addition, the expression profile of adhesion molecules in these selected cells was analyzed and their clinical relevance was examined.

2. Results

2.1. In Vitro Selection of Breast Cancer Cells through the Adhesion to the Brain Endothelium

Tumor cells must adhere to the brain endothelium before extravasating into the tissue to generate brain metastases. To mimic the interaction between the endothelium and CTCs, we developed an in vitro system (Figure 1a) [19–21], in which a monolayer of brain endothelium (hCMEC/D3) was grown in the microfluidic channel and tumor cells in suspension were circulated under a steady flow in the system. Since CTCs are very rare in vivo (1–10 CTCs per mL blood) and not commercially available yet, tumor cells in suspension have been widely utilized as an alternative model for CTC study [22–26]. In this study, suspended MDA-MB-231 cells (WT) were used to mimic breast CTCs, and they expressed high levels of the CTC marker EpCAM with high metastatic potential [27,28]. The endothelial monolayer was formed without the pre-treatment under shear stress before the perfusion of suspended tumor cells. The adhesion of tumor cells to the endothelium could be allowed in the blood flow within the veins (0.5–4 dyn/cm^2) while being prevented in the arteries (4–30 dyn/cm^2) [5]. The shear rate of blood flow in the brain post-capillary venules varies within a range of 1–6 dyn/cm^2 [29–31]. To determine the suitable fluid shear stress level, wide type MDA-MB-231 cells (WT) were perfused at different shear stress (5, 2.5, 1, 0.5, and 0.25 dyn/cm^2) into the flow chamber for 15 min and the adhesion between tumor cells and endothelial cells was analyzed. The results show that a large number of WT cells adhered to the brain endothelium under 0.25 dyn/cm^2 wall shear stress (Figure 1b,c). However, the adhered cells considerably decreased when the shear stress increased to 0.5, 1, 2.5, and 5 dyn/cm^2. There was no significant difference in the number of adhered WT cells among the larger shearing groups (Figure 1c). We speculated that there existed a threshold between 0.5 dyn/cm^2 and 0.25 dyn/cm^2, which was strong enough to counteract the adhesion between cancer cells and the endothelium. Therefore, 1 dyn/cm^2 shear stress was chosen for the in vitro brain-endothelium-adhesion-based selection. The selected cells were defined as the "Flow Adhesion Selected" (FAS) group. To explore the effect of shear stress on the endothelium-adhesion-based selection, WT cells were suspended in the flow chamber to interact with the brain endothelium for 30 min without exposure to shear stress. The isolated tumor cells were defined as the "Static Adhesion Selected" (SAS) group.

Notably, almost all of FAS and SAS cells expressed the green fluorescence, indicating that they were mainly green fluorescent protein (GFP)-labeled tumor cells and free of endothelial cells (Figure S1c). To compare the adhesion ability of FAS and SAS groups, both groups were mixed with WT and allowed to interact with the brain endothelium under 1 dyn/cm^2 wall shear stress. The results show that the FAS group exhibited enhanced adhesion to the brain endothelium compared to both the SAS and WT groups (Figure 1d,e). Meanwhile, there was no difference between the SAS and WT groups. These findings suggest that the in vitro endothelium adhesion can select a subpopulation of tumor cells.

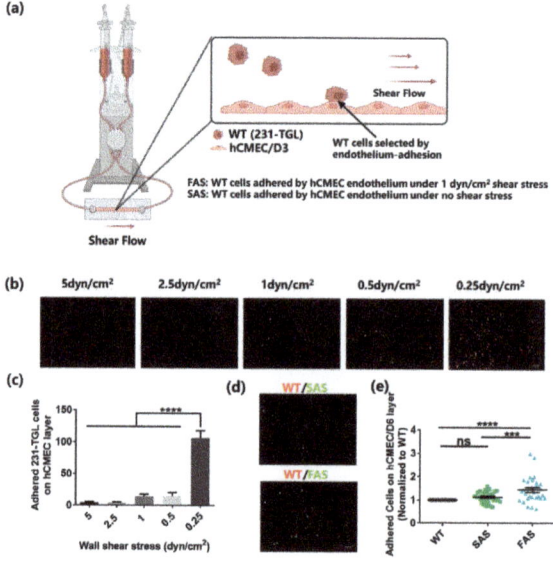

Figure 1. Brain-endothelium-adhesion-based selection of breast cancer cells. (a) The illustration of in vitro system for brain-endothelium-adhesion-based selection of breast cancer cells (created with biorender.com). (b,c) The adhesion of tumor cells to the brain endothelium under various levels of fluid shear stress. WT cells were labeled with red cell tracker and perfused into the flow chamber slides to adhere to the brain endothelium under 0.25, 0.5, 1, 2.5, and 5 dyn/cm^2 shear stress for 15 min. The number of adhered cells was counted under a fluorescence microscope. Scale bar = 100 μm. n = 3. (d,e) The FAS cells exhibited enhanced adhesion ability under shear stress. WT cells (labeled with red cell tracker) and FAS cells or SAS cells (labeled with green cell tracker) were equally mixed and perfused into the flow chamber at the rate of 1 dyn/cm^2 to adhere to the brain endothelium for 15 min. The number of adhered cells was counted under a microscope and normalized to the WT group. Scale bar = 100 μm. n = 3. All data are represented by mean ± SEM. The statistics among groups were analyzed using one-way ANOVA with the post hoc Bonferroni test. (ns: no significance, *** $p < 0.001$, **** $p < 0.0001$).

2.2. Adhesion-Selected Tumor Cells Up-Regulate Brain Metastasis Genes and Exhibit Enhanced Adhesion Strength and BBB Transmigration Ability

CTCs need to adhere to the endothelium before extravasation, indicating the importance of endothelial adhesion in tumor metastasis [32,33]. Therefore, we hypothesized that the endothelial adhesion might select a subpopulation of tumor cells with brain metastatic advantages. To test this hypothesis, the gene expression profile of the selected cells was examined. The results show that compared to both WT and SAS groups, the FAS group up-regulated *COX2*, *EREG*, *HBEGF*, *ITGAV*, *ITGB3*, *ANGPTL4*, *PIEZO2*, *SCNN1A*, and *LTBP1* (9 out of 11 genes; Figure 2a), which were reported to be highly expressed in breast cancer cells with brain metastasis ability [34–37]. In contrast to other metastasis sites, one special challenge for breast cancer cells in arriving at the brain is to cross the BBB, which can be mitigated by multiple key molecules, such as COX2, EREG, HBEGF, ITGAV, ITGB3, ST6GALNAC5, ANGPTL4, PIEZO2, and SCNN1A [34,35,37–40]. In particular, the FAS group showed higher expressions of *COX2* and *ITGB3*. COX2 can enhance the permeability of BBB by up-regulating the expression of matrix metallopeptidase 1 (MMP1) [37], and integrin αvβ3 (encoded by *ITGAV* and *ITGB3*) facilitates breast cancer brain metastasis by mediating the adhesion between cancer cells and brain endothelium and up-regulating the expressions of MMP2 and MMP9 [41,42]. We further examined the expressions of these two molecules at the protein level (Figures 2b–e and S1d,e). Compared to both WT and SAS groups, the FAS group enhanced the expressions of COX2 and integrin αvβ3. Note that there was no significant difference in the mRNA expressions of *ST6GALNAC5* and *SerpinB2* (Figure 2a). ST6GALNAC5 is a specific mediator of brain metastasis [34]. However, this gene might not be an indispensable biomarker for all breast cancer cells. It is found that overexpression of ST6GALNAC5 in MDA-MB-231 cells hinders their adhesion to the BBB [43], which may partially explain why MDA-MB-231 cells selected through adhesion (FAS) do not exhibit enhanced ST6GALNAC5 expression. ANGPTL4, PIEZO2, SCNN1A, LTBP1, and SerpinB2 promote the survival of breast cancer cells in the brain microenvironment [33–35,40]. All five genes except SerpinB2 were up-regulated in the FAS group (Figure 2a). Interestingly, FAS had enhanced protein expression of SerpinB2, indicating that the influence might be post-transcriptional. To elucidate the roles of individual molecules, it will be important to silence these genes and investigate the potential effects on the adhesion to the endothelium and the metastatic process.

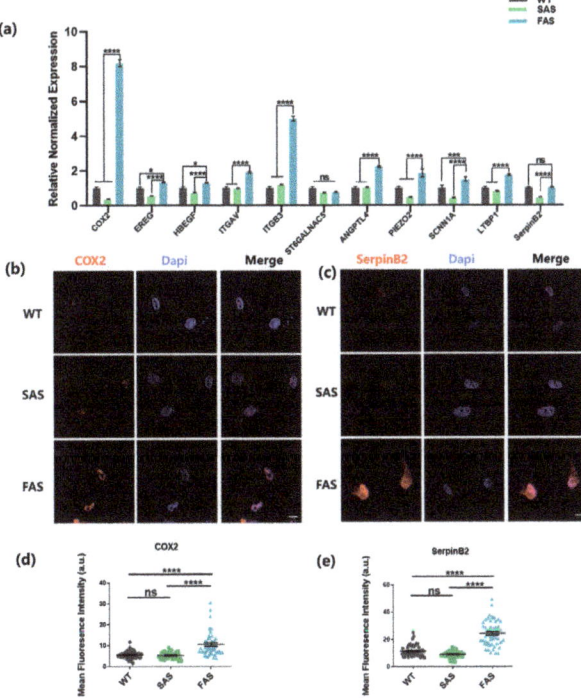

Figure 2. FAS cells up-regulate brain metastasis genes. (**a**) FAS cells showed enhanced expressions of brain metastasis-related genes. The gene expression was evaluated using qPCR. The statistics were calculated using two-way ANOVA with the post hoc Tukey test. $n = 3$. (**b,c**) The expressions of COX2 and SerpinB2 in all groups were tested using immunofluorescence staining. Scale bar = 10 μm. (**d,e**) Quantification of the fluorescence intensity of COX2 and SerpinB2 in (**b,c**). The statistics among three groups were calculated based on one-way ANOVA with the post hoc Bonferroni test. $n = 50$. (ns: no significance, * $p < 0.05$, *** $p < 0.001$, **** $p < 0.0001$).

Once captured by the endothelium, cancer cells begin to roll on it in the direction of the blood flow. During this process, complex molecules on tumor cells, such as integrins, can mediate strong adhesions to stabilize the interaction between tumor cells and endothelial cells [6]. To test whether FAS cells had advantages in establishing strong adhesion to the endothelium, different tumor cells were first co-cultured with the brain endothelium for 30 min to allow adhesion molecules to interact with each other. Then, different levels of shear stress were utilized to test the adhesion strength (Figures 3a,b and S1a,b) [44]. The results show that the number of adhered tumor cells decreased when the applied shear stress increased from 0.25 dyn/cm^2 to 2 dyn/cm^2 (Figure 3a,b). The FAS group had more tumor cells remaining on the brain endothelium compared to SAS and WT groups (Figure 3a,b). Interestingly, all three groups had a similar number of adhered tumor cells after exposure to wall shear stress higher than 2 dyn/cm^2 (Figure S1a,b). These results suggest that FAS cells can develop stable adhesion to the endothelium with enhanced strength.

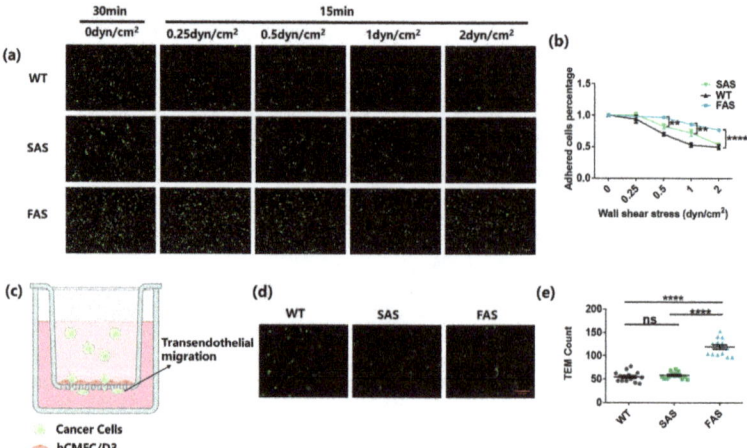

Figure 3. FAS cells exhibit enhanced adhesion to the brain endothelium and BBB transmigration ability. (**a**,**b**) FAS cells exhibited enhanced adhesion strength on the endothelium. All groups were labeled with green cell tracker and added into the flow chamber slides to adhere to the brain endothelium for 30 min. Then, 0.25, 0.5, 1, and 2 dyn/cm^2 wall shear stress were applied to each slide for 15 min, respectively. The cancer cells remaining on the brain endothelium after each treatment of fluid shear stress were counted using the fluorescence microscope. Two-way ANOVA along with post hoc Tukey test were used to calculate the statistics. Scale bar = 100 μm. n = 3. (**c**) The illustration of trans-endothelial migration assay (created by biorender.com). (**d**,**e**) FAS cells exhibited enhanced BBB transmigration ability. An hCMEC/D6 monolayer was cultured on the top of the insert membrane and cancer cells were added to transmigrate through the monolayer to the lower chamber. The transmigrated cancer cells (marked by the green cell tracker) were imaged and counted using fluorescence microscopy. One-way ANOVA along with post hoc Bonferroni test was performed to analyze the statistics. Scale bar = 50 μm. n = 3. All data are represented by mean ± SEM. (ns: no significance, ** $p < 0.01$, **** $p < 0.0001$).

After the stable attachment to brain endothelium, CTCs may start to transmigrate through the BBB and extravasate into the brain tissue [1,33]. To test the transendothelial migration, the brain endothelial monolayer was cultured on the top of the transwell membrane (Figure 3c) [45]. The number of tumor cells transmigrated to the bottom of the membrane was then counted. The results show that the FAS group had more than two times the number of tumor cells transmigrating through the hCMEC/D6 monolayer compared to the WT and SAS groups (Figure 3d,e), indicating the enhanced BBB transmigration ability of FAS cells. This might be related to the enhanced expressions of *COX2* and *ITGB3* (Figures 2a,b,d and S1d,e).

2.3. Adhesion-Selected Tumor Cells Exhibit Advantages in Cell Adhesion and Survival within the Soft Brain Microenvironment

After the transmigration across BBB, tumor cells invade into the brain tissue and need to adhere to the parenchyma and survive in the brain microenvironment. Soft polyacrylamide gels (0.6 kPa) coated with collagen I were utilized to mimic the soft environment of brain parenchyma [46]. To explore the adhesion ability of tumor cells on the soft tissue, WT cells were mixed with FAS or SAS cells equally. The cell mixture was then co-cultured on soft gels. After 15 min, the gels were rinsed to remove tumor cells with weak adhesion (Figure 4a). The results show that compared to both SAS and WT groups, the FAS group had around two-fold more cells remaining on the soft gels (Figure 4b,c), suggesting that FAS cells have advantages in adhering to the soft brain tissue.

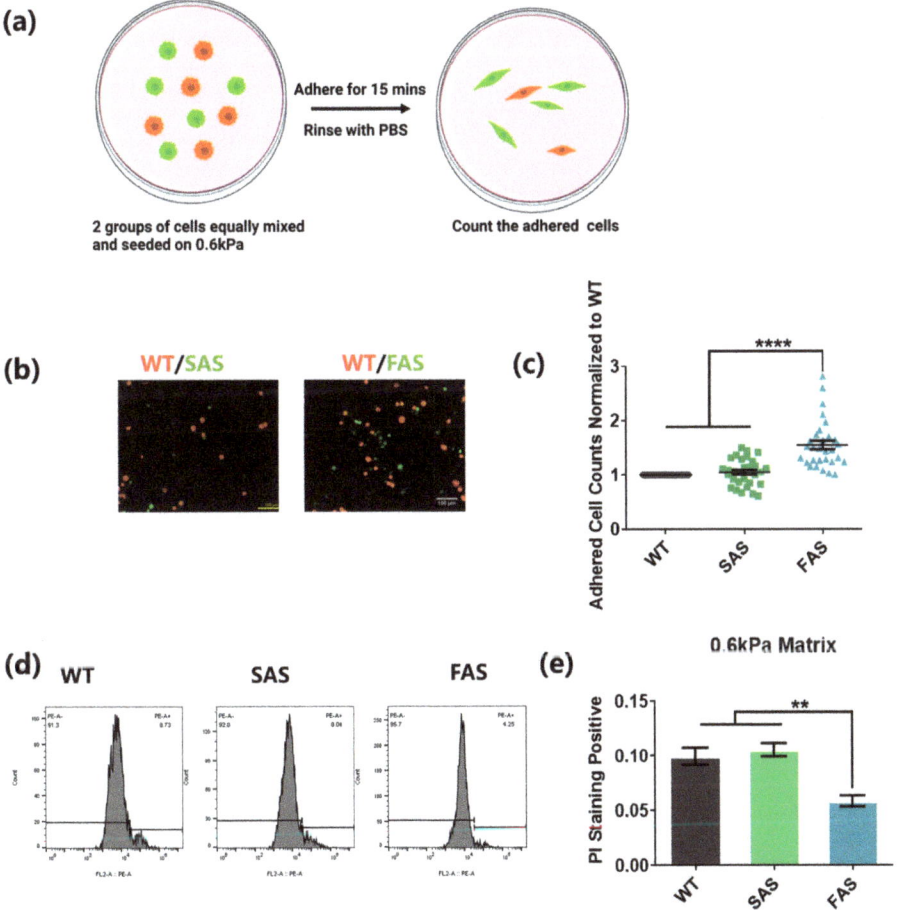

Figure 4. FAS cells exhibit advantages in cell adhesion and survival within the soft brain environment. (**a–c**) FAS cells had an enhanced adhesion ability on 0.6 kPa soft matrices. The same number of cancer cells from the WT group (labeled with red cell tracker) was mixed with cancer cells from the FAS group or the SAS group (labeled with green cell tracker) and seeded on the same 0.6 kPa polyacrylamide gels coated with collagen I. After 15 min, these cells were gently washed with PBS.

This illustration was created using Biorender.com (**a**). The remaining cells were imaged and counted under fluorescence microscope. Scale bar = 100 μm. n = 3. (**d,e**) FAS cells exhibited lower cell apoptosis on soft matrices. All groups were seeded on 0.6 kPa polyacrylamide gels coated with collagen I in low-FBS medium overnight. The dead cells were then marked with PI and tested through flow cytometry. n = 3. All data are represented by mean ± SEM. The statistics among three groups were calculated based on one-way ANOVA with the post hoc Bonferroni test. (** p < 0.01, **** p < 0.0001).

To test their ability to adapt and survive in a soft brain environment, the morphology, survival, and proliferation of tumor cells were examined. FAS cells showed a much lower cell apoptosis than WT and SAS cells in the soft microenvironment (Figure 4d,e), while there was no significant difference between the WT and SAS groups. This suggests that FAS cells may have survival advantages in the brain tissue after extravasation. The morphology analysis shows that there was no difference in cell-spreading area, circularity, and aspect ratio among the three groups when tumor cells were cultured on tissue culture plates (Figure S2a,b) [47,48]. On the soft gels, FAS exhibited a relatively larger spreading area than SAS cells but not larger than that of WT cells and a moderately higher aspect ratio than both the WT and SAS groups (Figure S2a,b). In addition, FAS cells proliferated faster than WT cells but not SAS cells and showed a similar migration ability on soft matrices compared to the other two groups (Figure S2c–f). All these results suggest that FAS cells may adapt better to soft microenvironments.

2.4. The Selected Tumor Cells Up-Regulate Adhesion Molecules That Are Correlated with Breast Cancer Brain Metastasis

To better understand the molecular mechanism underlying the enhanced brain metastatic abilities of the FAS cells, we examined the profile of the well-known adhesion molecules, which are reported to be highly involved in breast cancer brain metastasis [14,33,34]. The results show that compared to other groups, the FAS group had significantly higher mRNA expressions of multiple adhesion molecules, including MUC1, VCAM1, and VLA-4 (Integrin α4β1, encoded by *ITGA4* and *ITGB1*) (Figure 5a). The mRNA expressions of *MUC1* and *VCAM1* were notably up-regulated. We thus examined these two adhesion molecules at the protein level. Consistently, FAS cells expressed enhanced levels of both MUC1 and VCAM1 compared to WT and SAS group (Figure 5c–f). MUC1 interacts with selectins expressed on the surface of endothelial cells. Such adhesion forms fast and captures CTCs from the blood flow [8]. VCAM1 binds VLA-4 and both of them are expressed in tumor cells and brain endothelial cells. Their adhesion develops during the rolling of tumor cells on the endothelium and the binding strength is relatively strong [13,14]. This indicates that FAS cells might be selected by both transient capturing and the subsequent rolling process. In addition, the FAS cells up-regulated integrin β1 (encoded by *ITGB1*) that specifically interacts with collagen I, which might play a role in the enhanced adhesion and survival in soft microenvironments coated with collagen I (Figure 4). To further test the involvement of these adhesion molecules in breast cancer brain metastasis, we compared the expressions of these molecules using patient data collected from the Gene Expression Omnibus (GEO) database. The results show that *MUC1* and *ITGB1* tended to have higher expression levels in brain metastases compared with primary breast tumors (Figure 5b). Interestingly, the expression levels of *VCAM1* and *ITGA4* decreased in brain metastases compared to primary tumors. Together, these results indicate that the FAS cells up-regulate multiple adhesion molecules, which are clinically relevant in breast cancer brain metastasis.

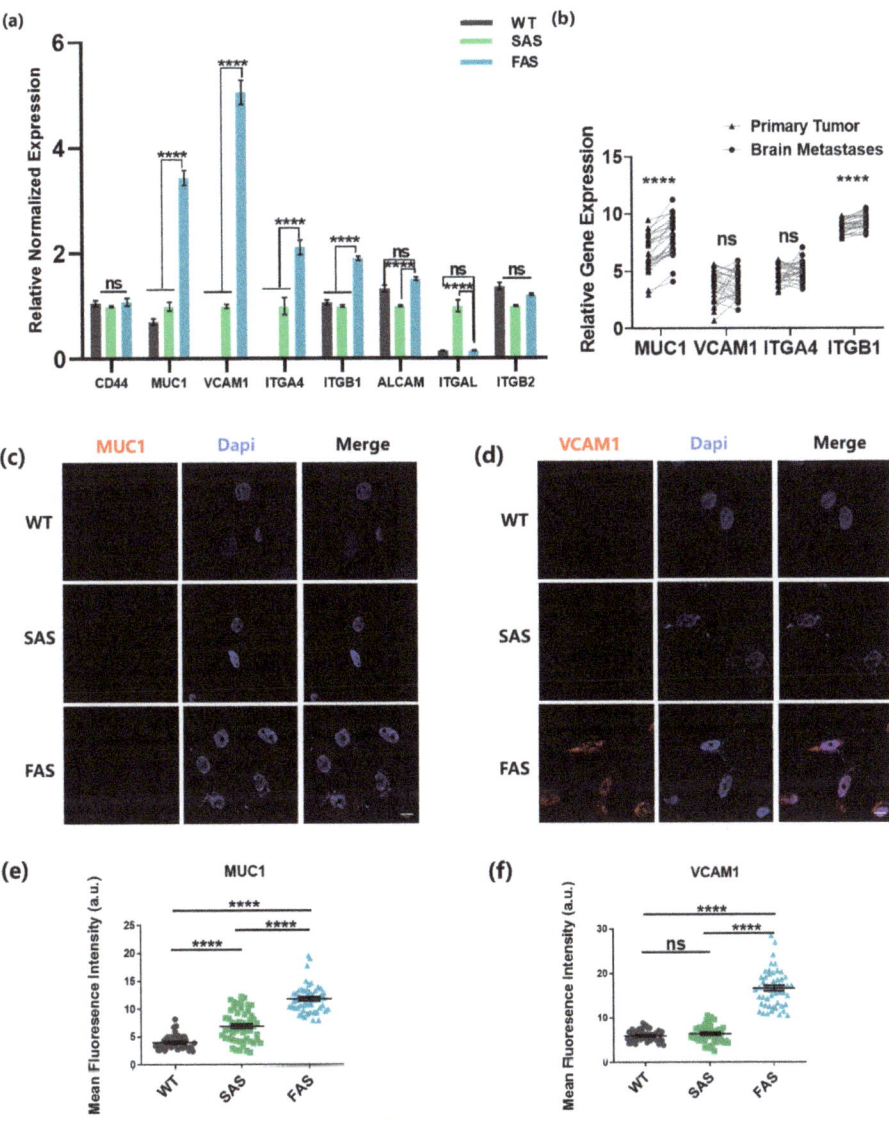

Figure 5. The FAS cells up-regulate multiple adhesion molecules that are clinically relevant to breast cancer brain metastasis. (a) FAS cells had enhanced expressions of brain metastasis-related adhesion genes. qPCR was conducted to test the gene expression. The statistics were analyzed based on two-way ANOVA along with the post hoc Tukey test. $n = 3$. (b) The highly expressed adhesion molecules in the FAS cells were clinically related to brain metastasis. The gene expression data were collected from the GEO public database (GSE173661, $n = 25$ samples). The paired comparison was analyzed using Graphpad. For each gene, a paired Student's t-test was used to analyze the difference between the statistics in two groups. (c,d) The expressions of MUC1 and VCAM1 were tested through immunofluorescence staining. Scale bar = 10 μm. (e,f) Quantification of the expressions of MUC1 and VCAM1 in (c,d). The statistics among three groups were calculated based on one-way ANOVA with the post hoc Bonferroni test. $n = 50$. All data are represented by mean ± SEM. (ns: no significance, **** $p < 0.0001$).

3. Discussion

Tumor cells disseminate to distant organs mainly through hematogenous metastasis, in which CTCs need to re-adhere to the endothelium before extravasation and the establishment of metastatic tumors. Therefore, it is rational to hypothesize that the adhesion to the endothelium can enrich a subpopulation of tumor cells with metastatic competence. This study focused on breast cancer brain metastasis, which has a relatively high incidence, poor prognosis, and short survival time [49–51]. A subpopulation of breast cancer cells were isolated based on their adhesion to the brain endothelium under shear stress. These selected breast cancer cells (FAS) up-regulated multiple genes related to brain metastasis and exhibited advantages of brain metastasis, including enhanced cell adhesion to the brain endothelium, elevated transmigration through BBB, and increased adhesion to soft brain tissue and reduced apoptosis within the soft brain microenvironment. It is known that fluid shear stress influences the expressions of adhesion molecules on endothelial cells [52–54]. In this study, the endothelial monolayer was not pre-treated under shear stress before tumor cells were perfused, which might affect the levels of surface adhesion proteins on endothelial cells and thus the selection of FAS cells. To better recapitulate the in vivo condition, brain endothelial cells will be pre-treated under shear stress mimicking the brain blood flow at different flow rates in the future (1–6 dyn/cm^2) [29]. The adhesion molecule profiles of the brain endothelial cells after shear treatment and the FAS cells selected by these pre-treated brain endothelial cells will be characterized.

Our findings suggest that the selected tumor cells (FAS) may be competent in generating brain metastases. Notably, breast cancer cells selected through adhesion to the brain endothelium without exposure to shear stress (SAS) did not obtain enhanced brain metastasis abilities and exhibited barely any difference from the wild-type cancer cells. This sheds light on the indispensable role of both cell adhesion and blood shear stress in metastasis. Further, several adhesion molecules were highly expressed in the FAS cells and clinically relevant to breast cancer brain metastasis. Therefore, this study provides the first piece of evidence to demonstrate that the re-attachment to the brain endothelium enriches CTCs with brain metastatic potential. This is consistent with the previous finding that cervical cancer cells selected through the adhesion to the endothelium under fluid shear stress for 48 h exhibit a high metastatic potential [20]. The influence of this selection process may involve the potential effect of long-time exposure (48 h) to fluid shear stress, while the short-time selection (15 min) in this study mainly reflects the influence of tumor cell–endothelium adhesion on brain metastasis ability. On the other hand, previous studies show that tumor cells strongly adhered to the underlying substrates are less migratory, while the cells with low adhesion strength have enhanced metastatic potential [55,56]. In addition, our results also support that the adhesion to the endothelium is an important rate-limiting factor in determining the metastasis inefficiency. Despite the adhesion molecules investigated in this study, many other proteins are known to involve in the adhesion of CTCs to the endothelium, such as VE-cadherin and N-cadherin [57,58]. In the future, proteomic analysis will be conducted to comprehensively characterize the adhesion molecule profile of FAS cells [59], which may identify the target adhesion molecules. Furthermore, the role of each target adhesion molecule will be elucidated by blunting its function using blocking antibodies and testing tumor cell adhesion and the functions of the selected cells, which can be further utilized to identify druggable targets. Thus, the results in this study suggest a possible therapeutic strategy—targeting the adhesion molecules on CTCs may prevent their re-adhesion and thus suppress tumor metastasis at the relatively early stage of tumor progression [13,15,49,60].

It is well-known that different types of cancer do not randomly metastasize to other organs; instead, they have preferred distant sites or exhibit organotropism [61]. For example, most prostate cancer metastasizes to bone and pancreatic cancer often metastasizes to the liver [62,63]. Breast cancer mainly disseminates to the bone, liver, brain, and lung [51]. The underlying mechanism remains unclear. In particular, it is largely unknown whether metastatic organotropism is related to the expression profiles of adhesion molecules in tumor cells and endothelial cells of specific organs. In this study, we report that the FAS cells highly express MUC1, VCAM1 and VLA-4. Meanwhile, the brain endothelial cells are reported to have relatively high expressions of their corresponding ligands: E-selectin, VLA-4, and VCAM1 [14,64]. These suggest that the adhesion molecule profiles of FAS cells and brain endothelial cells may match with each other. Previous studies show that endothelial cells from different organs are heterogeneous and distinct in their adhesion molecule profiles [17,18], which may arrest different subpopulations of CTCs with different organotropism. Further, the blood flow pattern of each metastasized organ varies, including the flow velocity and shear stress [5], which may affect the adhesion process between CTCs and the endothelium. Therefore, it is reasonable to assume that the endothelium-adhesion-based selection might contribute to organotropism. In the future, endothelial cells originating from bone, liver, lung, and brain will be used to select breast cancer cells under fluid shear stress mimicking the hematogenous pattern of the corresponding organ. In addition, proteomic analysis will be conducted to characterize the adhesion molecule profiles and gene signatures of the tumor cells selected through the adhesion to the endothelial cells of different organs. Finally, the metastatic preference and adhesion molecule profiles of selected breast cancer cells would be rigorously characterized.

4. Materials and Methods

4.1. Cell Culture

Human breast cancer cell line MDA-MB-231-TGL (WT) and human brain endothelial cell line hCMEC/D3 were purchased from Memorial Sloan Kettering Cancer Center and ATCC (Manassas, VA, USA), respectively. WT was a stable cell line by transfecting MDA-MB-231 cells with human herpesvirus 1 TK, EGFP, and firefly luciferase. WT cell line and its derivatives in this study were all cultured in Dulbecco's Modified Eagle Medium (DMEM; HyClone, Logan, UT, USA) with 10% fetal bovine serum (FBS; HyClone) and 1% penicillin/streptomycin (PS; Gibco, Dublin, Ireland) at 37 °C and 5% CO_2. hCMEC/D3 cells were cultured in Endothelial Cell Medium (ECM; ScienCell Research Laboratories, Carlsbad, CA, USA) with 10% FBS (ScienCell) and 1% PS (ScienCell), and 1% endothelial cell growth supplement (ScienCell) at 37 °C and 5% CO_2. All cell lines were passaged every 2 to 3 days using trypsin-EDTA solution (HyClone).

4.2. Isolation of Flow-Adhesion-Selected Cells (FAS) and Static-Adhesion-Selected Cells (SAS)

In the microfluidic system, a brain endothelial monolayer was cultured on flow chamber slides (μ-Slide I Luer, Cat. No: 80176, ibidi, Martinsried, Germany), while the shear stress was generated using a peristaltic pump (P-230, Harvard Instruments, Cambridge, MA, USA) to mimic the blood flow. In short, the slide channels were first coated with 0.2 mg/mL rat-tail collagen type I (Thermo Fisher Scientific, Waltham, MA, USA). Then, 100,000 hCMEC/D3 cells were seeded in the slide channel for 2 days to reach high confluency. For the selection of FAS cells, 100,000 WT cells (EGFP labeled) were suspended in DMEM full medium and then infused into the flow chamber under 1 dyn/cm^2 wall shear stress for 15 min following the protocol as previously described [20,21]. The adhered tumor cells and hCMEC/D3 cells were harvested and cocultured for 2 days in tissue culture plates (TCP). Since all tumor cells were labeled with EGFP, the selected tumor cells could be separated from the unlabeled hCMEC cells based on the fluorescence expression using BD FACSAria III Cell Sorter (BD Bioscience, San Jose, CA, USA). To test the purity of these selected cells, the expression of EGFP was then examined under a fluorescence microscope (Leica, Wetzlar, Germany). For the selection of SAS cells, an hCMEC/D3 monolayer was formed; then, 100,000 WT cells were added into the flow chamber and allowed to interact with the endothelium without exposure to shear stress for 30 min. Non-adhered tumor cells were removed via gentle washing with PBS (HyClone) and the adhered cells were isolated using the same method as FAS. The selected cells were cultured in TCP for at least three passages before experiments. They were detached using 0.2% EDTA solution (HyClone) and passaged every 3 days.

4.3. Quantitative RT-PCR Analysis

The total RNAs of each sample were extracted using Aurum Total RNA Mini Kit (Bio-Rad, Hercules, CA, USA). The complementary DNA was synthesized from the RNA samples using the RevertAid RT Reverse Transcription Kit (Thermo Fisher Scientific). The Forget-Me-Not qPCR Master Mix Kit (Biotium, Fremont, CA, USA) was used to prepare the PCR mixture for the quantitative analysis via the CFX96 Real-Time System (Bio-Rad). The relative expression of target genes was normalized to the expression of the housekeeper gene GAPDH. All primers used in the qPCR analysis were designed based on the National Center for Biotechnology Information database (NCBI; Bethesda, MD, USA) and listed in Supplementary Table S1.

4.4. Transendothelial Migration Assay

The transendothelial migration assays were carried out using Corning® Transwell® (Corning, NY, USA) with an 8 μm pore. In brief, the transwell inserts were coated with collagen type I (Thermo Fisher Scientific) and then seeded with 20,000 hCMEC/D3 cells. These cells were then incubated for at least two days to form a monolayer. A total of 50,000 cancer cells were then marked with cell tracker CytoTrace™ Green CMFDA (Thermo Fisher Scientific) following the manufacturer's instructions and added into the inserts containing low-serum medium (DMEM + 1%FBS + 1%PS). DMEM full medium was then added to the lower chamber, inducing the cancer cells to transmigrate through the brain endothelial monolayer. After 24 h, the transmigrated cancer cells were fixed with a 4% paraformaldehyde solution (Thermo Fisher Scientific) and counted under a fluorescence microscopy. For each insert, microscopic images of the transmigrated cells were taken at five random views, and then the number of cells was calculated using the particle analysis in ImageJ 1.46 (NIH, Bethesda, MD, USA).

4.5. Flow Adhesion Assay

An hCMEC monolayer was formed in the flow chamber slides (ibidi) as described above. Then, two groups of cancer cells were marked with two different colors of cell trackers CytoTrace™ Green CMFDA (Thermo Fisher Scientific) and PKH26 Red Fluorescent Cell Linker (Sigma-Aldrich, Saint Louis, MA, USA) following the manufacturer's instructions. In total, 100,000 cells from each group were mixed and then infused into one slide under 1 dyn/cm^2 shear stress for 15 min. The number of adhered cells was counted.

4.6. Static Adhesion Assay

The hCMEC monolayer was formed in the flow chamber slides (ibidi). A total of 300,000 cancer cells were marked with cell tracker CytoTrace™ Green CMFDA (Thermo Fisher Scientific) following the manufacturer's instructions and then added into the slides. The slides were then incubated at 37 °C for 30 min to allow cancer cells to interact with the endothelium. DMEM full medium was perfused through the slide, creating certain level of shear stress for 15 min. The number of adhered cancer cells was counted.

4.7. Preparation of Polyacrylamide Hydrogels (PA-Gel)

The PA-gels were synthesized and pretreated using methods reported elsewhere [46]. In brief, 40% acrylamide solution (Bio-Rad), 2% bis-acrylamide solution (Bio-Rad), and water were mixed in the ratio of 179:15:6 (0.6 kPa). Then, 1% v/v ammonium persulfate (APS, Thermo Fisher Scientific) and 0.1% v/v methylethylenediamine (TEMED, Thermo Fisher Scientific) were added into the mixture. An adequate amount of the mixed solution was added onto the chloro-silanated glass surface and then covered by amino-silanated coverslips. After solidification, the PA gels stuck to the coverslip were detached from the glass together for later use. Before assays, the PA gel was coated with 200 μg/mL collagen type I (Thermo Fisher Scientific) via the crosslinker sulfosuccinimidyl 6-(4′-azido-2′-nitrophenylamino) hexanoate (sulfo-SANPAH; Thermo Fisher Scientific).

4.8. Adhesion Assay on PA-Gel

A total of 100,000 cells of WT group marked with CytoTrace™ Green CMFDA (Thermo Fisher Scientific) according to the manufacturer's instruction and 100,000 cells of the SAS group or FAS group marked with PKH26 Red Fluorescent Cell Linker (Sigma-Aldrich) were mixed and seeded on the 0.6 kPa PA gel and then incubated at 37 °C for 15 min. The cells were rinsed gently with PBS. The adhered cells remaining on the gels were counted.

4.9. Cell Viability Assay

A total of 300,000 cancer cells were seeded on each 0.6 kPa PA-gel using low-serum medium (DMEM + 1%FBS + 1%PS) and then incubated at 37 °C for 24 h. All cells were collected and stained using propidium iodide (PI; Abcam, MA, USA) following the manufacturer's instructions. The ratio of dead cells to live cells was examined using a flow cytometer (BD Accuri C6, BD Bioscience). The data was analyzed using the software FlowJo_v 10.6.2 (BD Bioscience).

4.10. Cell Morphology Analysis

Cancer cells were either seeded on PA-gels or TCP and then incubated at 37 °C for 24 h. Microscopic images of cells were captured under bright fields. ImageJ (NIH, Bethesda) was used to mark the cell edge and measure the spreading area, aspect ratio and circularity of each cell. At least 100 cells were analyzed for each group.

4.11. Cell Proliferation Assay

A total of 300,000 cancer cells were seeded on 0.6 kPa PA-gels and then incubated overnight. The EdU (5-ethynyl-2′-deoxyuridine) kit (Beyotime, Shanghai, China) was used to mark and quantify the proliferating cells following the manufacturer's instructions. In brief, cancer cells were cultured in full medium containing 0.1% EdU for 2 h. These cells were then collected, fixed, and then permeabilized. A click reaction solution from the kit was then used to label the EdU with fluorescence dyes. The percentage of proliferating cells (EdU$^+$) was examined through flow cytometry.

4.12. Wound Healing Assay

A culture insert (Ibidi, Gräfelfing, Germany) with two separate chambers was first placed on 0.6 kPa PA-gel. A total of 200,000 cells were seeded into each chamber of the insert and incubated at 37 °C for 6 h to adhere to the PA gel. The culture insert was then removed, thus creating a wound between cells cultured in two chambers. These cells were incubated at 37 °C for another 24 h. In total, 4× microscopic images of the wound were recorded at 0 h and 24 h. ImageJ was used to mark and measure the wound area and then calculate the wound healing rate, which is the healed area (wound area at 0 h − wound area at 24 h) divided by the wound area at 0 h.

4.13. Immunofluorescence Staining

Cells were seeded on collagen-I-coated glass-bottom dishes (Ibidi) overnight. They were fixed with 4% formaldehyde (Sigma-Aldrich) for 15 min and then permeabilized with 0.1% Triton X-100 (Sigma-Aldrich) in PBS containing 1% BSA. Then, the cells were incubated with the primary antibodies (AbCam) of COX2 (ab188183), SerpinB2 (ab47742), VCAM1 (ab134047), MUC1(ab109185), and integrin αvβ3 (ab7166) at 4 °C overnight. After that, the cells were incubated with secondary antibodies (ab150079; ab150115) in the ark for one hour. Finally, the cells were stained with Dapi (Thermo Fisher Scientific) to counterstain the nuclei. For each group, at least 50 cells were imaged using fluorescence microscopy (Nikon, Tokyo, Japan). ImageJ was used to mark the cell boundary and measure the mean fluorescence intensity.

4.14. Statistical Analysis

All data in this project were shown as mean ± SEM. One-way ANOVA was adopted when there were three or more groups of samples. In the ANOVA tests, Tukey or Bonferroni tests were used to conduct the multi-comparison between every two paired groups with equal or unequal sample sizes. *, $p < 0.05$, **, $p < 0.01$, ***, $p < 0.001$, ****, $p < 0.0001$.

5. Conclusions

This study utilized an in vitro adhesion system to select a subpopulation of CTCs that had the ability to adhere to the brain endothelium under blood shear stress. The selected tumor cells expressed increased levels of brain metastasis genes and exhibited brain metastasis competency, including the ability to transmigrate through the blood–brain barrier and adhere/survive in the soft brain microenvironment. Further, the adhesion-selected cells up-regulated the adhesion molecules that were relevant to breast cancer brain metastasis. In summary, these findings demonstrated that the adhesion of CTCs to brain endothelium under shear stress could enrich a subpopulation of tumor cells with brain metastasis ability.

Supplementary Materials: The supporting information can be downloaded at: https://www.mdpi.com/article/10.3390/ijms24087087/s1.

Author Contributions: Y.T. conceived the project. Y.T., B.Z., X.L., K.T., Y.X., G.H., Y.Z., K.L. and C.Z. designed and conducted the experiments and analyzed data, Y.T. and B.Z. wrote the manuscript. All authors have read and agreed to the published version of the manuscript.

Funding: Y.T. acknowledges the support from the National Natural Science Foundation of China (Project no. 11972316), Shenzhen Science and Technology Innovation Commission (Project no. JCYJ20200109142001798, SGDX2020110309520303, and JCYJ20220531091002006), General Research Fund of Hong Kong Research Grant Council (PolyU 15214320), Health and Medical Research Fund (HMRF18191421), Germany/Hong Kong Joint Research Scheme (G-PolyU503/20), and an internal grant from the Hong Kong Polytechnic University Shenzhen Research Institute (I2021A002).

Institutional Review Board Statement: Not applicable.

Informed Consent Statement: Not applicable.

Data Availability Statement: The dataset (GSE173661) analyzed in our study was collected from the public GEO database. https://www.ncbi.nlm.nih.gov/geo/query/acc.cgi?acc=GSE173661.

Acknowledgments: We thank the University Life Science Facility in the Hong Kong Polytechnic University for providing flow cytometry and confocal microscope equipment.

Conflicts of Interest: The authors declare no conflict of interest.

Abbreviations

BBB	Blood–brain barrier
WT	Wild Type
FAS	Flow Adhesion Selected
SAS	Static Adhesion Selected
PA-gel	Polyacrylamide gel
GEO	Gene Expression Omnibus

References

1. Valastyan, S.; Weinberg, R.A. Tumor Metastasis: Molecular Insights and Evolving Paradigms. *Cell* **2011**, *147*, 275–292. [CrossRef]
2. Bakir, B.; Chiarella, A.M.; Pitarresi, J.R.; Rustgi, A.K. EMT, MET, Plasticity, and Tumor Metastasis. *Trends Cell Biol.* **2020**, *30*, 764–776. [CrossRef]
3. Gensbittel, V.; Kräter, M.; Harlepp, S.; Busnelli, I.; Guck, J.; Goetz, J.G. Mechanical Adaptability of Tumor Cells in Metastasis. *Dev. Cell* **2021**, *56*, 164–179. [CrossRef]
4. Turajlic, S.; Swanton, C. Metastasis as an evolutionary process. *Science* **2016**, *352*, 169–175. [CrossRef]

5. Follain, G.; Osmani, N.; Azevedo, A.S.; Allio, G.; Mercier, L.; Karreman, M.A.; Solecki, G.; Leòn, M.J.G.; Lefebvre, O.; Fekonja, N.; et al. Hemodynamic Forces Tune the Arrest, Adhesion, and Extravasation of Circulating Tumor Cells. *Dev. Cell* **2018**, *45*, 33–52.e12. [CrossRef]
6. Osmani, N.; Follain, G.; León, M.J.G.; Lefebvre, O.; Busnelli, I.; Larnicol, A.; Harlepp, S.; Goetz, J.G. Metastatic Tumor Cells Exploit Their Adhesion Repertoire to Counteract Shear Forces during Intravascular Arrest. *Cell Rep.* **2019**, *28*, 2491–2500.e5. [CrossRef]
7. Brodt, P.; Fallavollita, L.; Bresalier, R.S.; Meterissian, S.; Norton, C.R.; Wolitzky, B.A. Liver endothelial E-selectin mediates carcinoma cell adhesion and promotes liver metastasis. *Int. J. Cancer* **1997**, *71*, 612–619. [CrossRef]
8. Geng, Y.; Yeh, K.; Takatani, T.; King, M.R. Three to Tango: MUC1 as a Ligand for Both E-Selectin and ICAM-1 in the Breast Cancer Metastatic Cascade. *Front. Oncol.* **2012**, *2*, 76. [CrossRef] [PubMed]
9. Kang, S.-A.; Hasan, N.; Mann, A.P.; Zheng, W.; Zhao, L.; Morris, L.; Zhu, W.; Zhao, Y.D.; Suh, K.S.; Dooley, W.C.; et al. Blocking the Adhesion Cascade at the Premetastatic Niche for Prevention of Breast Cancer Metastasis. *Mol. Ther.* **2015**, *23*, 1044–1054. [CrossRef] [PubMed]
10. Fujisaki, T.; Tanaka, Y.; Fujii, K.; Mine, S.; Saito, K.; Yamada, S.; Yamashita, U.; Irimura, T.; Eto, S. CD44 stimulation induces integrin-mediated adhesion of colon cancer cell lines to endothelial cells by up-regulation of integrins and c-Met and activation of integrins. *Cancer Res.* **1999**, *59*, 4427–4434. [PubMed]
11. Su, C.-Y.; Li, J.-Q.; Zhang, L.-L.; Wang, H.; Wang, F.-H.; Tao, Y.-W.; Wang, Y.-Q.; Guo, Q.-R.; Li, J.-J.; Liu, Y.; et al. The Biological Functions and Clinical Applications of Integrins in Cancers. *Front. Pharmacol.* **2020**, *11*, 579068. [CrossRef]
12. Wang, H.-S.; Hung, Y.; Su, C.-H.; Peng, S.-T.; Guo, Y.-J.; Lai, M.-C.; Liu, C.-Y.; Hsu, J.-W. CD44 Cross-linking induces integrin-mediated adhesion and transendothelial migration in breast cancer cell line by up-regulation of LFA-1 ($\alpha L\beta 2$) and VLA-4 ($\alpha 4\beta 1$). *Exp. Cell Res.* **2005**, *304*, 116–126. [CrossRef]
13. Schlesinger, M.; Roblek, M.; Ortmann, K.; Naggi, A.; Torri, G.; Borsig, L.; Bendas, G. The role of VLA-4 binding for experimental melanoma metastasis and its inhibition by heparin. *Thromb. Res.* **2014**, *133*, 855–862. [CrossRef] [PubMed]
14. Soto, M.S.; Serres, S.; Anthony, D.C.; Sibson, N.R. Functional role of endothelial adhesion molecules in the early stages of brain metastasis. *Neuro-Oncology* **2014**, *16*, 540–551. [CrossRef] [PubMed]
15. Zhang, C.; Xu, Y.; Hao, Q.; Wang, S.; Li, H.; Li, J.; Gao, Y.; Li, M.; Li, W.; Xue, X.; et al. FOXP3 suppresses breast cancer metastasis through downregulation of CD44. *Int. J. Cancer* **2015**, *137*, 1279–1290. [CrossRef] [PubMed]
16. Chen, W.; Hoffmann, A.D.; Liu, H.; Liu, X. Organotropism: New insights into molecular mechanisms of breast cancer metastasis. *NPJ Precis. Oncol.* **2018**, *2*, 4. [CrossRef]
17. Feng, W.; Chen, L.; Nguyen, P.K.; Wu, S.M.; Li, G. Single Cell Analysis of Endothelial Cells Identified Organ-Specific Molecular Signatures and Heart-Specific Cell Populations and Molecular Features. *Front. Cardiovasc. Med.* **2019**, *6*, 165. [CrossRef]
18. Scott, D.W.; Patel, R.P. Endothelial heterogeneity and adhesion molecules N-glycosylation: Implications in leukocyte trafficking in inflammation. *Glycobiology* **2013**, *23*, 622–633. [CrossRef]
19. Shetty, S.; Weston, C.J.; Adams, D.H.; Lalor, P.F. A Flow Adhesion Assay to Study Leucocyte Recruitment to Human Hepatic Sinusoidal Endothelium Under Conditions of Shear Stress. *J. Vis. Exp. JoVE* **2014**, *85*, e51330.
20. Chuang, P.-C.; Lu, C.-W.; Tsai, C.-C.; Tseng, S.-H.; Su, W.-H. MicroRNA-128 Confers Anti-Endothelial Adhesion and Anti-Migration Properties to Counteract Highly Metastatic Cervical Cancer Cells' Migration in a Parallel-Plate Flow Chamber. *Int. J. Mol. Sci.* **2021**, *22*, 215. [CrossRef]
21. Clément-Demange, L.; Mulcrone, P.L.; Tabarestani, T.Q.; Sterling, J.A.; Elefteriou, F. β2ARs stimulation in osteoblasts promotes breast cancer cell adhesion to bone marrow endothelial cells in an IL-1β and selectin-dependent manner. *J. Bone Oncol.* **2018**, *13*, 1–10. [CrossRef]
22. Xin, Y.; Li, K.; Yang, M.; Tan, Y. Fluid Shear Stress Induces EMT of Circulating Tumor Cells via JNK Signaling in Favor of Their Survival during Hematogenous Dissemination. *Int. J. Mol. Sci.* **2020**, *21*, 8115. [CrossRef]
23. Xu, Z.; Li, K.; Xin, Y.; Tang, K.; Yang, M.; Wang, G.; Tan, Y. Fluid shear stress regulates the survival of circulating tumor cells via nuclear expansion. *J. Cell Sci.* **2022**, *135*, jcs259586. [CrossRef] [PubMed]
24. Xin, Y.; Chen, X.; Tang, X.; Li, K.; Yang, M.; Tai, W.C.-S.; Liu, Y.; Tan, Y. Mechanics and Actomyosin-Dependent Survival/Chemoresistance of Suspended Tumor Cells in Shear Flow. *Biophys. J.* **2019**, *116*, 1803–1814. [CrossRef]
25. Wei, R.-R.; Sun, D.-N.; Yang, H.; Yan, J.; Zhang, X.; Zheng, X.-L.; Fu, X.-H.; Geng, M.-Y.; Huang, X.; Ding, J. CTC clusters induced by heparanase enhance breast cancer metastasis. *Acta Pharmacol. Sin.* **2018**, *39*, 1326–1337. [CrossRef] [PubMed]
26. Fan, R.; Emery, T.; Zhang, Y.; Xia, Y.; Sun, J.; Wan, J. Circulatory shear flow alters the viability and proliferation of circulating colon cancer cells. *Sci. Rep.* **2016**, *6*, 27073. [CrossRef] [PubMed]
27. Brugnoli, F.; Grassilli, S.; Lanuti, P.; Marchisio, M.; Al-Qassab, Y.; Vezzali, F.; Capitani, S.; Bertagnolo, V. Up-modulation of PLC-β2 reduces the number and malignancy of triple-negative breast tumor cells with a CD133+/EpCAM+ phenotype: A promising target for preventing progression of TNBC. *BMC Cancer* **2017**, *17*, 617. [CrossRef]
28. Fiorillo, M.; Scatena, C.; Naccarato, A.G.; Sotgia, F.; Lisanti, M.P. Bedaquiline, an FDA-approved drug, inhibits mitochondrial ATP production and metastasis in vivo, by targeting the gamma subunit (ATP5F1C) of the ATP synthase. *Cell Death Differ.* **2021**, *28*, 2797–2817. [CrossRef]
29. Follain, G.; Herrmann, D.; Harlepp, S.; Hyenne, V.; Osmani, N.; Warren, S.C.; Timpson, P.; Goetz, J.G. Fluids and their mechanics in tumour transit: Shaping metastasis. *Nat. Rev. Cancer* **2020**, *20*, 107–124. [CrossRef]

30. Mehta, A.; Desai, A.; Rudd, D.; Siddiqui, G.; Nowell, C.J.; Tong, Z.; Creek, D.J.; Tayalia, P.; Gandhi, P.S.; Voelcker, N.H. Bio-Mimicking Brain Vasculature to Investigate the Role of Heterogeneous Shear Stress in Regulating Barrier Integrity. *Adv. Biol.* **2022**, *6*, e2200152. [CrossRef]
31. Wang, X.; Xu, B.; Xiang, M.; Yang, X.; Liu, Y.; Liu, X.; Shen, Y. Advances on fluid shear stress regulating blood-brain barrier. *Microvasc. Res.* **2020**, *128*, 103930. [CrossRef] [PubMed]
32. Bendas, G.; Borsig, L. Cancer Cell Adhesion and Metastasis: Selectins, Integrins, and the Inhibitory Potential of Heparins. *Int. J. Cell Biol.* **2012**, *2012*, 676731. [CrossRef]
33. Wang, Y.; Ye, F.; Liang, Y.; Yang, Q. Breast cancer brain metastasis: Insight into molecular mechanisms and therapeutic strategies. *Br. J. Cancer* **2021**, *125*, 1056–1067. [CrossRef] [PubMed]
34. Bos, P.D.; Zhang, X.H.F.; Nadal, C.; Shu, W.; Gomis, R.R.; Nguyen, D.X.; Minn, A.J.; van de Vijver, M.J.; Gerald, W.L.; Foekens, J.A.; et al. Genes that mediate breast cancer metastasis to the brain. *Nature* **2009**, *459*, 1005–1009. [CrossRef] [PubMed]
35. Pardo-Pastor, C.; Rubio-Moscardo, F.; Vogel-González, M.; Serra, S.A.; Afthinos, A.; Mrkonjic, S.; Destaing, O.; Abenza, J.F.; Fernández-Fernández, J.M.; Trepat, X.; et al. Piezo2 channel regulates RhoA and actin cytoskeleton to promote cell mechanobiological responses. *Proc. Natl. Acad. Sci. USA* **2018**, *115*, 1925–1930. [CrossRef]
36. Wikman, H.; Lamszus, K.; Detels, N.; Uslar, L.; Wrage, M.; Benner, C.; Hohensee, I.; Ylstra, B.; Eylmann, K.; Zapatka, M.; et al. Relevance of PTEN loss in brain metastasis formation in breast cancer patients. *Breast Cancer Res.* **2012**, *14*, R49. [CrossRef]
37. Wu, K.; Fukuda, K.; Xing, F.; Zhang, Y.; Sharma, S.; Liu, Y.; Chan, M.; Zhou, X.; Qasem, S.; Pochampally, R.; et al. Roles of the cyclooxygenase 2 matrix metalloproteinase 1 pathway in brain metastasis of breast cancer. *J. Biol. Chem.* **2015**, *290*, 9842–9854. [CrossRef]
38. Leslie, T.K.; Brackenbury, W.J. Sodium channels and the ionic microenvironment of breast tumours. *J. Physiol.* **2022**. ahead-of-print. [CrossRef]
39. Cai, Y.-C.; Yang, H.; Wang, K.-F.; Chen, T.-H.; Jiang, W.-Q.; Shi, Y.-X. ANGPTL4 overexpression inhibits tumor cell adhesion and migration and predicts favorable prognosis of triple-negative breast cancer. *BMC Cancer* **2020**, *20*, 878. [CrossRef]
40. Gong, X.; Hou, Z.; Endsley, M.P.; Gronseth, E.I.; Rarick, K.R.; Jorns, J.M.; Yang, Q.; Du, Z.; Yan, K.; Bordas, M.L.; et al. Interaction of tumor cells and astrocytes promotes breast cancer brain metastases through TGF-β2/ANGPTL4 axes. *npj Precis. Oncol.* **2019**, *3*, 24. [CrossRef]
41. Niland, S.; Eble, J.A. Hold on or Cut? Integrin- and MMP-Mediated Cell–Matrix Interactions in the Tumor Microenvironment. *Int. J. Mol. Sci.* **2021**, *22*, 238. [CrossRef] [PubMed]
42. Zhu, C.; Kong, Z.; Wang, B.; Cheng, W.; Wu, A.; Meng, X. ITGB3/CD61: A hub modulator and target in the tumor microenvironment. *Am. J. Transl. Res.* **2019**, *11*, 7195–7208. [PubMed]
43. Drolez, A.; Vandenhaute, E.; Delannoy, C.P.; Dewald, J.H.; Gosselet, F.; Cecchelli, R.; Julien, S.; Dehouck, M.-P.; Delannoy, P.; Mysiorek, C. ST6GALNAC5 Expression Decreases the Interactions between Breast Cancer Cells and the Human Blood-Brain Barrier. *Int. J. Mol. Sci.* **2016**, *17*, 1309. [CrossRef] [PubMed]
44. Erbeldinger, N.; Rapp, F.; Ktitareva, S.; Wendel, P.; Bothe, A.S.; Dettmering, T.; Durante, M.; Friedrich, T.; Bertulat, B.; Meyer, S.; et al. Measuring Leukocyte Adhesion to (Primary) Endothelial Cells after Photon and Charged Particle Exposure with a Dedicated Laminar Flow Chamber. *Front. Immunol.* **2017**, *8*, 627. [CrossRef] [PubMed]
45. Ding, H.; Sagar, V.; Agudelo, M.; Pilakka-Kanthikeel, S.; Subba Rao Atluri, V.; Raymond, A.; Samikkannu, T.; Nair, M.P. Enhanced blood–brain barrier transmigration using a novel transferrin embedded fluorescent magneto-liposome nanoformulation. *Nanotechnolgy* **2014**, *25*, 055101. [CrossRef] [PubMed]
46. Tang, K.; Xin, Y.; Li, K.; Chen, X.; Tan, Y. Cell Cytoskeleton and Stiffness Are Mechanical Indicators of Organotropism in Breast Cancer. *Biology* **2021**, *10*, 259. [CrossRef]
47. Chen, C.S.; Mrksich, M.; Huang, S.; Whitesides, G.M.; Ingber, D.E. Geometric Control of Cell Life and Death. *Science* **1997**, *276*, 1425–1428. [CrossRef]
48. Mogilner, A.; Keren, K. The Shape of Motile Cells. *Curr. Biol.* **2009**, *19*, R762–R771. [CrossRef]
49. Bailleux, C.; Eberst, L.; Bachelot, T. Treatment strategies for breast cancer brain metastases. *Br. J. Cancer* **2021**, *124*, 142–155. [CrossRef]
50. Corti, C.; Antonarelli, G.; Criscitiello, C.; Lin, N.U.; Carey, L.A.; Cortés, J.; Poortmans, P.; Curigliano, G. Targeting brain metastases in breast cancer. *Cancer Treat. Rev.* **2022**, *103*, 102324. [CrossRef]
51. Pedrosa, R.M.S.M.; Mustafa, D.A.; Soffietti, R.; Kros, J.M. Breast cancer brain metastasis: Molecular mechanisms and directions for treatment. *Neuro-Oncology* **2018**, *20*, 1439–1449. [CrossRef] [PubMed]
52. Chappell, D.C.; Varner, S.E.; Nerem, R.M.; Medford, R.M.; Alexander, R.W. Oscillatory Shear Stress Stimulates Adhesion Molecule Expression in Cultured Human Endothelium. *Circ. Res.* **1998**, *82*, 532–539. [CrossRef]
53. Ho, R.X.-Y.; Tahboub, R.; Amraei, R.; Meyer, R.D.; Varongchayakul, N.; Grinstaff, M.; Rahimi, N. The cell adhesion molecule IGPR-1 is activated by and regulates responses of endothelial cells to shear stress. *J. Biol. Chem.* **2019**, *294*, 13671–13680. [CrossRef] [PubMed]
54. Nagel, T.; Resnick, N.; Atkinson, W.J.; Dewey, C.F.; A Gimbrone, M. Shear stress selectively upregulates intercellular adhesion molecule-1 expression in cultured human vascular endothelial cells. *J. Clin. Investig.* **1994**, *94*, 885–889. [CrossRef] [PubMed]
55. Beri, P.; Popravko, A.; Yeoman, B.; Kumar, A.; Chen, K.; Hodzic, E.; Chiang, A.; Banisadr, A.; Placone, J.K.; Carter, H.; et al. Cell Adhesiveness Serves as a Biophysical Marker for Metastatic Potential. *Cancer Res.* **2020**, *80*, 901–911. [CrossRef] [PubMed]

56. Fuhrmann, A.; Banisadr, A.; Beri, P.; Tlsty, T.D.; Engler, A.J. Metastatic State of Cancer Cells May Be Indicated by Adhesion Strength. *Biophys. J.* **2017**, *112*, 736–745. [CrossRef] [PubMed]
57. Breier, G.; Grosser, M.; Rezaei, M. Endothelial cadherins in cancer. *Cell Tissue Res.* **2014**, *355*, 523–527. [CrossRef]
58. Brock, T.; Boudriot, E.; Klawitter, A.; Großer, M.; Nguyen, T.; Giebe, S.; Klapproth, E.; Temme, A.; El-Armouche, A.; Breier, G. The Influence of VE-Cadherin on Adhesion and Incorporation of Breast Cancer Cells into Vascular Endothelium. *Int. J. Mol. Sci.* **2021**, *22*, 6049. [CrossRef]
59. Zandian, A.; Forsström, B.; Häggmark-Månberg, A.; Schwenk, J.M.; Uhlén, M.; Nilsson, P.; Ayoglu, B. Whole-Proteome Peptide Microarrays for Profiling Autoantibody Repertoires within Multiple Sclerosis and Narcolepsy. *J. Proteome Res.* **2017**, *16*, 1300–1314. [CrossRef]
60. Ma, F.; Chen, D.; Chen, F.; Chi, Y.; Han, Z.; Feng, X.; Li, X.; Han, Z. Human Umbilical Cord Mesenchymal Stem Cells Promote Breast Cancer Metastasis by Interleukin-8- and Interleukin-6-Dependent Induction of $CD44^+/CD24^-$ Cells. *Cell Transplant.* **2015**, *24*, 2585–2599. [CrossRef]
61. Gao, Y.; Bado, I.; Wang, H.; Zhang, W.; Rosen, J.M.; Zhang, X.H.-F. Metastasis Organotropism: Redefining the Congenial Soil. *Dev. Cell* **2019**, *49*, 375–391. [CrossRef] [PubMed]
62. Peixoto, R.D.; Speers, C.; McGahan, C.E.; Renouf, D.J.; Schaeffer, D.F.; Kennecke, H.F. Prognostic factors and sites of metastasis in unresectable locally advanced pancreatic cancer. *Cancer Med.* **2015**, *4*, 1171–1177. [CrossRef] [PubMed]
63. Gandaglia, G.; Abdollah, F.; Schiffmann, J.; Trudeau, V.; Shariat, S.F.; Kim, S.P.; Perrotte, P.; Montorsi, F.; Briganti, A.; Trinh, Q.-D.; et al. Distribution of metastatic sites in patients with prostate cancer: A population-based analysis. *Prostate* **2014**, *74*, 210–216. [CrossRef] [PubMed]
64. Poller, B.; Gutmann, H.; Krähenbühl, S.; Weksler, B.; Romero, I.; Couraud, P.-O.; Tuffin, G.; Drewe, J.; Huwyler, J. The human brain endothelial cell line hCMEC/D3 as a human blood-brain barrier model for drug transport studies. *J. Neurochem.* **2008**, *107*, 1358–1368. [CrossRef] [PubMed]

Disclaimer/Publisher's Note: The statements, opinions and data contained in all publications are solely those of the individual author(s) and contributor(s) and not of MDPI and/or the editor(s). MDPI and/or the editor(s) disclaim responsibility for any injury to people or property resulting from any ideas, methods, instructions or products referred to in the content.

Article

Characterization and Clinical Relevance of Endometrial CAFs: Correlation between Post-Surgery Event and Resistance to Drugs

Raed Sulaiman [1,†], Pradip De [2,3,4,†], Jennifer C. Aske [2], Xiaoqian Lin [2], Adam Dale [2], Kris Gaster [5], Luis Rojas Espaillat [6], David Starks [6] and Nandini Dey [2,3,*]

[1] Department of Pathology, Avera Research Institute, Sioux Falls, SD 57105, USA
[2] Translational Oncology Laboratory, Avera Research Institute, Sioux Falls, SD 57105, USA
[3] Department of Internal Medicine, University of South Dakota SSOM, USD, Sioux Falls, SD 57105, USA
[4] Viecure, Greenwood Village, CO 80111, USA
[5] Assistant VP Outpatient Cancer Clinics, Avera Cancer Institute, Sioux Falls, SD 57105, USA
[6] Department of Gynecologic Oncology, Avera Cancer Institute, Sioux Falls, SD 57105, USA
* Correspondence: nandini.dey@avera.org
† These authors contributed equally to this work.

Abstract: Cancer-associated fibroblasts (CAFs) within a solid tumor can support the progression of cancer. We studied the identification and characterization of patient-derived endometrial CAFs in the context of their clinical relevance in endometrial cancers. We established patient-derived primary cultures of CAFs from surgically resected tumors (TCAF) and tumor-adjacent normal (NCAF) tissues in 53 consented patients with success rates of 97.7% and 75%, respectively. A passage of CAF was qualified by the (1) absence of CK 8,18,19, EpCAM, CD45, and CD31, and (2) presence of SMAalpha, S100A4, CD90, FAP, TE-7, CD155, PD-L1, TGFB, PDGFRA (qRT-PCR, flow cytometry, Western blot, ICC). Out of the 44 established CAFs, 31 were aggressive (having an early, i.e., 4–7 week, establishment time and/or >3 passages) compared to 13 which were non-aggressive. A post-surgery-event (PSE) was observed in 7 out of 31 patients bearing aggressive CAFs, 2 of whom were also positive for CTCs, while none of the 13 patients bearing non-aggressive CAFs had events. A positive correlation was found between patients with grade 3 (p = 0.025) as well as stage 3/4 diseases (p = 0.0106) bearing aggressive CAFs and the PSE. Finally, aggressive TCAFs from patients with PSE resisted the effects of paclitaxel and lenvatinib on the growth of HUVEC and endometrial tumor cells. Our study is the first to report *a correlation between the PSE and the aggressive nature of CAFs in endometrial cancers* and provides an undeniable reason to study the in-depth mechanism of CAF function towards the development of treatment resistance in endometrial cancers.

Keywords: CAFs primary culture; CAF markers; patient-specific aggressive CAFs; post-surgery event; drug resistance; endometrial cancers

1. Introduction

Statement of Translational Relevance

Endometrial cancer-associated fibroblasts (CAFs) can be designated as aggressive or non-aggressive within 4–7 weeks of the time of surgery based on the characteristics of the established primary ex vivo culture. An aggressive CAF from an individual patient with a high grade/stage of the disease can be tested for its role in the development of resistance to the same drugs (chemotherapy, targeted, and/or anti-angiogenic drugs) received as adjuvant treatment within eight weeks of the surgery. The designation of the aggressiveness of an established CAF will provide an opportunity for a personalized test to predict the development of a TME/CAF-mediated resistance to the adjuvant treatment in real-time, prior to clinically recorded events. Thus, the data obtained from our study present a not-so-far-reported *mode of personalized testing for the development of real-time resistance to adjuvant therapy in high-risk patients* with endometrial cancers well before it is clinically encountered in the patients.

Endometrial cancer is the most common gynecological malignancy in high-income countries, and the highest rate of this cancer is currently observed in North America [1,2]. The incidence in the United States in recent years is making uterine cancer the fourth most common cancer in women and the fifth most common cause of cancer death, and its five-year, age-adjusted survival has not improved [3]. With high-grade endometrial cancers that tend to recur, it is desirable to prevent recurrence as the prognosis for recurrence is dismal [1]. Extra-pelvic disease recurrence/recurred metastatic endometrial cancers have a limited response to standard therapies, and the overall survival for most patients with recurrent or metastatic disease is around one year [4,5], thus highlighting an unmet need for a novel treatment strategy.

Tumor cells within a progressive disease enforce several pro-metastatic event(s) via direct and indirect cross-talk with the immune, angiogenic, and mesenchymal stromal cells of TME. CAFs, the most abundant and influential cells of TME, have a significant contextual role in this progression by means of resisting the therapy and ensuing metastasis reported in colon, pancreas, breast, esophagus, and skin cancers. Understanding CAFs' undeniable role in the resistance to therapy and the progression of solid tumors has recently encouraged CAF-directed clinical trials as part of a next-generation cancer drug design and discovery innovation strategy [6–9].

Acknowledging the fact that CAFs partner with tumor cells and all components of TME in a solid tumor, working in favor of the progression in spite of therapy [8], and the specific role of endometrial CAFs in the disease progression [7], we hypothesize that endometrial tumors bear characteristic CAFs which have a clinical relevance from the viewpoint of post-surgery events (PSE). To test our hypothesis, we generated patient-derived CAFs from resected tumor tissues (TCAF) and tumor-adjacent normal tissues (NCAF) at the time of surgery. We identified aggressive CAFs as defined by an early establishment (within 4–7 weeks) and a higher passage number (>3 passages) of the primary culture. By designating the aggressiveness of the CAFs from each patient's tumor sample, we *tested the clinical relevance of the aggressiveness of CAFs and demonstrated that the post-surgery event in patients with a high grade and stage of the disease is directly correlated to the aggressive nature of the CAFs in endometrial cancers.* Aggressive endometrial CAFs from patients with PSE resisted the effects of paclitaxel and lenvatinib on the growth of endometrial cells and HUVEC cells in a hybrid-co-culture (HyCC), indicating an effect mediated via secretomic paracrine signals as compared to direct contact.

2. Results

Our cohort involved 53 patients with endometrial cancers who provided informed consent for the study. Table 1 presents information about the patients included in the study. CAFs were established from both T and TAN received at the time of surgery, depending on the availability of tissue samples. Out of 53 tissue samples, 9 samples were used for the standardization of the CAFs. A total of 44 tissue samples from patients were used for establishing CAF cultures. Out of 44 patient samples, 39 TCAFs (5 failed to grow) and 15 NCAFs (4 failed to grow) were established. In many instances, we received tumor tissue paired with tumor-adjacent normal tissue or Fallopian tubes as the normal tissue.

As we progressed from getting consent from the patients to characterizing the primary CAFs for our study, we passed through different steps, culminating in different CAF samples for the study. Sometimes, we received unpaired samples. Sometimes, we received Fallopian tube samples as the tumor-adjacent normal tissue from the pathology. In our experience, none of the Fallopian tube samples produced NCAFs in our way of creating a non-enzymatic culture from a feeder layer. Often, the paired samples did not grow equally to yield CAFs for passages and characterization by ICC, qRT-PCR, WB, and flow cytometry, which led to the differences in the number of CAF samples in the study.

Table 1. Information on patients included in the study.

All Patients	
Age group	# Patients (n = 53)
Below 55	7
Above 55	46
Pathology Parameters	
Histology	
Carcinosarcoma	3
Endometrioid Adenocarcinoma	42
Endometrial Cancer Not Otherwise Specified	1
High-Grade Papillary Serous Carcinoma	1
Mixed High-Grade Carcinoma	3
High-Grade Serous Carcinoma	3
Stage	
I	40
II	2
III	8
IV	3
Grade	
1	27
2	11
3	15
Myometrial Invasion %	
0–25	24
26–50	19
51–75	4
76–100	6
Lymphovascular Invasion	
+	13
−	39
Indeterminate	1
Lymph Node Positivity	
+	11
−	35
None Submitted	7

2.1. Characterization of Patient-Derived Endometrial CAFs

The purity and specificity of the CAFs were tested by (1) the negative expression of non-CAF markers, including CK 8,18,19, EpCAM, CD45, and CD31, and (2) the positive expression of SMAalpha, S100A4, CD90, FAP, TE-7, TGFB, FGFR1, and PDGFRA. The CAFs were characterized by testing the expression of mRNA using qRT-PCR as well as protein expression using flow cytometry and ICC. The CAFs established from the patient samples were characterized by a negative expression of cell type markers, including CD45, CD31, EpCAM, and CK 8,18, with a parallel positive expression of SMAalpha, FAP, S100A4, and TE-7. PD-L1 and CD44 were additionally used separately. The success rates of the establishment of CAFs were 97.7% (43 out of 44) and 80% (12 out of 15) for TCAFs and NCAFs, respectively. Figure 1A–D presents the expressions of mRNA for different positive (*SMA, FAP-A, S100A4, PDGFR-A, FGFR1, CD44, PD-L1, PD-L2*) and negative (*CD31, EpCAM, CD45*) markers of CAF in endometrial NCAFs and TCAFs as obtained using qRT-PCR. Figure 1D compares the qRT-PCR expression of certain markers with the protein expression of the same markers and more using the WB of a representative non-aggressive NCAF and TCAF pair, demonstrating the expression of SMA in both in contrast to a higher expression of PD-L1 in TCAF than in NCAF. Figure 2 presents the expression of the protein markers of CAF in endometrial NCAF and TCAF pairs using flow cytometry. It depicts the expression of the six positive (CD155, CD90, SMA-A, PD-L1, S100A4, FAP-A) and two negative (EpCAM and CD31) markers of CAF in the established aggressive TCAFs and NCAFs from the tumor samples of two patients (A, B and C, D) with endometrial cancers by flow cytometry. Although the expressions in TCAFs and NCAFs were found to be comparable in terms of both negative and positive markers, a general trend of a higher expression of FAP and CD90 was noted in TCAFs. Figure 2D demonstrates a flow cytometry expression of SMA and S100A4 comparable to the immunoblot expression from representative aggressive CAFs (TCAF and NCAF). The immunoblot expressions of PD-L1 and TGFB are also presented (Figure 2E; densitometric analyses from WB by Image J). A three-color scale-based conditional formatting of the pattern of the % expression of CAF markers of EpCAM, SMA-A, S100A4, and FAP in the pairs of cultured TCAFs and NCAFs (early passages) from 17 representative patients (as presented with color codes) is presented as a heatmap (Figure 2F) which shows a comparable pattern of % distribution of the markers in NCAFs and TCAFs. Figure 3 presents the representative photomicrographs (Figure 3A–C) of the subcellular expression of different positive (CAF-specific markers such as SMA, S100A4, and TE-7, and immune checkpoint marker PD-L1) and negative markers (CK 8,18,19 and EpCAM) of CAF obtained using ICC. The NCAFs and TCAFs are indistinguishable morphologically by H&E stain, although their expression patterns vary for marker proteins. We observed multinucleated TCAFs in a few patients' tumor samples (Figure 3A upper panel left). Figure 3B,C presents a representative photomicrograph of a positive expression of SMAalpha, S100A4, TE-7, and PD-L1 in the NCAFs and TCAFs of different patients, respectively. Noticeably, SMAalpha, S100A4, and PD-L1 were expressed in multinucleated NCAFs and TCAFs. Both epithelial markers, CK 8,18,19 and EpCAM, were negative in all TCAFs and NCAFs (Figure 2A lower panel). A heatmap of the % expression (Figure 3D) of the positive and negative marker proteins of TCAFs from 29 patients with endometrial cancers of different histologies, stages, and grades obtained using ICC is presented in Figure 3D. The heatmap of the expression of the markers in the TCAFs showed (1) a uniformity of the negative expression of the epithelial markers CK 8,18,19 and EpCAM in all TCAFs, (2) a uniformity of the expression of the fibroblast markers SMAalpha and TE-7 in all TCAFs, and (3) a differential expression of S100A4 and PD-L1 in TCAFs irrespective of the histologies, grades, and stages of the tumors. The inset shows a positive correlation between PD-L1 positive TCAFs and S100A4 positive TCAFs in 27 patients. Trendline demonstrated the Pearson correlation between S100A4 and PD-L1 as obtained by ICC in TCAFs from patients ($p = 0.0373$, $n = 27$; Created using GraphPad Prism Version 9.4.0).

2.2. Designation of Aggressive and Non-Aggressive Endometrial CAFs

We categorized CAFs based on criteria concerning their growth patterns, as mentioned in the method section. Out of established patient-derived CAFs, we obtained aggressive CAFs from 31 patients and non-aggressive CAFs from 13 patients. Interestingly, all 31 patients in the aggressive CAF category had TCAFs. We identified six patients with aggressive NCAFs. Two patients had aggressive TCAFs but non-aggressive NCAFs. Two out of seven aggressive CAF-bearing patients' blood was positive for CTC, as shown using a laboratory-friendly method of double-immunocytochemistry and triple-immunofluorescence as detailed elsewhere [10]. Table 2 presents the patients' information for the aggressive and non-aggressive CAFs.

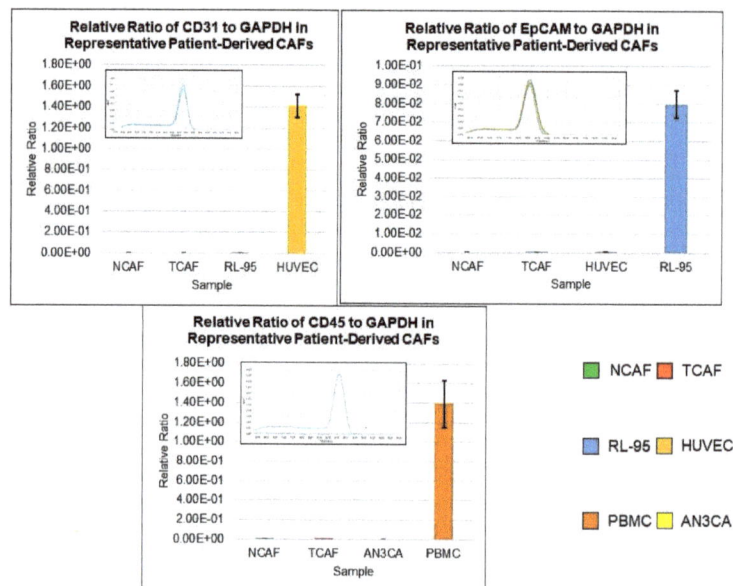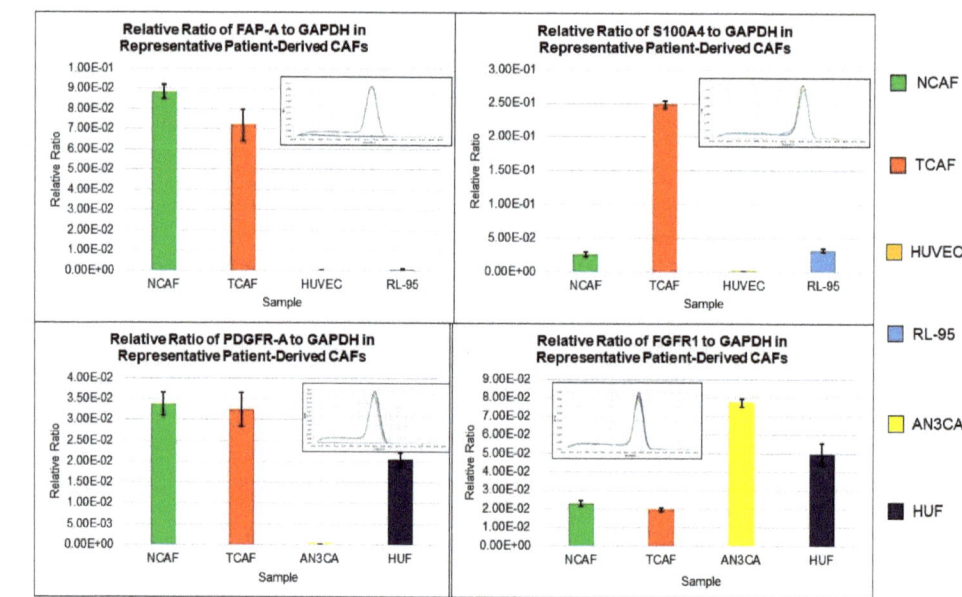

Figure 1. *Cont.*

C

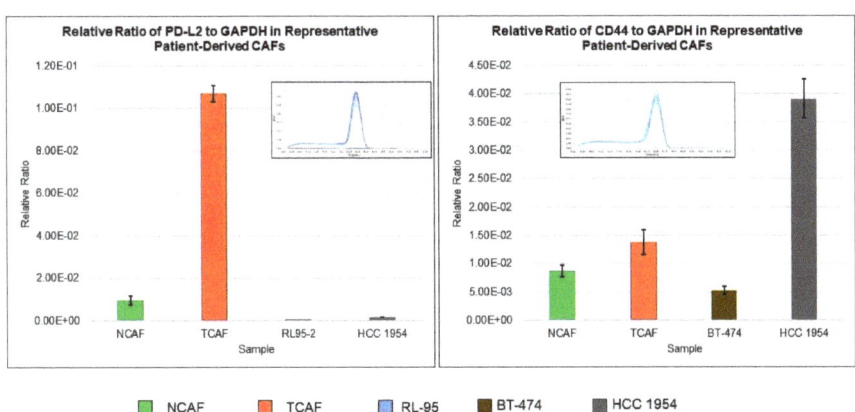

D

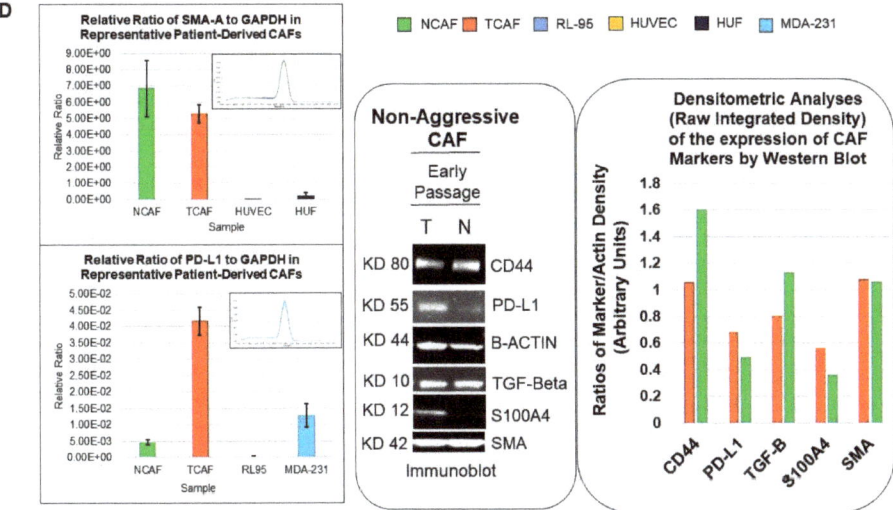

Figure 1. Expression of mRNA for marker proteins of CAF by qRT-PCR: Each graph represents the expression of a gene of interest presented as its relative ratio to *GAPDH*. Inset in each graph is a melting curve of the gene of interest. CAF samples and the positive and negative controls for each gene are color-coded and listed at the bottom of the figure. (**A**) presents expressions of three negative markers of CAF, endothelial marker (*CD31*), epithelial marker (*EpCAM*), and common leucocyte antigen (*CD45*), in NCAFs and TCAFs with respective validation controls. (**B**) presents expressions of four positive markers of CAF (*FAP-A*, *S100A4*, *PDGFR-A*, and *FGFR-1*) in NCAFs and TCAFs with respective validation controls. (**C**) presents expressions of *PD-L2* and *CD44* in NCAFs and TCAFs with respective validation controls. (**D**) compares the qRT-PCR expression of markers with the protein expression of the same markers (SMA and PD-L1) by WB (inset) from a representative non-aggressive early passage NCAF and TCAF. The expressions of CD44, TGFB, and S100A4 by WB are also presented with the corresponding densitometric analyses (bar diagrams) using Image J. Beta-actin was used as the loading control.

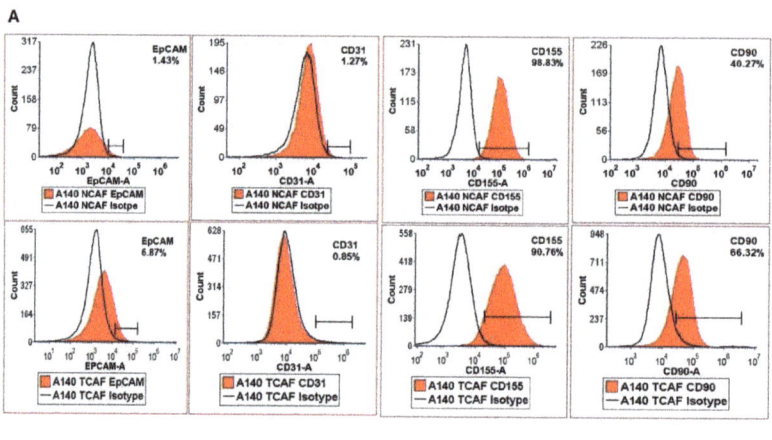
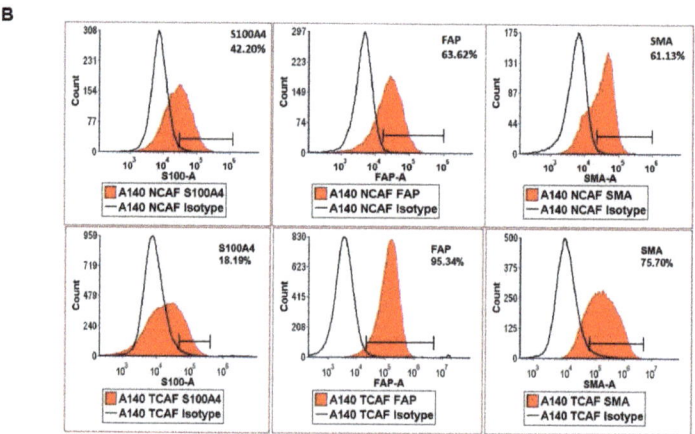
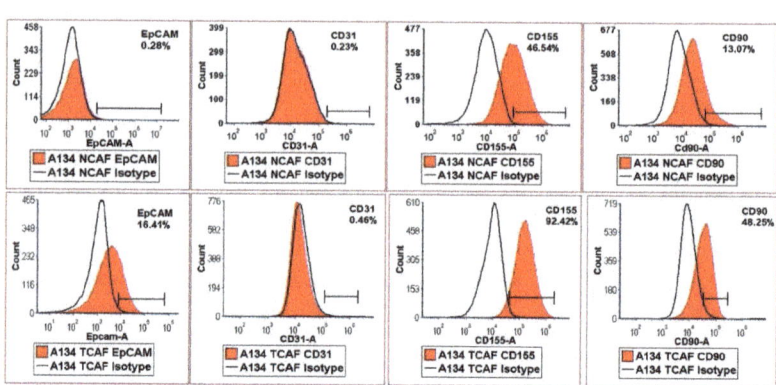

Figure 2. *Cont.*

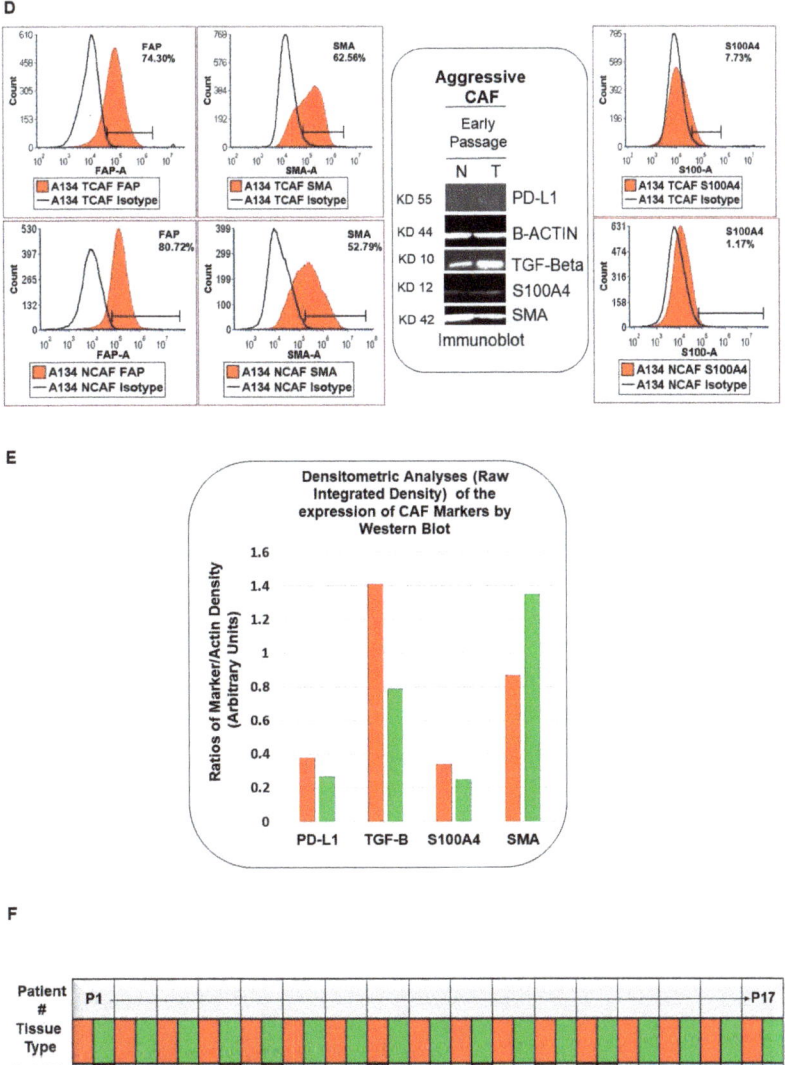

Figure 2. Expression of marker proteins of CAF by flow cytometry: Expression of the six positives (CD155, CD90, SMA-A, S100A4, FAP-A) and two negatives (EpCAM and CD31) markers of CAF in the established TCAFs and NCAFs from the T and TAN samples from two patients; one bearing non-aggressive CAFs (**A**,**B**) and bearing aggressive CAFs (**C**,**D**) by flow cytometry. (**D**) compares the

flow cytometry expression of SMA and S100A with the WB expression (inset) of SMA and S100A from the same representative aggressive CAFs (TCAF and NCAF). The expressions of PD-L1, and TGFB, by WB, are also presented. The protein expression of the same markers (SMA and S100A4) by WB (inset) from a representative aggressive early passage NCAF and TCAF. The expressions of other CAF markers, such as PD-L1 and TGFB, by WB are also presented with the corresponding densitometric analyses (bar diagrams) using Image J (**E**). Beta-actin was used as the loading control. The isotype controls for the respective protein are marked as open black lines compared to the red area under curves for the proteins. A three-color scale-based conditional formatting of the pattern of % expression of CAF markers of EpCAM, SMA-A, S100A4, and FAP (by flow cytometry) in the pairs of cultured TCAFs and NCAFs (early passages) from 17 representative patients (as presented with color codes) is presented as a heatmap (**F**). Red bars represent TCAFs and green bars represent NCAFs (**E**).

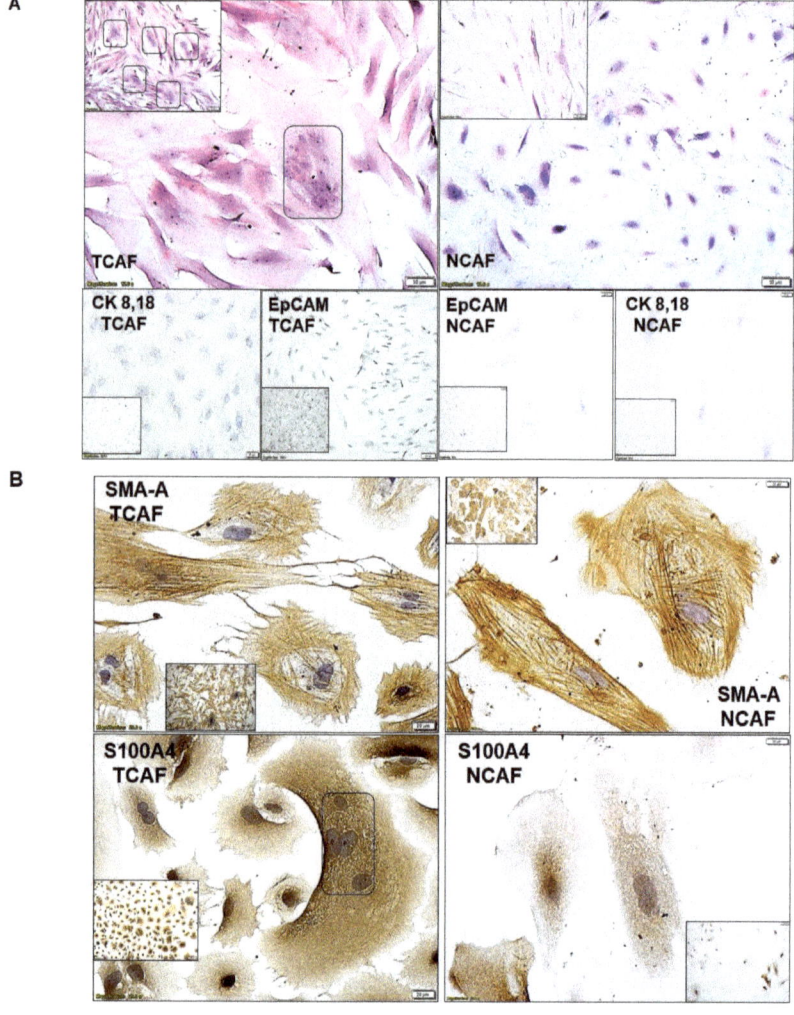

Figure 3. *Cont.*

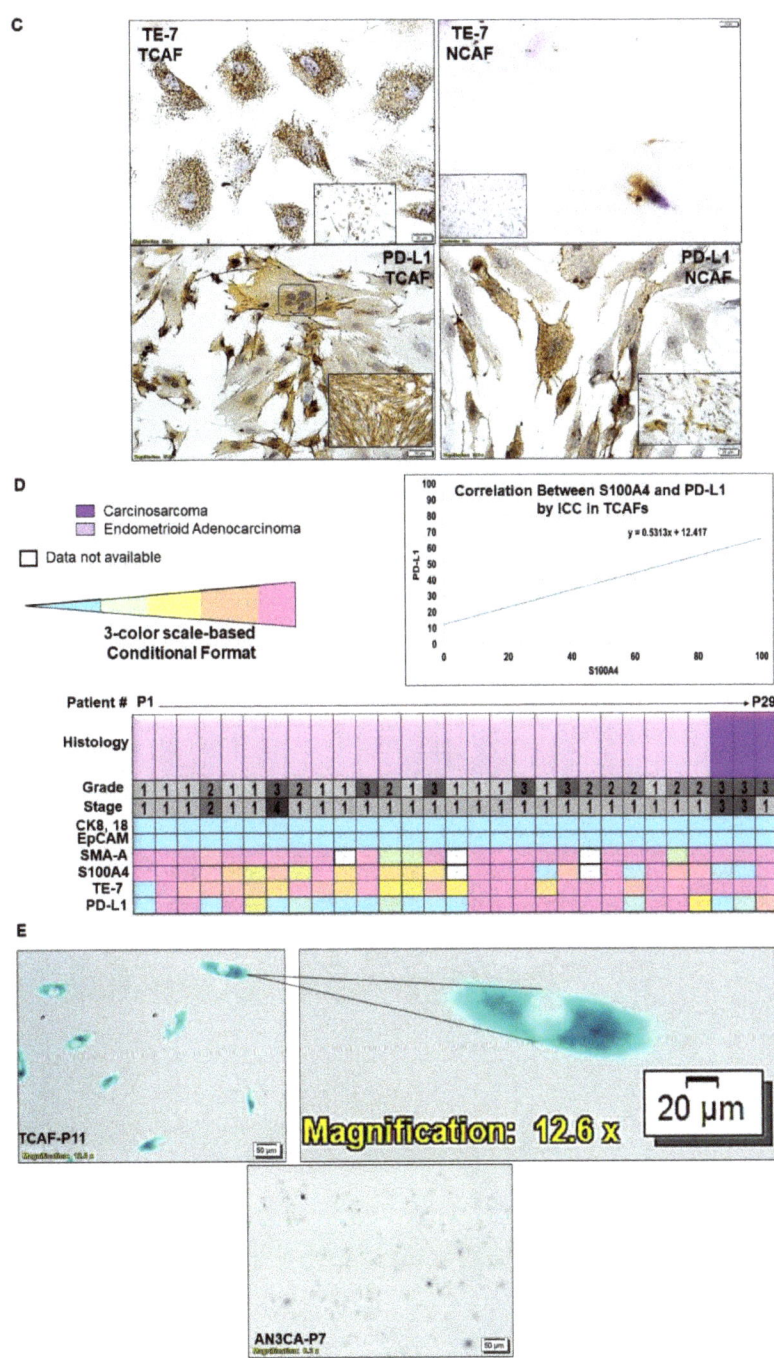

Figure 3. Representative photomicrographs showing subcellular localization of different positive markers of CAF, SMAalpha, S100A4, TE-7, and PD-L1, as well as negative markers, EpCAM and CK 8,18 of CAFs in established primary cultures of NCAF and TCAF samples from patients with

endometrial cancers by ICC: Images show H&E (**A**) upper panel, and negative markers such as CK 8,18 and EpCAM (**A**) lower panel as well as expressions of positive markers including SMAalpha (**B**) upper panel, S100A (**B**) lower panel, TE-7 (**C**) upper panel, and PD-L1 (**C**) lower panel in NCAFs and TCAFs of tumor samples from different patients. Hematoxylin was used as the counterstain. For ICC, pictures were taken at 20× and 40× dry-objective of Olympus BX43 Microscope using cellSens 1.18 LIFE SCIENCE IMAGING SOFTWARE (OLYMPUS CORPORATION). Photomicrographs show the information stamps for magnifications and the scale bars. A three-color scale-based conditional formatting of the pattern of % expression of CAF markers of CK 8,18, EpCAM, SMAalpha, S100A4, and TE-7 (by ICC) in cultured TCAFs (early passages) from 29 representative patients sorted on the basis of histology of the disease with corresponding grades/stages (as presented with color codes and numbers) is presented as a heatmap (**D**). The heatmap displays the % expression of CK 8,18, EpCAM, SMAalpha, S100A4, and TE-7 in TCAFs from 29 out of a total of 31 patients as evaluated by ICC. A very similar pattern of the expression of markers in TCAFs between different histologies, endometrioid adenocarcinoma, and carcinosarcomas of endometrial cancers was observed, although the number of patients for the latter was much smaller in the cohort. Inset shows a positive correlation between the % expressions of S100A4 and PD-L1 in TCAFs in these 27 patients out of a total of 29 patients. Trendline demonstrated the Pearson correlation between S100A4 and PD-L1 by ICC in TCAFs from patients. $p = 0.0373$, $n = 27$. Created using GraphPad Prism Version 9.4.1. The beta-galactosidase stain of endometrial CAFs of late passage (P11) as compared to epithelial tumor cell lines of endometrial AN3CA cell line (P7): Endometrial CAFs stain positive in contrast to AN3CA endometrial cancer cell line (**E**). Inset shows a higher magnification of CAF (the scale bar is computer-generated), showing the subcellular distribution of the stain.

Table 2. Pathological parameters of endometrial cancer patients bearing aggressive and non-aggressive CAFs.

Patients with Aggressive CAF	
Histology	
Carcinosarcoma	2
Endometrioid Adenocarcinoma	26
Endometrial Cancer Not Otherwise Specified	1
High-Grade Papillary Serous Carcinoma	0
Mixed High-Grade Carcinoma	1
High Grade Serous Carcinoma	1
Stage	
I	25
II	1
III	3
IV	2
Grade	
1	14
2	8
3	9
Myometrial Invasion %	
0–25	11
26–50	14
51–75	2
76–100	4

Table 2. *Cont.*

Patients with Aggressive CAF	
Lymphovascular Invasion	
+	5
−	25
Indeterminate	1
Lymph Node Positivity	
+	5
−	23
None Submitted	3
Patients with Non-Aggressive CAF	
Histology	
Carcinosarcoma	1
Endometrioid Adenocarcinoma	9
Endometrial Cancer Not Otherwise Specified	0
High Grade Papillary Serous Carcinoma	1
Mixed High-Grade Carcinoma	1
High-Grade Serous Carcinoma	1
Stage	
I	9
II	0
III	3
IV	1
Grade	
1	8
2	1
3	4
Myometrial Invasion %	
0–25	7
26–50	3
51–75	1
76–100	2
Lymphovascular Invasion	
+	5
−	8
Indeterminate	0
Lymph Node Positivity	
+	4
−	5
None Submitted	4

2.3. Clinical Relevance of Patient-Derived Aggressive Endometrial CAFs

To test the clinical relevance of CAFs, we have conducted follow-ups on the outcome data from patients bearing both aggressive and non-aggressive CAFs. We recorded that 7 patients out of the 31 patients bearing aggressive CAFs had PSEs, while zero events were observed among the patients with non-aggressive CAFs. We tested the correlation between a PSE during the follow-up period and the presence of aggressive CAFs in patients with both high grades and stages of the disease. Figure 4A presents the positive correlation between aggressive-CAF-bearing patients with a high grade (Grade 3) or stage (Stage III/IV) of the disease and post-surgery events. Trendline showed the Pearson correlation

(95% Confidence Intervals) between aggressive CAFs obtained from patients with high-grade (Grade 3) disease and a PSE. $p = 0.0250$ (Figure 4A(i)). Trendline showed the Pearson correlation (95% Confidence Intervals) between aggressive CAFs from patients with high-stage (Stage III/IV) disease and a PSE. $p = 0.0106$ (Figure 4A(ii)). The data present a positive correlation indicating a functional relationship between the aggressiveness of the CAFs at higher grades and stages of the disease and the occurence PSEs in endometrial cancers. Figure 4B presents a heatmap of the ICC % expression of the TCAF markers sorted by events and no-events for a total of 26 patients (Event = 7, Non-Event = 19). Interestingly, both patients with carcinosarcomas bearing a high grade and stage of the disease had events. Since we had a distribution pattern for PD-L1 and S100A4 expression in TCAFs with No-events ($n = 19$) and With-event ($n = 7$), we tested the correlation between the % expression PD-L1 (C(i)) and S100A4 (C(ii)) in TCAFs and events separately (Figure 4C). Trendline demonstrated the Pearson correlation between a post-surgery event and the PD-L1 expression levels as determined by ICC in aggressive TCAFs from 26 patients ($p = 0.0198$, $n = 26$), as created using GraphPad Prism Version 9.4.0. Similarly, Trendline demonstrated the Pearson correlation between a post-surgery event and the S100A4 expression levels as obtained by ICC (right panel) in aggressive TCAFs from 24 patients ($p = 0.0119$, $n = 24$), as created using GraphPad Prism Version 9.4.0. Table 3 presents the pathological parameters, the treatment history, and the details of the post-surgery events in seven patients bearing aggressive CAFs. Table 3 shows the central theme of the study: that, despite different pathological and genomic parameters of the tumors in patients, PSE was recorded in all cases bearing aggressive CAFs. By designating the aggressiveness of the CAFs from each patient's tumor sample, we tested the clinical relevance of the aggressiveness of CAFs and demonstrated that the occurrence of PSEs in patients with a high grade and stage of the disease is directly correlated with the aggressive nature of CAFs in endometrial cancers.

Figure 5A shows the heatmap for the alterations of genes from one of the aggressive CAFs between TCAF and NCAF compared to those of HUF cells. NCAF and TCAF were strikingly different in the alteration of genes from the HUF cells' heatmap expression, while the patterns of the alterations of genes were comparable between NCAFs and TCAFs. We report a comprehensive cancer panel on all exon coverage of 409 genes (most commonly altered in neoplasms) in aggressive TCAFs and NCAFs from the same patient (Figure 5A). The panel covered both somatic and germline mutations of genes.

Heatmaps depicting the genetic alterations identified in NCAF, TCAF, or HUF samples are presented. The coding regions of 409 cancer-related genes are sequenced. The variants with a significant allele frequency difference among the sample groups ($n = 3$) are summarized. The numbers of significantly altered variants of those genes are plotted. The data show a clear segregation of HUF expressions from both NCAFs and TCAFs, while the expression pattern of NCAF and TCAF pairs appears comparable. Interestingly, we observed a comparable pattern of the expression of CAF markers between NCAF and TCAF pairs, with a few exceptions, such as S100A4 and PD-L1 in some patients. In line with the above data, a heatmap generated out of the % expression as obtained by ICC in NCAF and TCAF (Figure 5B) demonstrated (1) a uniformity of the negative expression of the epithelial markers CK 8,18,19 and EpCAM in all NCAFs and TCAFs, irrespective of the stage, grade, and histology of the tumors, (2) a uniformity of the high expression of the CAF markers SMA and TE-7 in all NCAFs and TCAFs, irrespective of the stage, grade, and histology of the tumors, and (3) a differential expression of S100A4 and PD-L1 in NCAFs and TCAFs, irrespective of the stage, grade, and histology of the tumors. We studied the clinical relevance of the characterization of CAFs into aggressive and non-aggressive forms by evaluating the aggressiveness of the CAFs and PSE in the EMR. PSE was recorded in 23% of patients with aggressive CAFs, irrespective of the grade, stage, and genomic alterations. No PSE was recorded in our patients bearing non-aggressive CAFs. We observed an inverse correlation between PSE and S100A4 expression (the same is also true for the PD-L1 expression) (Figure 4C). In line with the above results, we observed (1) a higher expression of PD-L1 and S100A4 as determined by both qRT-PCR and

WB (Figure 1) in the non-aggressive-TCAF-bearing patient, and (2) a lower expression of PD-L1 and S100A4 as determined by both flow cytometry and WB (Figure 2D) in the aggressive-TCAF-bearing patient.

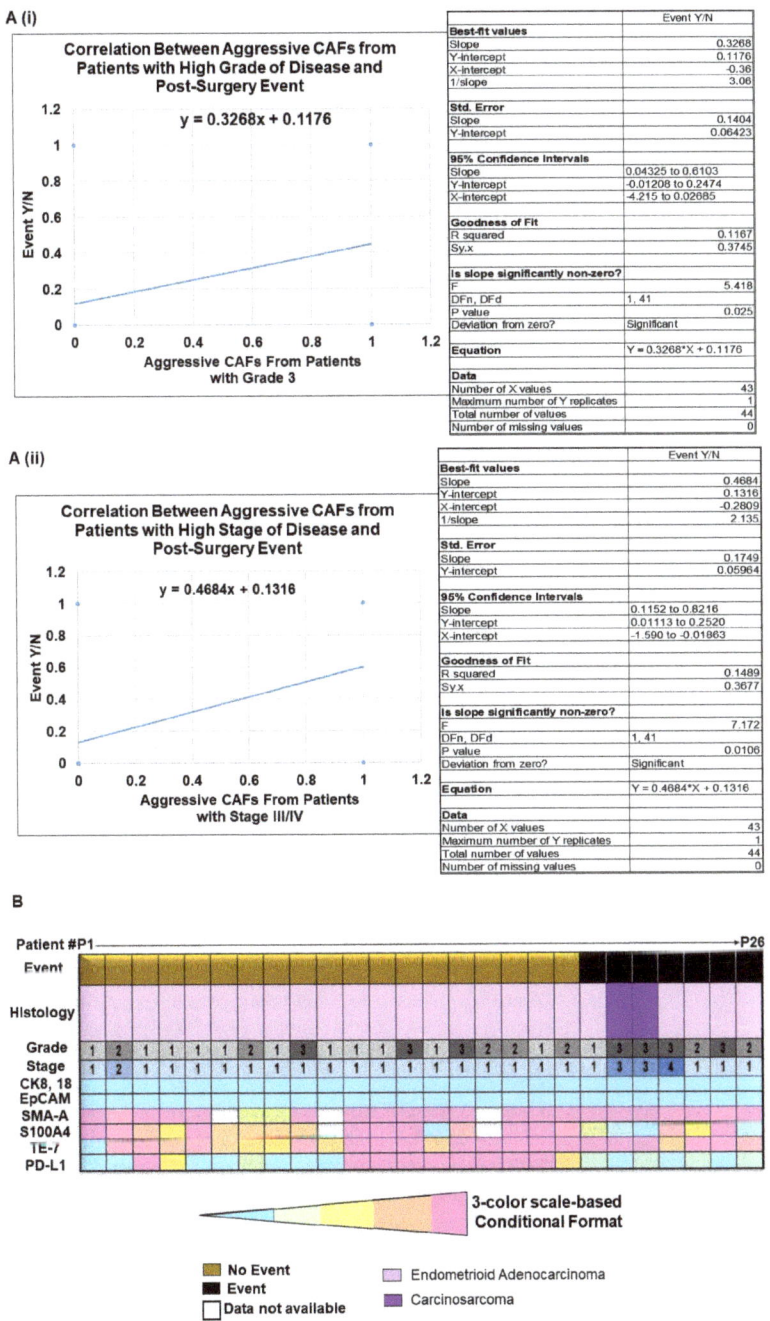

Figure 4. *Cont.*

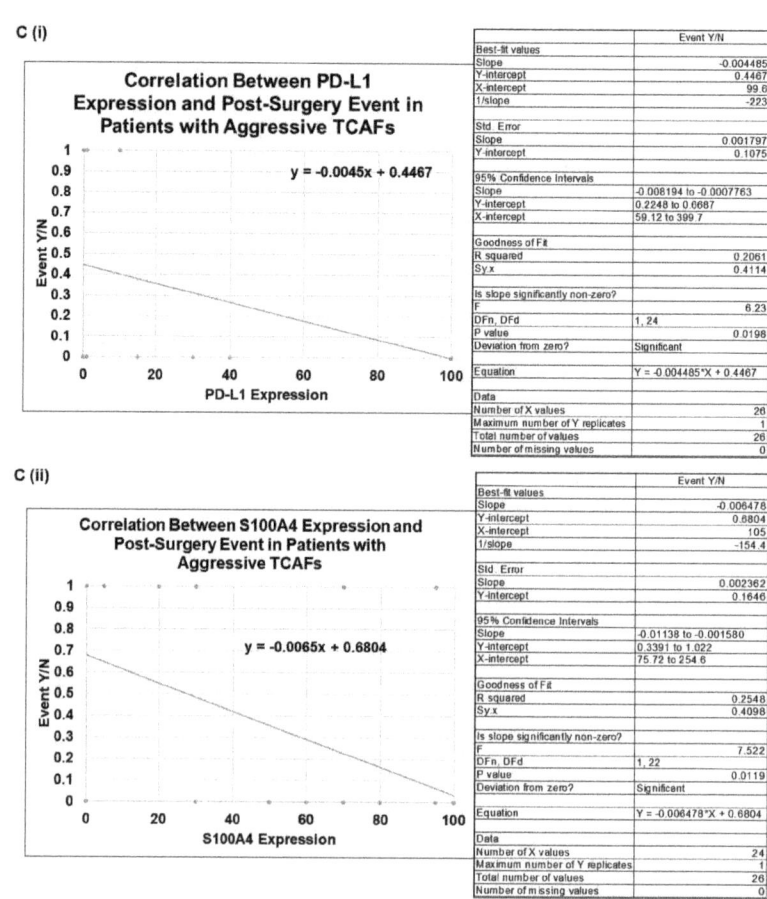

Figure 4. Pearson correlation between aggressive primary CAFs from patients with high grade (Grade 3)/high stage (Stage III/IV) of disease as well as the relationship between % of ICC expression of PD-L1 and S100A4 in aggressive TCAFs and a post-surgery event: Trendline showing the Pearson correlation (95% Confidence Intervals) between aggressive CAFs from patients with high grade (Grade 3) disease and a post-Surgery event. $p = 0.0250$ (**A(i)**). Trendline showing the Pearson correlation (95% Confidence Intervals) between aggressive CAFs from patients with higher stage (Stage III/IV) disease and a post-Surgery event. $p = 0.0106$ (**A(ii)**). $n = 43$ (Created using GraphPad Prism Version 9.4.1) (**A**). A heatmap of ICC % expression of TCAF markers of the aggressive CAFs sorted by events and no-events is presented (**B**). A correlation between % expression PD-L1 in the aggressive TCAFs and events (**C(i)**) and a correlation between % expression S100A4 in the aggressive TCAFs and events (**C(ii)**) are presented (**C**). We presented the aggressiveness by (present or absent) and the PSE by (PSE recorded by Sept. 2022 designated as "Yes" or No PSE recorded by Sept. 2022 as "No") to binary states. Several points overlap because the patient data are the same (for example, we have many patients who were both aggressive CAF and high grade, but it only shows one point on the graph because they share the exact same XY coordinate).

Table 3. Pathological parameters, genomic alterations, and treatment details of endometrial cancer patients bearing aggressive CAFs who presented with a post-surgery event.

#	Treatment		Post-Surgery Events			Pathology Parameter(s)	Genomics
	Neo-Adjuvant Treatment	Adjuvant Treatment	Specific Event	Time of Event (Months)	OS (Overall Survival in Months)		
1	None	None	Developed "squamous cell carcinoma of the skin"	25–37	NA	Endometrioid Adenocarcinoma; Grade 1; Stage I; LVI Negative; Lymph Node: Negative; Myometrial Invasion: 30%; Mismatch Repair Competent	Not performed
2	None	Carboplatinum + Paclitaxel x6. Radiation post-chemo.	Vaginal bleeding, biopsy demonstrated recurrent disease.	10	14	Carcino-sarcoma; Grade 3; Stage III; LVI Positive; Lymph Node: Positive; Myometrial Invasion: 72%; Mismatch Repair Competent	*PTEN*: p.L186fs Frameshift. Variant Allele Fraction 52.1%; *CTNNB1*: p.S33F Missense variant. Variant Allele Fraction 34.2%; *PTEN*: p.R130G Missense variant. Variant Allele Fraction 22.9%; *ARID1A*: p.P109fs Frameshift. Variant Allele Fraction 10.4%
3	None	Carboplatinum + Paclitaxel x6. Radiation post-chemo.	CT scan demonstrated lymph nodes concerning for metastatic disease, biopsy confirmed recurrence.	19	NA	Carcino-sarcoma; Grade 3; Stage III; LVI Negative; Lymph Node: Positive; Myometrial Invasion: 38%; Mismatch Repair Competent	*PIK3CA*: p.Q546K Missense Variant. Variant Allele Fraction 68%; *CDKN2A*: p.P81L Missense Variant. Variant Allele Fraction 4%; *TP53*: p.K132R Missense Variant. Variant Allele Fraction 53%; *FBXW7*: p.R505C Missense Variant. Variant Allele Fraction 20%; *KMT2D*: p.Q2811Sfs*40 Frameshift. Variant Allele
4	None	Carboplatinum + Paclitaxel x6.	Patient had worsening symptoms and was found to have moderate volume of ascites and peritoneal nodularity. CT scan showed changes concerning for progression.	10	13	Endome-trioid Adeno-carcinoma; Grade 3; Stage IV; LVI Negative; Lymph Node: None Submitted; Myometrial Invasion: 50%; Mismatch Repair Competent	*PIK3CA*: p.H1047R Missense variant (exon 20). Variant Allele Fraction 52.8%; *PTEN*: p.E288* Stop gain. Variant Allele Fraction 71%; *TP53*: p.Y234C Missense Variant. Variant Allele Fraction 68.5%; *NF1*: Copy Number Loss; *KHDRBS3-TP63*: Chromosomal rearrangement

Table 3. Cont.

Characteristics of Patients with Aggressive CAF and Post-Surgery Events

#	Treatment		Post-Surgery Events				
	Neo-Adjuvant Treatment	Adjuvant Treatment	Specific Event	Time of Event (Months)	OS (Overall Survival in Months)	Pathology Parameter(s)	Genomics
5	None	None	Patient presented with hematuria and worsening peripheral edema. There was concern for recurrent disease but patient passed away before definitive diagnosis.	11	11	Endome-trioid Adeno-carcinoma Grade 2 Stage I LVI Negative Lymph Node: Negative Myometrial Invasion: 44% Mismatch Repair Competent	Not performed
6	None	Radiation.	Patient presented with vaginal spotting, biopsy revealed recurrent disease. CT scan demonstrated widespread metastatic recurrence.	27	NA	Endome-trioid Adeno-carcinoma Grade 3 Stage I LVI Negative Lymph Node: Negative Myometrial Invasion: 95% Mismatch Repair Competent	*PIK3CA*: p.Q546H Missense variant. Variant Allele Fraction 55% 17q12q21.2 (Focal—*ERBB2*): Gain, Copies = 3.5 *SMARCA4*: p.L410Tfs*90 Frameshift. Variant Allele Fraction 28% *SPOP*: p.R121Q Missense variant. Variant Allele Fraction 59% *TP53*: pJ195F Missense variant. Variant Allele Fraction 46% 17p13.3q12 (whole 17p, part 17q—*TP53*): Loss, Copies = 1.5 10q23.1q23.32 (Sub-arm—*PTEN*): Loss, Copies = 1.5
7	None	Radiation.	Patient had lesion on top of vagina, biopsy revealed recurrent disease.	8	NA	Endome-trioid Adeno-carcinoma Grade 2 Stage I LVI Negative Lymph Node: Negative Myometrial Invasion: 6% Mismatch Repair Deficient	Invitae—Negative

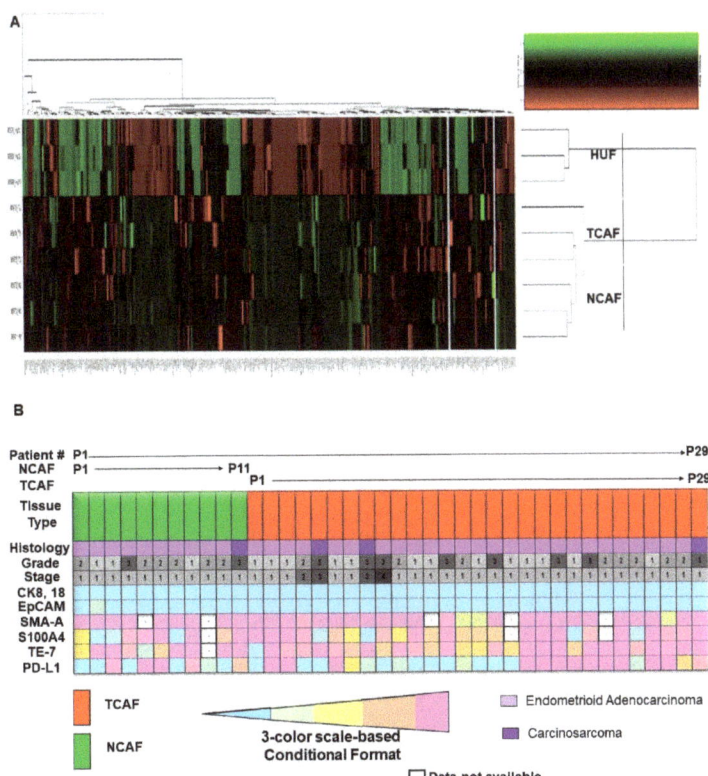

Figure 5. A comparison of NCAF with TCAF in terms of the alterations in genes and expression of protein markers: A comprehensive cancer panel report on all exon coverage on 409 genes involved in some of the most common known cancers shows the alteration of genes in aggressive TCAFs and NCAFs from the same patient compared to HUF cells (**A**). The pathway-based gene selection profiled the mutational spectrum in known cancer driver genes and drug targets along with signaling cascades, apoptosis genes, DNA repair genes, transcription regulators, inflammatory response genes, and growth factor genes. The panel covered both somatic and germline mutations. Heatmaps depicting the genetic alterations identified in NCAF, TCAF, or HUF samples are presented. The coding regions of 409 cancer-related genes are sequenced. Variants with a significant allele frequency difference between sample groups ($n = 3$) are summarized. Numbers of significantly altered variants of those genes are plotted. A 3-color scale-based conditionally formatted expression (% of expression by ICC) of CAF markers in aggressive TCAFs, and NCAFs from 29 representative patients sorted on the basis of CAFs with corresponding histologies, grades/stages (as presented with color codes, and numbers) is presented in a heatmap (**B**).

2.4. Growth Resistance Property of Aggressive Endometrial CAFs

Searching for the mechanistic explanation for the clinical relevance of patient-derived aggressive CAFs in endometrial cancers, we tested the role of aggressive CAFs in resisting the effects of the chemotherapy drug paclitaxel on endometrial tumor cells. Figure 6A presents the effects of the aggressive CAFs cultured from the tumor sample of a patient with a post-surgery event in resisting paclitaxel-mediated growth of endometrial tumor cells following seven days of treatment. The effect of paclitaxel on the clonogenic growth of endometrial tumor cells plated on the aggressive TCAFs derived from patients with PSEs using a HyCC was tested, and the results indicate that TCAF resisted the inhibitory effect of paclitaxel on the 3D clonogenic growth of AN3CA cells (Figure 6B). Neither the on-plate nor on-Top 3D matrigel clonogenic growth of the tumor cells was inhibited following treatment with paclitaxel in the presence of representative aggressive TCAFs at 48 h. At similar times, the same aggressive CAFs also blocked the effect of lenvatinib on HUVEC cells (Figure 6C). The parallel experiments showed the inhibitory effects of the same doses of paclitaxel and lenvatinib on the 3D growth of the test cells. Paclitaxel inhibited the growth of AN3CA. Thus, aggressive TCAFs from patients with PSEs resisted the effects of paclitaxel and lenvatinib on growth in a HyCC. Since paclitaxel failed to inhibit the growth of endometrial tumor cells in the presence of the aggressive CAFs, we tested the basic mode of action of the CAFs by physically isolating the 2 HyCC culture formats of "On-C-slip" HyCC and "On-Plate" HyCC (Figure 6D,E). Diagrammatic representation of the formats of "On-C-slip" HyCC and "On-Plate" HyCC are shown in Figure S1. "On-C-slip" HyCC included cover-slips pre-coated with DiO-stained CAFs on which DiI-stained AN3CA cells were plated. "On-Plate" HyCC did not include pre-coated (with DiO-stained CAFs) cover-slips and consisted only of DiI-stained AN3CA cells plated on the plate. Interestingly, we observed no change in the growth of the endometrial cells between the two formats following the application of paclitaxel, indicating that the growth-inhibition-resisting effect of CAFs does not require direct cell contact and can be mediated via secretomic signals in the media covering the cells in both the formats. Merged pictures of the DiO-stained aggressive CAF and DiI-stained AN3CA of the "On-C-slip" HyCC and "On-Plate" HyCC of the vehicle-treated cells from the same microscopic field showed on dramatic difference in the patterns of cell growth (Figure 6D; upper right panel vs. lower right panel pictures), which was also comparable to the paclitaxel-treated HyCC (Figure 6E; upper right panel vs. lower right panel pictures).

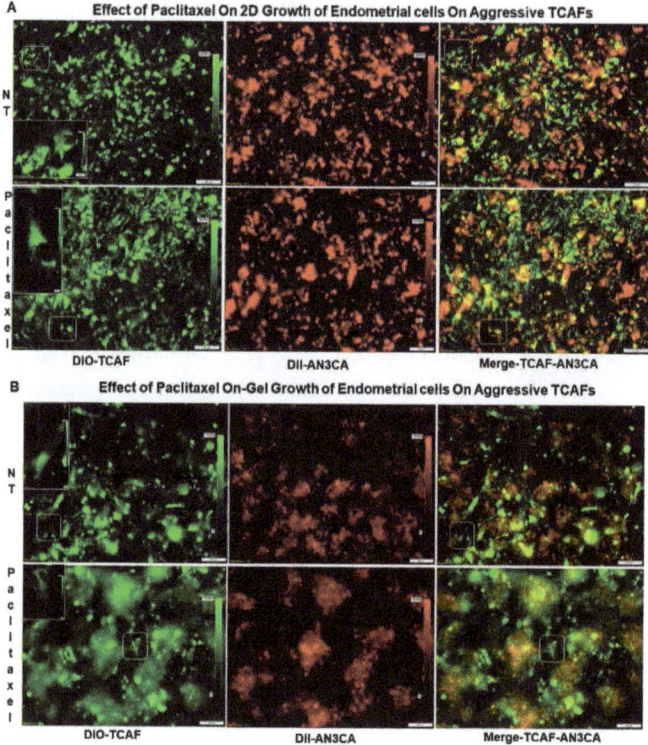

Figure 6. *Cont.*

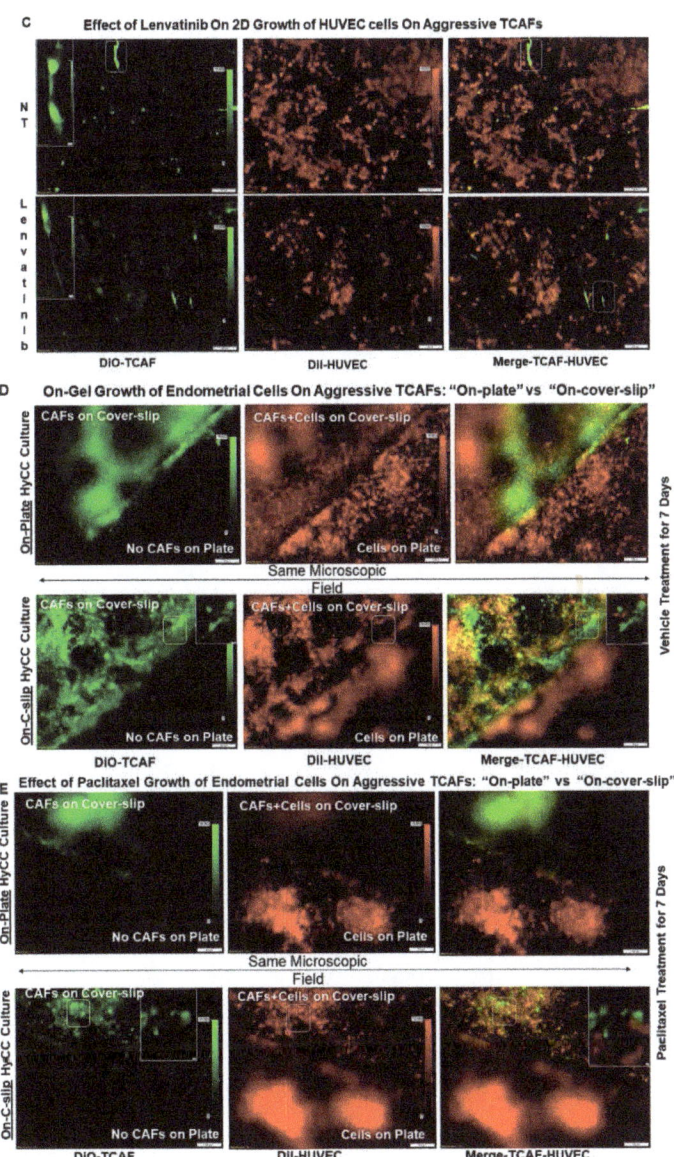

Figure 6. Effect of aggressive CAFs from a patient with PSE on the paclitaxel-mediated inhibition of clonogenic growth of endometrial cancer cells and on the lenvatinib-mediated inhibition of HUVEC growth in hybrid-co-cultures using different formats: Primary culture of aggressive CAFs (passage # 2) from a patient with PSE was stained with DiO and plated on matrigel 24 h before DiI stained AN3CA endometrial cells were added in the 3D On-Top clonogenic cultures with or without 20 nM paclitaxel. Parallel experiments were run without CAFs. Pictures were taken at 20X/40X dry-objective of Olympus BX43 Microscope using cellSens 1.18 LIFE SCIENCE IMAGING SOFTWARE (OLYMPUS CORPORATION). Photomicrographs show the information stamps for magnifications and the scale bars. Growth (2D) of DiI-stained AN3CA endometrial cells in HyCC without coverslips on DiO-stained aggressive endometrial TCAFs with PSE in the presence and absence of paclitaxel

for 7 days (**A**), on-gel growth of DiI-stained AN3CA endometrial cells in HyCC without coverslips on DiO-stained aggressive endometrial TCAFs with PSE in the presence and absence of paclitaxel for 7 days (**B**), no-gel-growth of DiI-stained HUVEC (passage #3) in HyCC on DiO-stained aggressive endometrial TCAFs with PSE in the presence and absence of lenvatinib for 7 days (**C**), and "On-plate" (upper panel) versus "On-C-slip" (lower panel) matrigel On-Top growth in HyCC in the absence (**D**) or presence of paclitaxel (**E**) for 7 days are presented as marked in the photomicrographs. Bar represents 100 micron (μM).

3. Discussion

We explored the functional association between aggressive CAFs and resistance to the growth-inhibitory effect of chemo/targeted drugs in endometrial cancers. CAFs are the most abundant stromal cells and play a significant contextual role in shaping tumor initiation, disease progression on therapy, and metastasis. Understanding the function of CAFs, the most influentially abundant stromal cells with a significant contextual role in the progression and metastasis of several solid tumors [6,8], has encouraged CAF-directed clinical trials as part of a next-generation cancer drug design and discovery innovation strategy. CAFs have therefore begun to emerge as a part of future cancer management.

Contrary to the standard system of isolating/growing CAFs based on enzyme digestion, our system is not artificial, as we derived the CAFs from surgically resected tumors and tumor-adjacent normal tissues following pathological grossing in each case. Considering the diversity of the tumor sample received in each case, their grossing pattern, the histology, and the stage/grade of the tumor, we designed a unique way to set up the primary CAF culture from the feeder layer, NOT from the standard enzyme digested model which can give the number of cell counts. To the best of our knowledge, this is the first report of a non-enzymatic isolation of CAFs and their clinical relevance in endometrial cancers. We believe that this novel way of using a natural method of isolating the CAFs has empowered us to identify two different populations of CAFs. Keeping in mind the nature of the heterogeneity of samples, we have performed "viability of the tissues, as well as tumor/stroma ratio of each and every sample".

As we were aware of the intratumoral heterogeneity of the CAF population within the tumor sample, we used the feeder layer from the entire tissue sample provided to us to set up the primary culture. It is possible that the CAF population we cultured may have more than one subpopulation of CAFs, since the tumor sample(s) we received may well represent a heterogeneous TME (CAF) ecosystem, as reported in PDACs. Hence, our CAFs from a single patient could be a heterogeneous mixture (of different CAF subpopulations), as heterogeneous as it could possibly be in the original tumor sample from that particular patient. Indeed, our data demonstrated that some cells of the CAFs from the same patient bear multiple nuclei, while other CAFs in the same population are mono-nuclear in the same passage (Figure 3A–C).

We identified and characterized CAFs in the context of their clinical relevance in endometrial cancers to demonstrate that a direct correlation exists between the presence of aggressive CAFs and PSE in patients with endometrial cancers. In establishing a patient-derived primary culture of endometrial CAFs, we tested every passage of the NCAFs and TCAFs as per their availability in the primary culture. We considered both the properties of a primary culture of the CAFs, the time of establishment, and the number of passages before the cells show the senescence markers to designate them as aggressive and non-aggressive CAFs. No uniform CAF marker is expressed in the same pattern across all cancers [6]. We demonstrated that the endometrial CAF markers in our cohort bear a characteristic expression pattern in terms of mRNA (Figure 1), protein (Figure 2), and their subcellular localization (Figure 3). Although we observed a comparable % expression of markers between NCAFs and TCAFs as obtained by parallel qRT-PCR, flow cytometry, and ICC, we observed a trend in the classification of CAFs based on the differential expression of S100A4 and PD-L1. Interestingly, we observed a significant positive correlation between the expressions of S100A4 and PD-L1 in CAFs ($p = 0.0373$) (Figure 3D). However, the clinical significance of such observation remains unclear. We observed that NCAFs are lesser in number than TCAFs in our cohort. Delving into the reason for this observation (that NCAFs are lesser in number than TCAFs) in our cohort, we found that fewer TAN tissues were received compared to the tumor tissues obtained. Furthermore, sometimes we received samples from the Fallopian tubes as the TAN samples, and no NCAFs could be established from those samples. This fact indicates that NCAFs are fundamentally different from the fibroblasts of the distant normal tissues of a different organ (more so in the case when the disease is of low grade and stage and is strictly localized). In support of the above logic, we have also observed no difference in the sequencing data for alterations in genes between NACFs and TCAFs of one of our patients bearing aggressive NCAFs and TCAFs as compared to HUF cells as presented in the heatmap (Figure 5). We used HUF only as an internal technical control, since the background data on endometrial CAF are very limited in the literature.

Additionally, we did not find an overall difference in the expression of markers between NACFs and TCAFs in ICC (Figure 5B).

To test whether the aggressiveness of the patient-derived CAF can be regarded as a companion marker with a high grade/stage of the disease for a PSE, we tested the correlation between events and the high-grade/stage patients bearing aggressive CAFs. Interestingly, all 31 patients of the "aggressive CAF category" had TCAFs. Our study has limitations. Our cohort consists of 53 patients with endometrial cancers over a period of 5 years, with most of the patients bearing stage I and grade 1 disease, which naturally caused a skewness towards low grades/stages in the population of patients. Our CTC data provided very limited information, as we did not receive a longitudinal CTC. Since non-aggressive CAFs did not grow beyond passages 2–3 over a longer period (more than a month), the yields of non-aggressive CAFs were practically too low for any experiments other than the testing of markers by ICC. Hence, we could not compare the aggressive and non-aggressive CAFs for the HyCC. One of the limitations of our study is that, due to the nature of the growth patterns of the endometrial sample-derived primary CAFs, the number of samples for some experiments was low for certain categories of experiments. Using single-cell sequencing of tissue-purified CAFs, a subpopulation of aggressive CAFs with a specific gene expression pattern has been defined in several works of the organ-type cancers whose CAFs are already characterized, such as PDAC, melanoma, colorectal, breast, lung, and ovarian cancers. In all of these organ types, the CAF was found to have clinical relevance. Unfortunately, the work has not progressed to that level in the case of endometrial cancers yet. Since the markers of the CAF subpopulations are organ-type specific, we will need more information before we start to tease out subpopulations of CAFs in endometrial cancers per se. In this study, we first sought to find the clinical relevance of the CAFs in endometrial cancers.

We show a positive correlation between patients with grade 3 disease bearing aggressive CAFs and a PSE ($p = 0.0106$), as well as between patients with stage 3/4 disease bearing aggressive CAFs and a PSE ($p = 0.0017$). Considering the fact that the probability of PSE is a function of time in patients and is closely associated with disease progression, we observed PSE in most of the earlier patients in our cohort. It should be noted that the PSEs currently observed in 7 out of 31 patients bearing aggressive CAFs may increase over time, and we may observe a greater number of patients with PSEs within the cohort of aggressive CAFs. Our study is the first to report a correlation between the post-surgery event and the aggressive nature of CAFs in endometrial cancers and provides an undeniable reason to study the in-depth mechanism of CAF function in the context of tumor cells and the rest of the stromal cells of a TME, immune cells, and angiogenic cells. We chose to test the effects of paclitaxel and lenvatinib because these are among the drugs which are clinically used in treating endometrial cancers [11,12]. We provide the first experimental evidence of the mode in which the role of CAF is proved in resisting the growth-inhibitory effect of both paclitaxel and lenvatinib via secretomic signals (Figure 6). Ascertaining direct experimental proof of the role of CAFs in developing resistance to anti-tumor drugs will provide an opportunity to investigate new drugs for counteracting CAF-mediated treatment resistance and thus normalizing TME in aggressive endometrial cancers.

Thus, our tumor cell-on-CAF ex vivo HyCC model provides a unique opportunity for the personalized testing of anti-tumor drugs, positively predicting the development of future resistance well before it is clinically encountered in patients. Including anti-stromal therapy to normalize the TME will broaden drug design and target the TME. Our data prove that the horizon ought to be widened to include the molecular targets contributed by the oncologic signals from CAFs. As we progress on the roadmap for effective cancer management in the coming decade, we are optimistic about *a CAF-inclusive target*.

Currently, the management of the disease in endometrial cancers is targeted at the tumor and the immune compartment of the TME [7,13,14]. Our data can initiate a different perspective to include CAF-directed therapy, especially in the high-risk patients undergoing surgery with high grade and stage of the diseases. Together, our study sheds light on the development of resistant conditions in the presence of aggressive CAFs and suggests a new therapeutics potential for CAF-targeted therapy to enhance the response to cancer treatment and outcome.

4. Methods and Materials

4.1. Tissue Collection at the Time of Surgery

All experimental protocols were approved by the institutional and/or licensing committee(s). Informed consent (IRB approved: Protocol Number Study: 2017.053-100399_ExVivo001) was obtained from 53 patients and/or their legal guardian(s). The resected tumor (T) and tumor-adjacent normal (TAN) tissues were collected during surgery in designated collection media as per the guidelines and relevant regulations and provided by the pathologist, depending upon the availability of the tissue on a case-to-case basis. We included samples from consecutive consented patients with endometrial tumors at any stage/grade of the disease undergoing surgery (Table 1) with or without pre-treatment/history of any previous carcinoma.

4.2. Cell Lines and Reagents

Human uterine fibroblasts (HUF; Primary Uterine Fibroblasts, Cat # PCS-460-010), HUVEC cells (cat # PCS-100-013), endometrial cells (RL-95-2 and AN3CA), MCF7 cells, and NCI-H441 cells were procured from ATCC (USA) and were cultured according to the standard cell culture procedures as per ATCC (USA). MCF-7 cells were cultured in atmospheric 95% air, 5% CO_2. The complete medium consisted of the base medium for this cell line (ATCC-formulated Eagle's Minimum Essential Medium, Catalog No. 30-2003). To make the complete growth medium, we added 0.01 mg/mL of human recombinant insulin and fetal bovine serum to a final concentration of 10% and 1% pen/strep (Temperature 37 °C). RL-95-2 cells were cultured at 37 °C and 5% CO_2. The complete medium for this cell consisted of the base medium (ATCC-formulated DMEM: F12 Medium Catalog No. 30-2006). To make the complete growth medium, we add 0.005 mg/mL of insulin fetal bovine serum to a final concentration of 10% and 1% pen/strep. The subculturing procedure included the removing the growth medium, rinsing with 0.25% trypsin, 0.03% EDTA solution, and allowing the flask to sit at room temperature (or at 37 °C) until the cells detached. Fresh culture medium was added into new culture flasks in a subcultivation ratio of 1:2 to 1:5. AN3CA cells were cultured in atmospheric 95% air, 5% CO_2. The complete medium consisted of the base medium for this cell line (ATCC-formulated Eagle's Minimum Essential Medium, Catalog No. 30-2003). To make the complete growth medium, we added a fetal bovine serum to a final concentration of 10% and 1% pen/strep (a subcultivation ratio of 1:3 to 1:6). The NCI-H441 cells' complete medium consisted of the base medium for this cell line (ATCC-formulated RPMI-1640 Medium, ATCC 30-2001). To make the complete growth medium, we added fetal bovine serum (ATCC 30-2020) to a final concentration of 10% and 1% pen/strep (a subcultivation ratio of 1:3 to 1:8). HUVEC cells were cultured in a vascular cell basal medium (ATCC PCS-100-030) supplemented with endothelial cell growth kit BBE (ATCC PCS-100-040) at 37 °C and 5% CO_2. The complete medium for HUF cells (Primary Uterine Fibroblasts; Normal, Human (ATCC PCS-460-010) consisted of the components of the growth kit with the Fibroblast Basal Medium-Low Serum (ATCC PCS-201-041), which were added to the basal medium (ATCC PCS-201-030). The Fibroblast Growth Kit-Low Serum components are rh FGF b, 0.5 mL (final concentration 5 ng/mL), L-glutamine, 18.75 mL (final concentration 7.5 mM), Ascorbic acid, 0.5 mL (final concentration 50 μg/mL), Hydrocortisone Hemisuccinate, 0.5 mL (final concentration 1 μg/mL), rh Insulin, 0.5 mL (final concentration 5 μg/mL), and Fetal Bovine Serum, 10.0 mL (final concentration 2%) at 37 °C, and 5% CO_2. Cells were seeded in a new flask during subculturing at a density of 2500 to 5000 cells per cm^2. We bought HUF cells from ATCC (ATCC PCS-460-010). These primary uterine fibroblast cells are bipolar, refractile, and adherent spindle-shaped, were isolated from the wall of a donor's uterus, and were obtained from the non-decidualized uterus via hysterectomy. We have also used primary fibroblasts from the tumor-adjacent normal tissue of the patient as provided to us by the pathologist after grossing following the surgical resection.

Other cell lines for qRT-PCR were procured from ATCC. Antibodies for ICC were bought from Cell Marque (Rocklin, CA, USA), NOVUS (Centennial, CO, USA), Abcam (Waltham, MA, USA), Agilent-Dako (Santa Clara, CA, USA), and Cell Signaling (Danvers, MA, USA). All cells were used within 7–8 passages and tested negative for mycoplasma.

4.3. Patient-Derived Primary Culture of Endometrial CAFs

The primary culture of CAFs (TCAF and NCAF) from the endometrial T and TAN samples was set up from the feeder layer. The initial seeding of cells was cultured in media containing DMEM/F-12 + Glutamax. The primary culture of CAFs (TCAF and NCAF) was set up from the resected endometrial tumors and tumor-adjacent normal tissue samples. Since the culture was initiated from a feeder layer, the entire process was independent of any enzymatic digestion of the tissue. The initial seeding of cells was cultured in media containing DMEM/F-12 + Glutamax. Once the fibroblast cells grew in passage zero, they were passaged thereof by differential trypsinization (trypsin-sensitive and trypsin-resistant cells). We monitored the purity and the extent of the epithelial cell contamination of the cultures by testing (1) the negative expression of non-CAF markers, including epithelial cell markers CK 8,18,19, and EpCAM, leucocyte common antigen CD45, and endothelial cell marker CD31, and (2) the positive expression of fibroblast/CAF markers, including SMAalpha, S100A4, CD90, FAP, CD155, TE-7, PDGFRA, and FGFR. Expression of the stem cell marker CD44 and the immune checkpoint marker PD-L1 are also monitored. The expression of fibroblast markers was monitored throughout each of the following passages until the late passages tested positive for beta-galactosidase stain. We employed qRT-PCR, Western blot, flow cytometry, and ICC to test the expression of the above-mentioned markers (mRNAs and proteins). The purity and the extent of epithelial cell contamination of the cultures were monitored by testing the expression of mRNA by qRT-PCR as well as the protein expression by flow and ICC. Acknowledging the fact of CAF heterogeneity and the fact that very limited data have been published in the literature on the endometrial CAFs, we examined three categories of CAF markers in our study, (1) negative markers, (2) positive markers, and (3) other associated immune markers and stem cell markers in paired NCAF and TCAF samples. We have also presented a simultaneous readout by using two testing methods for multiple markers (positive) in the paired samples for clarity.

A passage of the primary culture of CAF is qualified by (1) the negative expression of non-CAF markers, including epithelial cell markers CK 8,18,19, and EpCAM, leucocyte common antigen CD45, and endothelial cells marker CD31, and (2) the positive expression of fibroblast/CAF markers, including SMAalpha, S100A4, CD90, FAP, CD155, TE-7, PDGFRA, and FGFR. The expression of stem cell marker CD44 and immune checkpoint marker PD-L1 were also monitored. The expression of fibroblast markers was monitored throughout each passage. The first 3 passages are designated as early, followed by mid and late passages. Depending upon the viability and expression of the markers, the late passage CAFs have been tested for senescence (beta-galactosidase assays), as presented in Figure 3E.

4.4. Comprehensive Cancer Panel for Aggressive CAF Pair and HUF

We obtained a Comprehensive Cancer Panel (CCP) for an aggressive NCAF-TCAF pair and compared it with HUF cells (a 'commercially available DNA sequencing CCP' from PrimBio Research Institute, Exton, PA, USA). The CCP (DNA sequencing panel) consisted of the most comprehensive cancer panel available, with all exon coverage on the 409 genes involved in the most common known cancers. The workflow consisted of the construction of the library using the Ampliseq 2.0 Library Kit (Cat. no. 4475345), preparation of the template, sequencing (An Ion S5 sequencer using an Ion Torrent Amplieseq CCP DNA run plan), and data analysis (Ion Torrent Browser Plugin for VCF generation and data uploaded to Ion Reporter). Approximately 50 pM of pooled libraries were used for templating using the Life Technologies Ion Chef S5 kit (Cat# 4488377) and the manufacturer's recommended protocol.

4.5. Expression of mRNA for CAF Markers by qRT-PCR

As mentioned earlier, the expressions of CAF markers were tested by qRT-PCR. RNA extraction and qRT-PCR were performed as mentioned elsewhere [15]. In short, RNA extraction was performed using the Qiagen RNEasy MiniKit and Qiashredder Kit according to the manufacturer's protocol. RNA was extracted from lysed cell pellets (Qiashredder system) and converted to cDNA using iScript Reverse Transcription Supermix. qRT-PCR was performed using the Roche LightCycler96 platform. Appropriate primers (Integrated DNA Technologies, Coralville, IA, USA) for each gene of interest were mixed with Roche FastStart Essential DNA Green Master Mix and run in triplicate. FastStart Essential DNA Green Master (Roche, Basel, Switzerland) was used for product detection (Relative Ratio of the gene of interest to GAPDH) using the Roche LightCycler 96 Software version 1.1. The list of primers used is presented in Table 4.

4.6. Expression of Protein Markers of CAF by Flow-Cytometer and Western Blot

Flow cytometry was performed using SMA-FITC, FAP-PE, S100A4-PerCP, EpCAM-APC, CD31-FITC, CD155-PE, CD90 PE-Vio615, and PD-L1- APC. For flow cytometry, cells were trypsin released and rinsed in FACS Buffer (phenol red-free RPMI with 1% FBS). Cells were stained for 15 min with cell surface antibodies (CD31 Miltenyi, CD155 Miltenyi, CD90 Miltenyi, FAP R&D systems, PD-L1 Miltyeni) or corresponding isotype control antibodies (Miltenyi). Cells were fixed using the kit from Miltenyi for 30 min, followed by re-suspension in a permeabilization buffer from the same kit. Cells were stained for intracellular antibodies (SMA and S100A4, both from Novus biologicals) for 30 min. Stained cells were run on a BD Accuri C6 flow cytometer and analyzed using FCS express (DeNovo software, 7.16.0047). TCAFs and NCAFs from all patients with established CAFs at every passage of the primary culture were tested for the expression of markers to confirm specificity and aggressiveness. The list of antibody conjugates used in the study is presented in Table 4. The expression of a few CAF markers (qRT-PCR, flow cytometry, and Western blot) in representative aggressive and non-aggressive CAFs, one each, was shown (Figures 1 and 2).

4.7. Cellular Localization of CAF Markers by ICC

For ICC, CAFs were cultured on coverslips. Both NCAFs and TCAFs from each passage were stained for EpCAM, CK 8,18, SMAalpha, S100A4, TE-7, and PD-L1 to confirm the specificity and aggressive nature of the CAFs. ICC for protein was first validated and evaluated by a pathologist and then was run with corresponding positive and negative controls. Endometrial tumor cells (RL-95-2 and AN3CA) were used as the positive control for EpCAM and CK 8,18 and as the negative control for SMAalpha, S100A4, and TE-7. HUF was used as the positive control for SMAalpha, S100A4, TE-7, and the negative control for

EpCAM and CK 8,18. HUVEC cells were used as the negative control for EpCAM, CK 8,18, SMAalpha, S100A4, and TE-7 and as the positive control for CD31. NCI-H441 cells were used as the positive control for PD-L1, and MCF7 cells were used as the negative control for PD-L1. Hematoxylin was used as the counterstain. For ICC, pictures were taken at 20× and 40× dry-objectives of an Olympus BX43 Microscope using cellSens 1.18 LIFE SCIENCE IMAGING SOFTWARE (OLYMPUS CORPORATION). The list of antibodies used is presented in Table 4.

Table 4. Detailed information on the antibodies and primers used in the study.

Primers for qRT-PCR	
Gene	Primer Sequence (Sequences Listed 5′–3′)
ACTA-2/SMA	F: CGT TAC TAC TGC TGA GCG TGA
	R: GCC CAT CAG GCA ACT CGT AA
CD31	F: ATT GCA GTG GTT ATC ATC GGA GTG
	R: CTG GTT GTT GGA GTT CAG AAG TGG
CD44	F: AGC ACT TCA GGA GGT TAC ATC T
	R: CTT GCC TCT TGG TTG CTG TCT
CD45	F: CTTCAGTGGTCCCATTGTGGTG
	R: CCACTTTGTTCTCGGCTTCCAG
CD90/THY1	F: GAAGGTCCTCTACTTATCCGCC
	R: TGATGCCCTCACACTTGACCAG
EpCAM	F: AGC GAG TGA GAA CCT ACT GGA
	R: CGC GTT GTG ATC TCC TTC TGA
FAP-A	F: GGA AGT GCC TGT TCC AGC AAT G
	R: TGT CTG CCA GTC TTC CCT GAA G
GAPDH	F: TCA AGG CTG AGA ACG GGA AG
	R: CGC CCC ACT TGA TTT TGG AG
FGFR1	F: GAC ACC ACC TAC TTC TCC GTC AA
	R: CAA TAT GGA GCT ACG GGC ATA CG
PDGFRA	F: TGG CAG TAC CCC ATG TCT GAA
	R: CCA AGA CCG TCA CAA AAA GGC
PD-L1	F: ACC TAC TGG CAT TTG CTG AAC G
	R: ATA GAC AAT TAG TGC AGC CAG GT
S100A4	F: CAG AAC TAA AGG AGC TGC TGA CC
	R: CTT GGA AGT CCA CCT CGT TGT C
Antibodies for ICC		
---	---	---
Antibody	Manufacturer	Cat.#
Cytokeratin 8 & 18 (B22.1 & B23.1)	Cell Marque	818M-94
Ep-CAM/Epithelial Specific Antigen (Ber-EP4)	Cell Marque	248M-94
Fibroblasts Antibody (TE-7)	NOVUS	NBP2-50082
Actin, Smooth Muscle (1A4)	Cell Marque	202M-94
Recombinant Anti-S100A4 Antibody	Abcam (Waltham, MA, USA)	ab124805
PD-L1 [Clone 22C3]	Agilent-Dako	M365329-1
PD-L2 (D7U8C)	Cell Signaling	82723
CD31	Cell Signaling	3528
Vimentin (SP20) Rabbit Monoclonal Antibody	Cell Marque	347R-14
Antibodies for Flow Cytometry		
Antibody	Manufacturer	Cat.#
CD31-FITC	Miltenyi (Waltham, MA, USA)	130-117-390
CD155-PE	Miltenyi	130-105-846
CD90 PE-Vio615	Miltenyi	130-114-909
S100A4-PerCP	NOVUS	NBP2-36431APCCY7
SMA-FITC	NOVUS	NBP2-34522F
FAP-PE	R&D Systems (McKinley Place NE, Minneapolis, MN, USA)	FAB3715P-025
PD-L1-APC	Miltenyi	130-122-816
EpCAM-APC	Miltenyi	130-133-260

4.8. Categorization of the Aggressiveness of CAFs

CAFs were categorized according to their pattern of growth. The CAFs from each patient's tumor sample were designated as either aggressive or non-aggressive CAFs based on two criteria of aggressiveness: first, the time required for establishing the CAF culture in weeks, and second, the number of passages the cultured CAFs were passaged before becoming positive for the senescence stain. An aggressive CAF is defined by an early establishment (within 4–7 weeks) and a higher passage number (>3 passages) of the primary culture. A non-aggressive CAF is defined by a late establishment (more than 4–7 weeks) and a lower passage number (<3 passages) of the primary culture. Each passage of the CAFs was tested for the expression of the negative and positive markers as well as for the senescence markers.

4.9. Post-Surgery Events in Endometrial Patients

To test the clinical relevance of the aggressiveness of the CAFs derived from the tumor samples of patients with a high grade/stage of the disease, we tested the correlation between the occurrence of a PSE and the high-grade/stage of the disease in patients bearing aggressive CAFs. The occurrence of PSEs in endometrial patients bearing both aggressive and non-aggressive CAFs was obtained from the EMR in accordance with the IRB approval of the Avera Cancer Institute. Clinical events during the post-surgery follow-up period were recorded along with the treatment(s) received by the patient. The PSEs included (1) the metastatic recurrence of the disease as evidenced by a CT scan or biopsy following vaginal spotting/bleeding, (2) worsening symptoms with a moderate volume of ascites and peritoneal nodularity, (3) death following hematuria and worsening peripheral edema.

4.10. Testing the Effect of Aggressive TCAFs from Patients with Post-Surgery Events on the 3D Clonogenic Growth of Endometrial Tumor Cells

We tested the effect of an aggressive TCAF derived from the tumor sample of a patient who presented a PSE. For example, a patient with lymph node-negative grade 1, stage I disease developed squamous cell carcinoma of the skin. A HyCC was performed using a dual format (96-well plates and 48-well plates) of the patient-derived primary TCAF (Figure S1). The endometrial tumor cells, AN3CA, were used as the tumor cell co-cultured on the TCAF. The growth patterns of the TCAF and tumor cells were also separately compared in the presence or absence of paclitaxel in parallel to the HyCC. TCAFs and tumor cells were characterized by flow cytometry before the set-up of the HyCC culture and after DiI and Di-O stains. The HyCC of the patient-derived primary DiO-stained TCAF was set up with DiI-stained endometrial tumor cells, AN3CA, in the presence or absence of 20 nM of paclitaxel. DiO-stained TCAFs were plated on growth factor-reduced phenol-red free matrigel. After 24 h, DiI-stained tumor cells with or without paclitaxel were plated on the DiO-stained TCAFs, and a 3D-on-Top culture was carried out. Paclitaxel's effect on the 3D growth of tumor cell colonies was recorded over 7 days in the presence or absence of TCAF. Photomicrographs of 3D colonies were captured using Olympus Fluorescence Microscope and cellSens Dimensions.

4.11. Statistical Evaluation

We determined the Pearson's correlation (95% Confidence Intervals) between the existence of aggressive primary CAFs taken from patients with a high grade (Grade 3)/high stage (Stage III/IV) of disease and the occurrence of a PSE by measuring both the strength and direction of the linear relationship between the two continuous variables.

Supplementary Materials: The following supporting information can be downloaded at: https://www.mdpi.com/article/10.3390/ijms24076449/s1.

Author Contributions: R.S.: The pathologist who provided the confirmatory evaluation of IHC staining and reviewed MS; P.D.: Senior Scientist, helped in CAFs characterization and in writing the MS; J.C.A.: Laboratory Supervisor, standardized and performed flow-cytometry; X.L.: Research Associate Lead, standardized, and performed double ICC and IHC of CAFs; A.D.: Research Associate, obtained consent from patients, performed qRT-PCR, tested correlation, and provided technical assistance in record keeping and management; K.G.: Provided insight into the overall logistical management of the study; L.R.E.: The surgeon provided clinical insight into endometrial and ovarian tumors corresponding to the blood samples; D.S.: The surgeon provided clinical insight into endometrial and ovarian tumors corresponding to the blood samples; N.D.: Senior Scientist, conceptualized and supervised the study, analyzed the data and wrote the MS. All authors have read and agreed to the published version of the manuscript.

Funding: The study was funded by Avera Cancer Institute.

Institutional Review Board Statement: Anonymized tissue samples were collected at surgery from patients with endometrial cancers following their Informed (IRB approved: Protocol Number Study: 2017.053-100399_ExVivo001) consent.

Informed Consent Statement: Informed (IRB approved: Protocol Number Study: 2017.053-100399_ExVivo001) consents for receiving resected tissue were obtained from 72 enrolled patients with endometrial cancers.

Data Availability Statement: Not applicable.

Acknowledgments: We acknowledge Avera Cancer Institute for funding the entire study. We acknowledge every patient and their family for their participation in the ex vivo study at the Avera Cancer Institute. We acknowledge PrimBio, Garnet Valley, PA, USA.

Conflicts of Interest: The authors declare no conflict of interest.

References

1. Koskas, M.; Amant, F.; Mirza, M.R.; Creutzberg, C.L. Cancer of the corpus uteri: 2021 update. *Int. J. Gynaecol. Obstet.* **2021**, *155* (Suppl. S1), 45–60. [CrossRef] [PubMed]
2. Crosbie, E.J.; Kitson, S.J.; McAlpine, J.N.; Mukhopadhyay, A.; Powell, M.E.; Singh, N. Endometrial cancer. *Lancet* **2022**, *399*, 1412–1428. [CrossRef] [PubMed]
3. Brooks, R.A.; Fleming, G.F.; Lastra, R.R.; Lee, N.K.; Moroney, J.W.; Son, C.H.; Tatebe, K.; Veneris, J.L. Current recommendations and recent progress in endometrial cancer. *CA Cancer J. Clin.* **2019**, *69*, 258–279. [CrossRef] [PubMed]
4. Connor, E.V.; Rose, P.G. Management Strategies for Recurrent Endometrial Cancer. *Expert Rev. Anticancer Ther.* **2018**, *18*, 873–885. [CrossRef] [PubMed]
5. Winterhoff, B.; Konecny, G.E. Targeting fibroblast growth factor pathways in endometrial cancer. *Curr. Probl. Cancer* **2017**, *41*, 37–47. [CrossRef] [PubMed]
6. Dzobo, K.; Dandara, C. Broadening Drug Design and Targets to Tumor Microenvironment? Cancer-Associated Fibroblast Marker Expression in Cancers and Relevance for Survival Outcomes. *OMICS* **2020**, *24*, 340–351. [CrossRef] [PubMed]
7. Pradip, D.; Jennifer, A.; Nandini, D. Cancer-Associated Fibroblasts in Conversation with Tumor Cells in Endometrial Cancers: A Partner in Crime. *Int. J. Mol. Sci.* **2021**, *22*, 9121. [CrossRef] [PubMed]
8. De, P.; Aske, J.; Dey, N. Cancer-Associated Fibroblast Functions as a Road-Block in Cancer Therapy. *Cancers* **2021**, *13*, 5246. [CrossRef] [PubMed]
9. Kennel, K.B.; Bozlar, M.; De Valk, A.F.; Greten, F.R. Cancer-associated fibroblasts in inflammation and anti-tumor immunity. *Clin. Cancer Res.* **2022**, *29*, 1009–1016. [CrossRef] [PubMed]
10. Sulaiman, R.; De, P.; Aske, J.C.; Lin, X.; Dale, A.; Vaselaar, E.; Koirala, N.; Ageton, C.; Gaster, K.; Plorde, J.; et al. A Laboratory-Friendly CTC Identification: Comparable Double-Immunocytochemistry with Triple-Immunofluorescence. *Cancers* **2022**, *14*, 2871. [CrossRef] [PubMed]
11. Hoskins, P.J.; Swenerton, K.D.; Pike, J.A.; Wong, F.; Lim, P.; Acquino-Parsons, C.; Lee, N. Paclitaxel and carboplatin, alone or with irradiation, in advanced or recurrent endometrial cancer: A phase II study. *J. Clin. Oncol.* **2001**, *19*, 4048–4053. [CrossRef] [PubMed]
12. Makker, V.; Colombo, N.; Casado Herraez, A.; Santin, A.D.; Colomba, E.; Miller, D.S.; Fujiwara, K.; Pignata, S.; Baron-Hay, S.; Ray-Coquard, I.; et al. Lenvatinib plus Pembrolizumab for Advanced Endometrial Cancer. *N. Engl. J. Med.* **2022**, *386*, 437–448. [CrossRef] [PubMed]
13. Kailasam, A.; Langstraat, C. Contemporary Use of Hormonal Therapy in Endometrial Cancer: A Literature Review. *Curr. Treat. Options Oncol.* **2022**, *23*, 1818–1828. [CrossRef] [PubMed]

14. Kalampokas, E.; Giannis, G.; Kalampokas, T.; Papathanasiou, A.A.; Mitsopoulou, D.; Tsironi, E.; Triantafyllidou, O.; Gurumurthy, M.; Parkin, D.E.; Cairns, M.; et al. Current Approaches to the Management of Patients with Endometrial Cancer. *Cancers* **2022**, *14*, 4500. [CrossRef] [PubMed]
15. De, P.; Carlson, J.H.; Jepperson, T.; Willis, S.; Leyland-Jones, B.; Dey, N. RAC1 GTP-ase signals Wnt-beta-catenin pathway mediated integrin-directed metastasis-associated tumor cell phenotypes in triple negative breast cancers. *Oncotarget* **2017**, *8*, 3072–3103. [CrossRef] [PubMed]

Disclaimer/Publisher's Note: The statements, opinions and data contained in all publications are solely those of the individual author(s) and contributor(s) and not of MDPI and/or the editor(s). MDPI and/or the editor(s) disclaim responsibility for any injury to people or property resulting from any ideas, methods, instructions or products referred to in the content.

Review

Association between Hepatocellular Carcinoma Recurrence and Graft Size in Living Donor Liver Transplantation: A Systematic Review

Alessandro Parente [1,2], Hwui-Dong Cho [2], Ki-Hun Kim [2] and Andrea Schlegel [3,4,*]

1. HPB and Transplant Unit, Department of Surgical Science, University of Rome Tor Vergata, 00133 Rome, Italy
2. Division of Hepatobiliary and Liver Transplantation, Department of Surgery, Asan Medical Center, University of Ulsan College of Medicine, Seoul 05505, Republic of Korea
3. Fondazione IRCCS Ca' Granda, Ospedale Maggiore Policlinico, Centre of Preclinical Research, 20122 Milan, Italy
4. Department of Surgery and Transplantation, Swiss HPB Centre, University Hospital Zurich, 8091 Zurich, Switzerland
* Correspondence: schlegel.andrea@outlook.de

Abstract: The aim of this work was to assess the association between graft-to-recipient weight ratio (GRWR) in adult-to-adult living donor liver transplantation (LDLT) and hepatocellular carcinoma (HCC) recurrence. A search of the MEDLINE and EMBASE databases was performed until December 2022 for studies comparing different GRWRs in the prognosis of HCC recipients in LDLT. Data were pooled to evaluate 1- and 3-year survival rates. We identified three studies, including a total of 782 patients (168 GRWR < 0.8 vs. 614 GRWR ≥ 0.8%). The pooled overall survival was 85% and 77% at one year and 90% and 83% at three years for GRWR < 0.8 and GRWR ≥ 0.8, respectively. The largest series found that, in patients within Milan criteria, the GRWR was not associated with lower oncological outcomes. However, patients with HCC outside the Milan criteria with a GRWR < 0.8% had lower survival and higher tumor recurrence rates. The GRWR < 0.8% appears to be associated with lower survival rates in HCC recipients, particularly for candidates with tumors outside established HCC criteria. Although the data are scarce, the results of this study suggest that considering the individual GRWR not only as risk factor for small-for-size-syndrome but also as contributor to HCC recurrence in patients undergoing LDLT would be beneficial. Novel perfusion technologies and pharmacological interventions may contribute to improving outcomes.

Keywords: living donor liver transplantation; hepatocellular carcinoma; graft-to-recipient weight ratio; graft size; tumor recurrence

1. Introduction

Hepatocellular carcinoma (HCC) is the most common primary liver cancer with increasing incidence and estimated death rate of 55% until 2040 [1,2]. Since the development of the Milan criteria in 1996, liver transplantation (LT) has been accepted as main curative liver cancer treatment [3]. In countries with no or limited access to a deceased donor organ pool for various reasons, living donor liver transplantation (LDLT) has been developed with excellent outcomes for HCC recipients.

Some early reports demonstrated advantages for LDLT over deceased donation, when considering patients with HCC, findings that were further supported by several prospective intention-to-treat analyses, where the overall recipient survival was comparable between living and deceased liver recipients [4,5]. A few other studies made such findings with even better recipient survival rates after LDLT, indicating that liver regeneration processes might be of limited relevance for HCC recurrence [6,7].

Another inherent advantage appears with the shorter waiting times for candidates where a living donor is available. Goldaracena et al. demonstrated that patients who had a potential live donor at the time of listing had a higher survival rate [6]. Interestingly, the authors found that waiting times of 9–12 months or ≥12 months were predictors of death [6]. In another study, Lai et al. showed that that having a potential live donor graft could decrease the intention-to-treat risk of death in patients with HCC who are on a waiting list for a liver transplant, due to fewer dropouts from the waiting list even in centers where both live donation and decease donation options are equally available [8].

Despite such results and the general opinion that LDLT imposes a lower risk and better outcomes in liver recipients, some other authors have also reported controversial data. Fisher et al. presented inferior oncological outcomes after LDLT compared to deceased donor transplantation, as patients undergoing LDLT had higher HCC recurrence rates within 3 years [9]. Such findings were later supported by another analysis, where authors observed a higher recurrence after LDLT, which could have been due to different tumor characteristics and HCC management before transplantation [10]. In 2013, Gant et al. provided a systematic literature review and meta-analysis and reported an inferior disease-free survival (DFS) after LDLT when compared to deceased donation [11]. The literature underlines the need for and improved study design and reporting in future trials to understand if the observed DFS difference might be attributed to a study bias or are the result of other contributors, as seen in context of LDLT [11].

The overall controversial clinical results also point to confounders, other than recipient cancer risk, that may have impact on recurrence rates. Donor risk factors (i.e., prolonged warm or cold ischemia), surgical parameters and stress (duration of transplantation, medical management) and the recipient risk (medical fitness, lab MELD score) are just a few adjunct variables, which add to the already evident recurrence risk conveyed by the candidate's liver cancer status and biology [12,13]. Such parameters may directly impact on the liver tissue quality and subsequently on the ability of the new liver to recover and regenerate after transplantation [14].

Despite the generally better quality of living donor liver grafts, compared to the deceased donor pool, the overall strategy is to protect the donors and to avoid the development of small-for-size-syndrome (SFSS). Per definition the development of a SFSS represents a situation where a small graft shows a primary dysfunction within the first postoperative week after transplantation without any sign of other pathologies (i.e., vascular complications, bile leak, sepsis) [15]. While the majority believe in the distinct clinical entity of a SFSS, the ongoing debate on underlying mechanisms involves portal hyper-perfusion along with a form of outflow obstruction [16]. This terminology appears to be somewhat misleading because the graft does not necessarily need to be small; SFSS-features can also occur when the liver tissue quality is impaired (i.e., steatosis) or when a partial liver is exposed to portal hypertension, as often seen in candidates with advanced liver disease staged as Child–Pugh grade C [16].

Additional risk factors may contribute to the observed SFSS, including the transplanted liver graft volume and the cytokine release triggered by both liver transection during donation surgery and after reperfusion [17]. The mechanistic link between an advanced hepatic ischemia-reperfusion injury (IRI) and liver tumor regrowth and metastasis was previously demonstrated in 2007 by Man et al., who described higher HCC recurrence rates and more lung metastases when a small liver remnant was evident [17,18].

Based on the above-mentioned concerns, the aim of this study was to perform a systematic literature review of all studies that reported data on oncological outcomes after LDLT for HCC and provide details on the association between liver graft rate recipient weight ratio (GRWR) and oncological outcomes.

2. Results

2.1. Baseline Characteristics and Demographics

A total of 45 papers were identified and screened. In 16 studies, the vast majority of authors reported their survival and SFSS-rates after LDLT using a cut-off for GRWR at <0.8% vs. ≥0.8% (Table 1) [19–34]. Most studies report a lower DFS when the liver volume is lower (i.e., GRWR < 0.8) compared to higher volumes. Centers from specific countries in Asia, including Japan, Hong Kong and India (Table 1), report the routine transplantation of living donor grafts with a smaller GRWR of <0.7 or even <0.6 in a very selective recipient cohort avoiding any additional risk factor, due to the higher risk of developing a SFSS. Despite the good overall pool of available data on the topic explored here, the number of studies transparently demonstrating the HCC risk factors together with liver GRWRs and acceptable follow-up duration after LDLT is very few.

The included studies reported a total of 782 patients; 168 underwent LDLT with a GRWR of <0.8%. Conversely, 614 were recipients, who received a liver with a GRWR of ≥0.8%. All patients were diagnosed with an HCC.

The mean donor age ranged between 26 and 33 years. The recipient age was found to be between 47 and 54 years, with the majority being male (range 76–98%), in the overall cohort. The baseline characteristics are summarized in Table 2.

From this pool of literature, three studies met the inclusion criteria (Figure 1). Such studies were all retrospective and published between 2007 and 2018 [35–37].

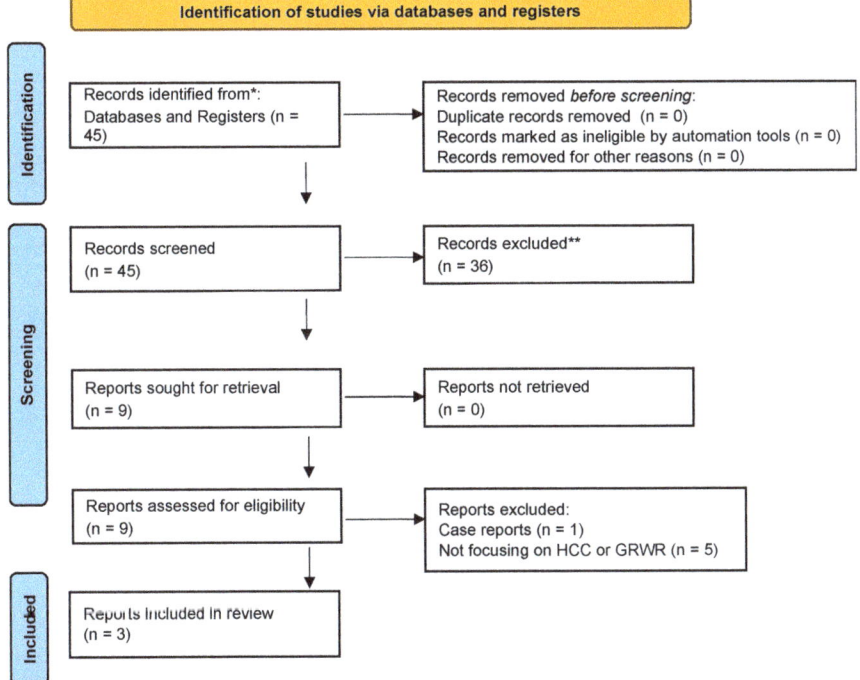

Figure 1. PRISMA flowchart. * MEDLINE, EMBASE and Web of Science; ** Sixteen studies were used to give an overview of survival and SFSS rates after LDLT using a cut-off for GRWR at <0.8 % vs. ≥0.8 %. SFSS: small for size syndrome.

2.2. Graft Characteristics

Three studies reported the graft types [35–37]. In the GRWR ≥ 0.8% cohort, there were 554 right lobes, 19 left lobes and 30 dual grafts. Conversely, there were 141 right lobes, 23 left lobes and 3 dual lobes in the GRWR < 0.8 group. The mean graft weight was reported only by one study, and it resulted in 680 (615–743) gram and 555 (500–607.5) gram liver graft weight for the two groups, GRWR ≥ 0.8% and GRWR < 0.8%, respectively [36]. Most studies [35,36] did not specify which liver lobe was used for LDLT (left: LLG (H1234) or right: RLG (H5678) [38]. The strategy for the management of the middle hepatic vein is not discussed; therefore, we have added the new terminology using LLG (H1234) or right: RLG (H5678) [38].

Lee et al. [37] exclusively transplanted right liver lobes (RLG: H5678) and reported that any middle hepatic vein branches of >5 mm in diameter were saved and underwent reconstruction. If the inferior right hepatic vein was >5 mm in size measured after right hepatic vein anastomosis, the inferior right hepatic vein was also anastomosed to the inferior vena cava.

2.3. Tumor Characteristics

One study [36] reported the overall number of locoregional treatments before transplantation, reporting a higher rate (n = 108/239, 45%) for GRWR $\geq$ 0.8% in contrast to GRWR < 0.8% (n = 16/56, 28%). Vascular invasion was present in the majority of patients in both groups, as reported by two studies and ranged between 42–78%. The mean number of tumor nodules was one in two studies (Table 3). Preoperative alpha-feto protein (AFP) was reported by the same two studies and ranged between 13 and 231 ng/mL in the entire cohort. The accepted tumor criteria for liver transplantation were also different. Two studies [35,37] used Milan criteria and reported a number of 256/375 (68.2%) and 75/112 (66.9%) recipients inside Milan criteria for GRWR $\geq$ 0.8% and GRWR < 0.8%, respectively. The third study [36] focused on the Hangzhou criteria and the authors reported 134 (65.6%) and 17 (50%) patients within criteria for GRWR $\geq$ 0.8% and GRWR < 0.8%, respectively. The graft and tumor characteristics are summarized in Table 3.

2.4. Risk of Bias Assessment

The risk of bias assessment of the included studies is demonstrated in Table 4. Overall, the two studies were deemed to be of good quality given the scores from NOS system. On the other hand, one study was deemed to be fair quality based on a score of six from the NOS system.

Table 1. Literature overview on outcomes according to accepted liver volume cutoffs for living donor liver transplantation.

Authors, Reference	Year	Country	Study Type	No. of Livers	Lowest Liver Volume				Intermediate Range				Highest Liver Volume			
					GRWR (%)	SFSS Rate (%)	Portal Inflow Modulation	Graft Loss (%)	GRWR (%)	SFSS Rate (%)	Portal Inflow Modulation	Graft Loss (%)	GRWR (%)	SFSS Rate (%)	Portal Inflow Modulation	Graft Loss (%)
Lee et al. [19]	2004	South Korea	Retrospective	79	<0.8	NA	NA	45.4%[3]	-	-	-	-	≥0.8	NA	NA	22.1%[4]
Troisi et al. [20]	2005	Belgium	Retrospective	13	<0.8		HPCS	NA ##	-	-	-	-	≥0.8			NA ##
Selzner et al. [21]	2009	Canada	Retrospective	337	0.59–0.79	9%	Splenectomy	11[1]	-	-	-	-	≥0.8	2.5%	Splenectomy	9[2]
Moon et al. [22]	2010	South Korea	Retrospective	427	<0.8	5.77%	NA	12.2 *	-	-	-	-	≥0.8	3.6%	NA	9.3 §
Chen et al. [23]	2014	Taiwan	Retrospective	196	<0.8	15.5%	NA	17.8%	-	-	-	-	≥0.8	5.9%	NA	18.6%
Vasavada et al. [24]	2014	Taiwan	Retrospective	156	<0.8	NA	SAL/Splenectomy	NA	-	-	-	-	≥0.8	NA	No	NA
Alim et al. [25]	2016	Turkey	Retrospective	649	<0.8	7%	SAL/Splenectomy	7%	-	-	-	-	>0.8	11%	None	11
Shoreem et al. [26]	2017	Egypt	Retrospective	174	<0.8	9.8%	Splenectomy	NA	≥0.8–<1	25%	Splenectomy	NA	≥1.0	25%	Splenectomy	NA
Kim et al. [27]	2018	South Korea	Retrospective	60	<0.8	0%	NA	15%	-	-	-	-	≥0.8	8.3%	NA	8.7%
Goja et al. [28]	2018	India	Retrospective	665	0.55–0.69	10.2%	HPCS/SAL	NA	0.7–0.8	5.4%	SAL	NA	>0.8	1.2%	No	NA
Sethi et al. [29]	2018	India	Retrospective	200	<0.8	12.1%	NA	15.5%[5]	-	-	-	-	≥0.8	7.04%	NA	22.8%[5]
Bell et al. [30]	2018	NA	Meta-analysis	1833	<0.8	10%	NA	NA	-	-	-	-	≥0.8	5%	NA	NA
Soin et al. [31]	2019	India	Retrospective	1321	0.54–0.69	4.2%	HPCS/SAL	18%	0.7–0.79	2.3%	HPCS/SAL	17%	≥0.8	2.4%	No	14%
Kusakabe et al. [32]	2021	Japan	Retrospective	417	<0.6	20%	Splenectomy	46%[6]	0.6–≤0.8	20.4%	Splenectomy	21.2%[6]	>0.8	10.7%	Splenectomy	22.8%[6]
Jo et al. [33]	2022	South Korea	Retrospective	118	<0.8	27.8%	SAL	13.8%	-	-	-	-	≥0.8	2%	SA ligation	4%
Wong et al. [34]	2022	Hong Kong	Retrospective	545	≤0.6	12.8%			0.6–≤0.8	13.2%			>0.8	0%		

Troisi et al. compared GRWR 0.8 with or without porto-caval hemi-transposition, no clear values were therefore available. [1]—this is extrapolated from "Graft survival at one year 89%"; [2]—this is extrapolated from "Graft survival at one year 91%"; * this is extrapolated from "Graft survival at one year 87.8%"; § this is extrapolated from "Graft survival at one year 90.7%"; [3]—this is extrapolated from "Graft survival at one year 54.6%"; [4]—this is extrapolated from "Graft survival at one year 77.9%"; [5]—please note this is 90 days mortality, one year not reported.; [6]—extrapolated from the KM curve and censored; in experienced centers the GRWR can go down to 0.6% and/or GV/SLV ≥ 30% in recipients with low MELD and no significant portal hypertension. GRWR: graft recipient weight ratio; HPCS: hemi-portocaval shunt NA: not available; SAL: splenic artery ligation. SFSS: small-for-size.

Table 2. Characteristics of the included studies and patients.

Study	Year	Country	Study Design	Study Period	Number of Patients	Number of Patients with GRWR ≥ 0.8/GRWR < 0.8	Donor Mean Age		Recipient Mean Age		Female n (%)		Lab MELD	
							GRWR ≥ 0.8	GRWR < 0.8	GRWR ≥ 0.8	GRWR < 0.8	GRWR ≥ 0.8	GRWR < 0.8	GRWR ≥ 0.8	GRWR < 0.8
Hwang et al. [35]	2007	South Korea	Retrospective	August 1992–December 2004	159	129/30	NA	NA	51.5 ± 6.1	50.3 ± 6.8	25 (19.3)	7 (23.3)	17.7 ± 8.8	16.5 ± 6.4
Hu et al. [36]	2016	China	Retrospective	January 2007–December 2009	295	239/56	26 (23.3–36.5)	28.4 (22.9–41.3)	48.6 (43–54.4)	47.3 (42.7–52.5)	23 (9.6)	1 (1.79)	NA	NA
Lee et al. [37]	2018	South Korea	Retrospective	January 2005–December 2015	328	246/82	33.4 ± 11.7	33.4 ± 12.6	54.7 ± 7.7	53.9 ± 7.6	48 (19.5)	10 (12.2)	11 (9–16)	11 (9–14)

Values are reported as mean ± standard deviation or median (interquartile range) as reported in the original manuscript. NA: not available.

Table 3. Graft and tumor characteristics.

Study (First Author [Ref])	Previous Treatment °		Vascular Invasion		Tumor Size (cm)		No. of Nodules		T Stage		Preop. AFP		Graft Type		Graft Weight (g)		Operation Time *		Blood Loss (mL)	
	GRWR ≥ 0.8	GRWR < 0.8	GRWR ≥ 0.8	GRWR < 0.8	GRWR ≥ 0.8	GRWR < 0.8	GRWR ≥ 0.8	GRWR < 0.8	GRWR ≥ 0.8	GRWR < 0.8	GRWR ≥ 0.8	GRWR < 0.8	GRWR ≥ 0.8	GRWR < 0.8	GRWR ≥ 0.8	GRWR < 0.8	GRWR ≥ 0.8	GRWR < 0.8	GRWR ≥ 0.8	GRWR < 0.8
Hwang et al. [35]	NA	NA	NA	NA	NA	NA	NA	NA	NA	NA	NA	NA	Right Lobe: 85 (H5678) Left lobe: 15 (H1234) Dual: 29	Right Lobe: 7 (H5678) Left lobe: 20 (H1234) Dual: 3	NA	NA	NA	NA	NA	NA
Hu et al. [36]	108 (45%)	16 (28%)	184 (77%)	44 (78.5%)	3.5 (2.4–5.5)	4 (2.6–5)	1 (1–3)	1 (1–2.5)	NA	NA	107.7 (9.2–1000)	231.6 (9.7–1210)	Right Lobe: 223 (H5678) Left lobe: 4 (H1234) Dual: 1	Right Lobe: 52 (H5678) Left lobe: 3 (H1234) Dual: 0	680 (615–743)	555 (500–607.5)	10.3 (8–12.5)	10.8 (9.5–12.85)	1800 (1000–3000)	2000 (1000–3000)
Lee et al. [37]	NA	NA	104 (42.2%)	41 (50%)	2.2 (1.5–3.5)	2.3 (1.6–3.5)	1 (1–3)	1 (1–2)	I—62 (25.2) II—141 (57.3) III—39 (15.9) IV—4 (1.6)	I—15 (18.3) II—50 (61.0) III—15 (18.3) IV—2 (2.4)	13 (5.1–117.8)	9.6 (3.7–150.4)	Right Lobe: 246 (H5678)	Right Lobe: 82 (H5678)	NA	NA	447.8 ± 105.1	446.8 ± 121.2	1550 (800–3500)	1500 (800–3000)

Values are reported as mean ± standard deviation or median with interquartile range as reported in the original manuscript. ° Includes previous partial hepatectomy (9 cases for GRWR ≥ 0.8; 1 case for GRWR < 0.8), RFA (14 cases for GRWR ≥ 0.8; 2 cases for GRWR < 0.8), trans-arterial chemoembolization (64 cases for GRWR ≥ 0.8; 2 cases for GRWR < 0.8), percutaneous ethanol injection (1 case for GRWR ≥ 0.8; 11 cases for GRWR < 0.8), combined treatments (17 cases for GRWR ≥ 0.8; 2 cases for GRWR < 0.8); * Operation time is reported in hours for Hu et al. and in minutes for Lee et al. AFP: alpha-fetoprotein. NA: not available.

Table 4. Summary of risk of bias assessment using NOS system.

Study (Year)	Number of Stars			
	Selection *	Comparability #	Outcome °	Overall
Hwang 2007	3	2	2	7/9
Hu 2016	3	1	2	6/9
Lee 2018	3	2	2	7/9

* Maximum 4 stars; # Maximum 2 stars; ° Maximum 3 stars.

2.5. Outcomes Analysis

2.5.1. Overall Survival

One study [37] reported the 1-, 3- and 5-year overall survival (OS) rates, which were 87.8%, 80.3% and 78.7%, respectively, for patients with GRWR < 0.8%, and 93.5%, 87.1% and 84.1%, respectively, for patients with GRWR ≥ 0.8%. The other survival rates were extrapolated and are merged in Figure 2.

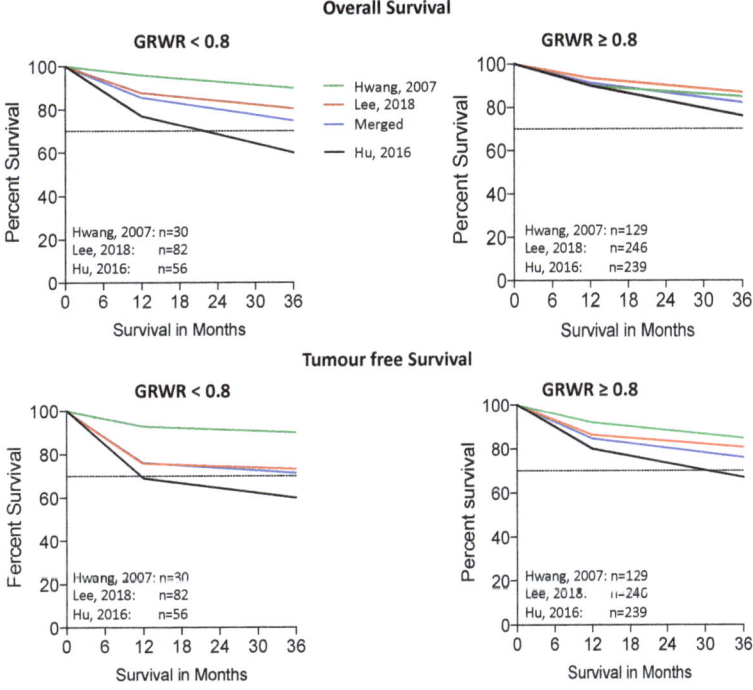

Figure 2. Composite Kaplan–Meier curve plot of overall and HCC-recurrence free survival.

2.5.2. Disease-Free Survival

One study reported [37] the 1-, 3- and 5-year disease-free survival rates which were 75.9%, 73.3% and 71.7%, respectively, for patients with GRWR < 0.8%, and 86.4%, 80.8% and 77.9%, respectively, for patients with GRWR ≥ 0.8%. The other survival rates were extrapolated and are merged in Figure 2.

3. Discussion

This systematic review focuses on patients undergoing living donor liver transplantation (LDLT) for hepatocellular carcinoma (HCC). The overall number of suitable studies that transparently report the graft volume together with the candidate's HCC cancer status with enough follow-up is limited to three. LDLT using small grafts with a graft recipient weight ratio (GRWR) of <0.8% was associated with lower overall and particularly tumor-free survival rates in patients with higher tumor burden outside Milan criteria. Although there were only three studies with a relatively small number of cases, this systematic review points to an outcome benefit considering donor liver volume in HCC recipients, particularly in candidates outside standard tumor criteria.

LDLT has emerged as standard treatment, especially in most Asian countries where the deceased donor transplant program is limited or not active [39]. LDLT offers the advantages of an elective procedure which can be planned in advance, and a promptly available graft, thus decreasing waitlist times, a key factor for better survival, particularly in candidates with liver cancers or metastases. However, some authors have shown that LDLT may increase recurrence rates with impaired disease-free survivals compared DDLT [40]. In fact, reducing waitlist times does not allow the prolonged evaluation of the tumor biology and thus prevent the selection of those patients, who would drop out in the waiting list. Patients that receive a selective LDLT may therefore have a more aggressive tumor biology, which could be the reason why in some studies LDLT was found to induce impaired survival and early HCC recurrence rates. In cohort studies, where LDLT and deceased donor transplants are well-matched for recipient tumor features and overall risk, comparable results are described in centers with experience in both techniques [41]. Additionally, shorter waiting times in context of LDLT are often beneficial in preventing the risk of rapid tumor progression with better overall and DFS rates.

Available intention-to-treat analyses have demonstrated a survival advantage for living donor liver recipients over DDLT [5,8,42,43]. Our study confirmed that most of the candidates benefit from LDLT, provided their HCC criteria are within standard profiles. However, a thorough graft selection for patients with an advanced tumor burden should be considered, particularly in cases where the calculated GRWR is expected to be <0.8%, because this sub-cohort is exposed to an elevated risk for impaired overall and tumor-free recipient survival rates as shown in the largest report [37]. Survival rates were substantially different comparing the here included studies. One possible explanation could be that most living donor grafts in the study by Hwang et al. [35] that had a GRWR of <0.8% were left lobes (H1234). In contrast, the majority of livers in the other studies were right lobes (H5678) [36,37]. Available studies assessing the impact of right or left living donor lobes on the development of SFSS or HCC recurrence are scarce. Additionally, it should be highlighted that the study of Hwang et al. was conducted between 1992 and 2004; therefore, the authors might have been careful in selecting donors and recipients; the medical management was probably different and surgical techniques have advanced since.

Although a GRWR of >0.8% is preferable, levels GRWR between of $\geq 0.6\%$ and 0.8% can be acceptable if an alternative living donor is lacking, for primary transplants of recipients with maximal one organ failure. For this approach, both the donor and recipient should consent to accept the higher risk [34] and an independent, multi-disciplinary team for LDLT should be involved. The calculation of the GRWR is generally based on liver graft weight and recipient weight, which can be challenging due to the known recipient edema, ascites and pleural fluid in contexts of advanced liver cirrhosis. Such clinical features may lead to a false high body weight, requiring a false high liver volume, protecting the recipient and potentially exposing the donor to a higher risk [44]. Despite the findings in this systematic review that suggest the importance of being cautious with low liver volumes to avoid higher HCC recurrence rates, donor safety remains the highest priority. Various factors have been proposed to contribute to the development of SFSS, including the venous congestion, portal hypertension along with arterial spasms and hypoperfusion. In addition, an insufficient liver mass and metabolically impaired tissue quality contribute further [16]. Small grafts (GRWR < 0.8) demonstrated significantly higher portal venous pressures (PVP) in different case series [45,46]. In such series, SFSS-grafts demonstrated elevated PVP for up to 2 weeks after transplantation, which was in contrast to liver grafts with appropriate volumes. The PVP of >20 mmHg was found to be associated with an impaired graft survival after LDLT. Also of interest are experimental studies with the transplantation of small partial liver grafts [47–49]. An elevated PVP in the early posttransplant phase correlated with rapid liver hypertrophy along with low portal blood levels of VEGF and elevated peripheral HGF values. These features were predictive for graft dysfunction (with coagulopathy, ascites, hyperbilirubinemia) and an overall poor outcome (Figure 3). These studies did, however, also demonstrate that an adequate elevation of portal venous pressure and flow is beneficial to trigger liver regeneration.

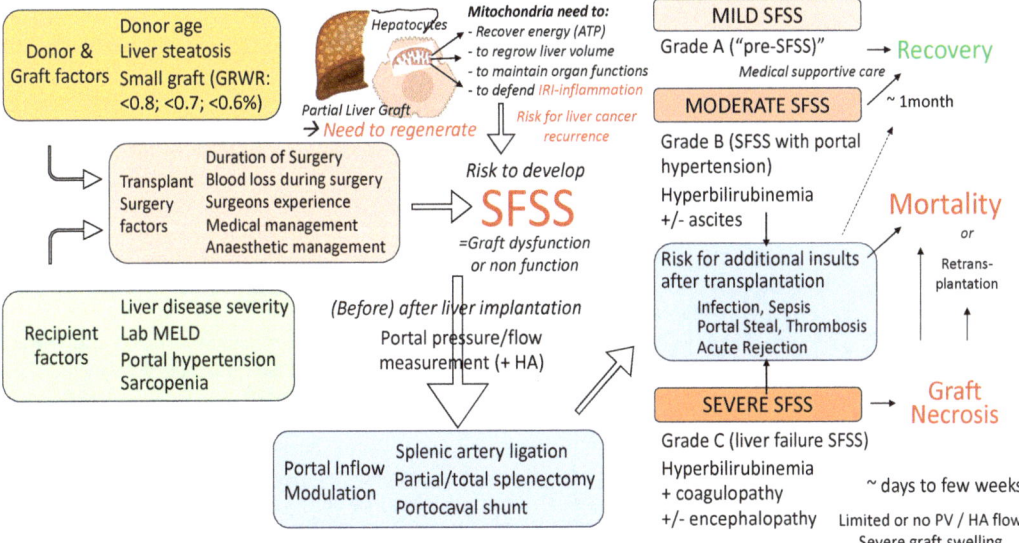

Figure 3. Multiple factors contributing to the development of a small for size syndrome (SFSS) after living donor liver transplantation: Overview on development, risk factors and outcomes of SFSS. In experienced centers, the GRWR can go down to 0.6% and/or GV/SLV $\geq$ 30% in recipients with low MELD and no significant portal hypertension, related to liver tissue quality. Disturbed blood flows along with elevated levels of IRI lead to an ongoing inflammation, creating a microenvironment that promotes the resettlement and replication of circulating cancer cells, further pronounced by pro-regenerative molecules released to promote liver repair and regeneration.

The relevance of blood flow and shear stress is well-described as trigger of endothelial cell dysfunction in different vessels. Disturbed blood flows with subsequent reduced shear stress at the endothelial cells, as seen with relevant portal hypertension in small liver grafts, may significantly impact mitochondrial function and increase mitochondrial ROS release with inflammation, thereby blocking mitochondria from their essential task to produce ATP—the key molecule needed for the full metabolic function of a liver, which can quickly regenerate without severe SFSS [50].

Based on the increasing surgical experience during the past few decades, various surgical techniques to modulate the liver inflow now exist, including splenic artery ligation, splenectomy or portocaval shunts. Such surgical steps are often required when the GRWR is below 0.7%. Intraoperative decision making is based on the portal vein and hepatic artery flow and pressure measurements (Figure 3). Other factors including the level of portal hypertension play an additional role intraoperatively, but also during the planning of the initial strategy in context for the estimated GRWR when donor and recipient are matched before transplantation [51]. In addition to the best possible inflow modulation, the optimal liver outflow reconstruction includes the creation of large anastomoses to the cava vein directly or through interposition grafts. Most surgeons consider the diameter of >5 mm for hepatic vein branches as relevant to perform an additional reconstruction.

The Cleveland group has recently demonstrated that an augmented graft outflow reconstruction (i.e., all three recipient hepatic veins used for the outflow of LLG-LDLT) together with meticulous PVP/flow measurement and inflow modulation whenever appropriate (before or after graft implantation) can lead to excellent and comparable results after LDLT using small grafts (GRWR < 0.7) [51].

Although living donor grafts are carefully selected and of optimal tissue quality, cells are subjected to a few more minutes warm ischemia at the time of recovery. Considering that such grafts undergo several blood-inflow clamping cycles (Pringle maneuver) during the transection of the donor liver, which activates the IRI-cascade, there is always a level of IRI seen in the recipient after transplantation. In fact, any intervention with subsequent tissue trauma, including surgery may trigger hypoxia and inflammation. The period of hypoxia (i.e., during pringle in an entire lobe or more regional during liver positioning for a specific preparation during partial liver resection) leads to an anaerobic metabolism in mitochondria with dys- or nonfunction of the respiratory chain and subsequent accumulation of succinate and other metabolites [52]. At reperfusion with the reintroduction of oxygen (i.e., after liver graft implantation, or removal of the pringle or repositioning of the lobe) in mitochondrial complex I, reactive oxygen species (ROS) are released, triggered by the level of accumulated succinate.

These ROS release initiates further downstream sterile inflammation and with damage of the mitochondrial and cellular wall thereby releasing further molecules including danger-associated-molecular patters (DAMPS, i.e., Hmgb1, mtDNA, ATP) and various pro-inflammatory cytokines (i.e., TNF-α, IL-1b, IL-18) [12]. An ongoing sterile tissue inflammation also involves the macrophages and endothelial cells, which trigger another wave of inflammation when in contact with patient or recipient blood components such as neutrophils. Further ROS, DAMPS and cytokines are released and the entire cascade of IRI inflammation is maintained. An increasing body of literature exists demonstrating the link between mitochondrial injury and function, the release of ROS and mitochondrial DNA and features of ongoing inflammation in the liver periphery. Particularly, this microenvironment was found to promote the resettling, migration and growth of circulating cancer cells from candidates with typical liver tumors, including HCC and cholangiocarcinoma. A favorable tumor biology with less active and less aggressive HCC-types are therefore beneficial and results in better survivals together with the best possible liver preservation and surgical techniques that are performed by experienced centers triggering the least possible inflammation [12].

The ROS–DAMPS–cytokine cascade instigates further downstream effects in the liver microenvironment, enabling the migration of circulating tumor cells through the sinusoidal endothelial cells (SEC) barrier. In fact, during inflammation, SECs can swell and lose their integrity, thus increasing their permeability. Consequently, tumor cells can migrate, invade and grow. In addition to these mechanisms, hypoxia triggers the release of the hypoxia-inducible-factor, which promotes tumor cell proliferation, migration and angiogenesis, with secretion of vascular endothelial growth factor. Such molecular pathways are of high relevance in liver transplantation, in particular when additional donor and recipient risk factors are involved, such as marginal donors or smaller grafts.

Next to the maintenance of liver function and recovery from IRI-injury, mitochondria are essential components to produce enough energy for liver regeneration. Impaired mitochondria trigger delays in liver regeneration particularly described in patients suffering from high inflammation after liver resection or partial transplantation. Delayed hepatocyte cell cycle passages and delayed entrance into the cell cycle were described.

The ongoing mitochondrial ROS release and chronic inflammation can also damage healthy cells and their mitochondrial DNA, which may lead to new DNA mutations, and upregulate proto-oncogenes together with a downregulation of tumor-suppressor genes, and hence, the development of new cancers [12].

Centre experience, logistics, medical recipient fitness and the duration of cold ischemia are additional factors with effect on the IRI-severity, the subsequent liver function, regeneration, and recipient recovery. Triggered by an elevated IRI-associated inflammation and related repair, tumor cell spread and growth can also be enhanced during liver transection in the donor and later regeneration [18]. In fact, IRI-based inflammation triggers local changes with the development of a favorable microenvironment for tumor cells to invade, migrate and grow.

Moreover, it has been demonstrated in experimental models that the process of liver splitting induces the release of inflammatory molecules as a result of the parenchyma transection, as observed in models of accelerated liver regeneration such as the Associating Liver Partition and Portal vein Ligation for Staged hepatectomy (ALPPS) technique [53,54]. Herein, the authors performed a quantitative polymerase chain reaction (PCR) for interleukin-6 (IL-6) tumor necrosis factor α (TNF-α) (mRNA) in these tissue samples in order to evaluate inflammatory response after ALPPS. Ultimately, these authors demonstrated that liver parenchymal transection dramatically increased the expression of early instigators of regeneration (IL-6 and TNF-α) [55].

Similarly, to the earlier concept of adding a small auxiliary liver graft, known as the concept of APOLT (auxiliary partial orthotopic liver transplantation), a technique that was initially performed based on the concept that the residual native recipient liver would support the overall function of the implanted graft until the graft that is, individually, too small would have grown enough [56].

Another interesting surgical technique to discuss is the living donor (LD) resection and partial liver segment 2–3 transplantation with delayed total hepatectomy (RAPID) procedure initially proposed by Königsrainer et al. [57]. Although experience in HCC candidates is scarce [58], an interesting review article describes 8 LD-RAPID procedures performed in Europe [59]. A clinical bi-institutional prospective study is currently ongoing in Germany to evaluate this technique further for patients with non-resectable colorectal liver metastases. The primary endpoint is a 36-month overall survival (NCT03488953) [59]. This and all other surgical techniques described above would be interesting to be tested in the context of the complex interplay between portal hypertension, liver volume, IRI and cancer recurrence.

Metabolically less favorable livers are often not easily detected just by quantifying the levels of steatosis or considering the age of the donor. Based on this, there is always a risk of developing significant IRI, contributing to the development of SFSS with impaired liver function [60]. The regeneration of the graft might be impaired despite optimal selection and LDLT conditions. This might be the reason why such complications are also observed in donor livers beyond all suggested volume thresholds (GRWR: >0.8–1%) [61–65].

Mitochondrial dysfunction and advanced inflammation in the recipient are the underlying mechanisms of and ongoing inflammation and downstream microvascular tumor spread and recurrence [12,66]. Sometimes, even a low level of ROS released from mitochondrial complex I can be sufficient to trigger this cascade [52]. The direct activation of other cells, including macrophages and endothelial cells, is the next consequence [67].

Provided that the recipient risk is given with the acceptance of the candidate for transplantation and the same is valid for an available living donor, the remaining "wheel to adjust" appears with the preservation of the organ. Early studies have shown that novel dynamic preservation technologies may reduce IRI, reprogram mitochondria and downregulate the recipient's immune response [68–70]. Two main approaches are of interest here: First, the reduction in additional injury, through a replacement of additional cold storage with normothermic machine perfusion (NMP). This technique induces an ex vivo liver IRI and inflammation which is used for viability assessment [71–74]. The optimal NMP technique is performed from the donor to the recipient center, bridging long cold storage times and the transport [74]. This technically challenging and labor-intensive procedure is tested in various clinical trials with the focus on the utilization of livers with advanced donor risk profiles [71,72,74,75].

The second main technique is the hypothermic oxygenated perfusion (HOPE). Precooled (4–10 °C) artificial fluids are recirculated in a pressure controlled system. Interestingly, the HOPE approach delivering high-perfusate oxygen levels under cold conditions was found to reprogram mitochondria and to prevent later IRI-associated inflammation after liver implantation. Various mechanistic and clinical studies exist together with four randomized controlled trials [62,76–82]. Based on the underlying mechanism of protection, this perfusion technique reduces posttransplant complications, improves graft survival and reduces re-transplantation rates [76,82]. The concept of improved peripheral sinusoidal flow and microvascular protection through the HOPE concept is also interesting [83,84]. In the context of LDLT, existing experimental studies demonstrated an improved growth and earlier entrance into the cell-cycle for quicker liver regeneration after hypothermic perfusion [85–87]. The combination of microvascular protection and better mitochondrial metabolism with more energy (ATP) levels may serve as an additional protective effect inducing better liver function and accelerated regeneration.

Although there are currently no clinical studies on the use of machine perfusion for living donor liver grafts, available experimental studies have triggered the enthusiasm to further assess the HOPE concept. Additional evidence comes from living donor kidney transplantation, where injury occurs during graft procurement which could be reversed with hypothermic perfusion, resulting in better early kidney graft function compared to standard cold storage [87]. Recently, a novel concept of "ischemia-free" organ transplantation (IFOT) has been developed by some authors [88]. The perfusion device is connected to the liver in the donor during procurement and entirely bridges the time spent between the donor and recipient, allowing the ischemic time to be minimal and avoiding cold flush and cold storage. In the first non-randomized clinical trial, the authors from China demonstrated good outcomes with low rates of allograft dysfunction [89]. This concept reduces IRI, inflammation and mitochondrial ROS and DAMPS release with subsequent potential reduction in HCC recurrence. The group from China has assessed the role of IFOT on cancer recurrence and described a reduction compared to standard cold storage [90]. Although such clinical findings support underlying mechanisms, prospective studies are required to gain more information on the role of novel preservation techniques in cancer recurrence and different scenarios with various risk factors.

The present study also highlights the importance of a careful donor selection when considering HCC recipients. Machine perfusion could potentially lead to better graft function, less IRI and downregulate immune response, which is directly linked to cancer recurrence [12,34]. The HOPE-tool could therefore increase the number of available donors for LDLT, considering the high number of steatotic livers in Asian and Arab countries, which mainly rely on LDLT for their candidates, with an ever increasing number of liver cancers.

The present study has several limitations, which should be considered when interpreting the results. The current literature lacks randomized controlled trials, the gold standard study design for comparative studies to provide high-quality evidence. Thus, the best available evidence is based on retrospective observational studies, which are inherently subject to selection bias. These studies consisted of relatively small cohorts and included patients that were treated over a prolonged period of time with more than 20 years. The long period may have introduced additional bias in the context of patient management and modified indications and surgical techniques over time. In addition, donor characteristics (e.g., family history or relation) were not reported in the included studies, which could have caused a bias when interpreting the results. Next, it is also important to highlight that two of the included studies did not report the locoregional liver cancer treatment prior to LDLT. Of note, the identification of prognostic factors appeared to be very challenging due to the lack of standardized parameter reporting. We were therefore not able to develop a formal meta-analysis model.

4. Conclusions

The present systematic review has demonstrated that in the setting of LDLT, smaller grafts (i.e., GRWR < 0.8%) could be potentially associated with lower survival rates in HCC recipients, in particular in candidates with tumors outside widely used acceptance criteria for transplantation. A careful donor-recipient selection considering the graft size and liver disease severity with level of portal hypertension for HCC recipients could improve outcomes further.

Future research should target IRI-associated inflammation to understand the relationship between graft size and quality, the ability to regenerate and to handle the relatively high portal flow and subsequent tumor recurrence in the setting of LDLT. Novel dynamic preservation techniques could be of benefit and their role should be explored further.

5. Material and Methods

5.1. Design and Study Selection

The eligibility criteria, methodology and investigated outcome parameters of the current study were underlined in a review protocol. The latter was registered at the International Prospective Register of Systematic Reviews (registration number: CRD42022384456). The methodology used in the present study respected the standards of Preferred Reporting Items for Systematic Reviews and Meta-Analyses (PRISMA) statement [91].

5.2. Eligibility Criteria and Types of Participants

All studies evaluating the outcomes of LDLT comparing the different GRWRs for any adult recipient (aged 18 or older), transplanted for HCC, were considered. Letters, expert opinions, congress abstracts, case series and case reports were excluded. In addition, pediatric transplantation for HCC cases were excluded.

5.3. Intervention and Comparison of Interest

LDLT in patients with HCC was considered as the intervention of interest and the different liver GRWRs were evaluated.

5.4. Outcomes

The inclusion criteria were studies that reported at least one of the following outcomes: 1-, 3- and 5-year recipient survival and DFS rates (oncological outcomes).

5.5. Literature Search Strategy

A comprehensive search strategy was undertaken based on thesaurus headings, search operators and limits in MEDLINE, EMBASE and Web of Science, and was conducted by two independent authors (A.P., A.S.). The last literature search was conducted on December 15th, 2022. A search algorithm that included the terms "liver transplantation" OR "liver transplant" AND "graft recipient weight ratio" OR "graft size" OR "GRWR" AND "hepatocellular carcinoma" OR "liver cancer" OR "HCC" was performed. No limits were set for the year of the publication, but the language was limited to English. Each study identified and included was searched for additional references to identify any further reports or studies of interest that might have been missed.

5.6. Selection of Studies

The assessment of the selected studies using title and abstract was conducted by two reviewers, independently (A.P., A.S.). The full texts of relevant articles were collected and evaluated with the eligibility criteria of this study.

5.7. Data Extraction and Management

An electronic data extraction spreadsheet according to Cochrane's recommendations for intervention reviews was created and was pilot-tested in randomly selected articles and adjusted accordingly. The following information was extracted from each of the included studies by two independent reviewers (A.P., A.S.) to ensure the homogeneity of data collection and to rule out any subjective influence in data collection:

- study-related data (first author, publication year, country of origin of the corresponding author, journal in which the study was published, study design, procedure performed and sample size of patients in each group);
- baseline demographic and clinical information of the study populations (donor age, gender, recipient age, recipient MELD, graft type, graft size, graft weight, previous surgical and/or oncological treatment other than transplant, criteria used for transplantation, tumor vascular invasion and GRWR);
- outcome data.

Disagreements between the two investigators during this process were resolved following iteration, discussion and consultation with a third, independent senior author (K-H.K.). Complete concordance for all variables was achieved.

5.8. Assessment of Risk of Bias

The quality of included studies was assessed using the Newcastle–Ottawa Scale (NOS) based on selection (four items), comparability (one item) and outcome (three items) [92]. A nine-star rating system (ranging from 0 to 9) in NOS was used for assessing the quality of observational studies; a study with seven or more stars was regarded as good quality. Conversely, a study with three or more stars but fewer than six was regarded as being of fair quality, whereas two or fewer stars indicated poor quality. Two investigators (A.P., A.S.) reviewed the publications, assessed the quality and extracted the data independently. Disagreements were resolved by discussion and consensus between the two investigators. If no agreement could be reached, a third independent senior author was consulted (K-H.K.). Ultimately, complete concordance was achieved.

5.9. Summary Measures, Synthesis, and Statistical Analysis

The potential role of GRWR in the prognosis of patients following LDLT was investigated using 0.8% as the cut-off value based on the reports among available studies. One reviewer (A.P.) independently entered the extracted data into the Review Manager 5.4 software for data synthesis. The entered data were subsequently checked independently by another senior author (A.S.).

Author Contributions: Conceptualization, A.P. and A.S.; methodology, A.P. and A.S.; software, A.P. and A.S.; validation, A.P. and A.S.; formal analysis, A.P. and A.S.; investigation, A.P. and A.S.; resources, A.P. and A.S.; data curation, A.P. and A.S.; writing—original draft preparation, A.P. and A.S.; writing—review and editing, A.P., H.-D.C., K.-H.K. and A.S.; visualization, A.P. and A.S.; supervision, A.P. and A.S.; project administration, A.P. and A.S. All authors have read and agreed to the published version of the manuscript.

Funding: This research received no external funding.

Institutional Review Board Statement: Considering the design of our study, ethical approval and consent were not required.

Informed Consent Statement: Not applicable.

Data Availability Statement: The data presented in this study are openly available.

Conflicts of Interest: The authors declare no conflict of interest.

Abbreviations

ALPPS: associating liver partition and portal vein ligation for staged hepatectomy; DAMPS: danger-associated-molecular patters; GRWR: graft recipient weight ratio; HCC: hepatocellular carcinoma; HMP: hypothermic machine perfusion; HOPE: hypothermic oxygenated perfusion; IRI: ischemia-reperfusion injury; LDLT: living donor liver transplantation; LT: liver transplantation; MELD: model end-stage liver disease; ROS: reactive oxygen species; SEC: sinusoidal endothelial cells. SFSS: small for size syndrome.

References

1. Rumgay, H.; Arnold, M.; Ferlay, J.; Lesi, O.; Cabasag, C.J.; Vignat, J.; Laversanne, M.; McGlynn, K.A.; Soerjomataram, I. Global Burden of Primary Liver Cancer in 2020 and Predictions to 2040. *J. Hepatol.* **2022**, *77*, 1598–1606. [CrossRef] [PubMed]
2. Siegel, R.L.; Miller, K.D.; Fuchs, H.E.; Jemal, A. Cancer Statistics, 2022. *CA Cancer J. Clin.* **2022**, *72*, 7–33. [CrossRef]
3. Mazzaferro, V.; Regalia, E.; Doci, R.; Andreola, S.; Pulvirenti, A.; Bozzetti, F.; Montalto, F.; Ammatuna, M.; Morabito, A.; Gennari, L. Liver Transplantation for the Treatment of Small Hepatocellular Carcinomas in Patients with Cirrhosis. *N. Engl. J. Med.* **1996**, *334*, 693–699. [CrossRef] [PubMed]
4. Shah, S.A.; Levy, G.A.; Greig, P.D.; Smith, R.; McGilvray, I.D.; Lilly, L.B.; Girgrah, N.; Cattral, M.S.; Grant, D.R. Reduced Mortality with Right-Lobe Living Donor Compared to Deceased-Donor Liver Transplantation When Analyzed from the Time of Listing. *Am. J. Transplant.* **2007**, *7*, 998–1002. [CrossRef] [PubMed]
5. Azoulay, D.; Audureau, E.; Bhangui, P.; Belghiti, J.; Boillot, O.; Andreani, P.; Castaing, D.; Cherqui, D.; Irtan, S.; Calmus, Y.; et al. Living or Brain-Dead Donor Liver Transplantation for Hepatocellular Carcinoma: A Multicenter, Western, Intent-to-Treat Cohort Study. *Ann. Surg.* **2017**, *266*, 1035–1044. [CrossRef]
6. Goldaracena, N.; Gorgen, A.; Doyle, A.; Hansen, B.E.; Tomiyama, K.; Zhang, W.; Ghanekar, A.; Lilly, L.; Cattral, M.; Galvin, Z.; et al. Live Donor Liver Transplantation for Patients with Hepatocellular Carcinoma Offers Increased Survival vs. Deceased Donation. *J. Hepatol.* **2019**, *70*, 666–673. [CrossRef]
7. Humar, A.; Ganesh, S.; Jorgensen, D.; Tevar, A.; Ganoza, A.; Molinari, M.; Hughes, C. Adult Living Donor Versus Deceased Donor Liver Transplant (LDLT Versus DDLT) at a Single Center: Time to Change Our Paradigm for Liver Transplant. *Ann. Surg.* **2019**, *270*, 444–451. [CrossRef]
8. Lai, Q.; Sapisochin, G.; Gorgen, A.; Vitale, A.; Halazun, K.J.; Iesari, S.; Schaefer, B.; Bhangui, P.; Mennini, G.; Wong, T.C.L.; et al. Evaluation of the Intention-to-Treat Benefit of Living Donation in Patients with Hepatocellular Carcinoma Awaiting a Liver Transplant. *JAMA Surg.* **2021**, *156*, e213112. [CrossRef]
9. Fisher, R.A.; Kulik, L.M.; Freise, C.E.; Lok, A.S.F.; Shearon, T.H.; Brown, R.S.; Ghobrial, R.M.; Fair, J.H.; Olthoff, K.M.; Kam, I.; et al. Hepatocellular Carcinoma Recurrence and Death Following Living and Deceased Donor Liver Transplantation. *Am. J. Transplant.* **2007**, *7*, 1601–1608. [CrossRef]
10. Kulik, L.M.; Fisher, R.A.; Rodrigo, D.R.; Brown, R.S.; Freise, C.E.; Shaked, A.; Everhart, J.E.; Everson, G.T.; Hong, J.C.; Hayashi, P.H.; et al. Outcomes of Living and Deceased Donor Liver Transplant Recipients with Hepatocellular Carcinoma: Results of the A2ALL Cohort. *Am. J. Transplant.* **2012**, *12*, 2997–3007. [CrossRef]
11. Grant, R.C.; Sandhu, L.; Dixon, P.R.; Greig, P.D.; Grant, D.R.; Mcgilvray, I.D. Living vs. Deceased Donor Liver Transplantation for Hepatocellular Carcinoma: A Systematic Review and Meta-Analysis. *Clin. Transplant.* **2013**, *27*, 140–147. [CrossRef] [PubMed]
12. Parente, A.; Flores Carvalho, M.; Eden, J.; Dutkowski, P.; Schlegel, A. Mitochondria and Cancer Recurrence after Liver Transplantation—What Is the Benefit of Machine Perfusion? *Int. J. Mol. Sci.* **2022**, *23*, 9747. [CrossRef] [PubMed]
13. Lai, Q.; Lesari, S.; Lerut, J.P. The Impact of Biological Features for a Better Prediction of Posttransplant Hepatocellular Cancer Recurrence. *Curr. Opin. Organ. Transplant.* **2022**, *27*, 305–311. [CrossRef]
14. Taketomi, A.; Shirabe, K.; Toshima, T.; Morita, K.; Hashimoto, N.; Kayashima, H.; Ikegami, T.; Yoshizumi, T.; Soejima, Y.; Maehara, Y. The Long-Term Outcomes of Patients with Hepatocellular Carcinoma after Living Donor Liver Transplantation: A Comparison of Right and Left Lobe Grafts. *Surg. Today* **2012**, *42*, 559–564. [CrossRef]
15. Dahm, F.; Georgiev, P.; Clavien, P.A. Small-for-size syndrome after partial liver transplantation: Definition, mechanisms of disease and clinical implications. *Am. J. Transplant.* **2005**, *5*, 2605–2610. [CrossRef]
16. Gonzalez, H.D.; Liu, Z.W.; Cashman, S.; Fusai, G.K. Small for size syndrome following living donor and split liver transplantation. *World J. Gastrointest. Surg.* **2010**, *27*, 389–394. [CrossRef]
17. Man, K.; Ng, K.T.; Lo, C.M.; Ho, J.W.; Sun, B.S.; Sun, C.K.; Lee, T.K.; Poon, R.T.P.; Fan, S.T. Ischemia-Reperfusion of Small Liver Remnant Promotes Liver Tumor Growth and Metastases—Activation of Cell Invasion and Migration Pathways. *Liver Transpl.* **2007**, *13*, 1669–1677. [CrossRef]
18. Man, K.; Lo, C.M.; Xiao, J.W.; Ng, K.T.; Sun, B.S.; Ng, I.O.; Cheng, Q.; Sun, C.K.; Fan, S.T. The Significance of Acute Phase Small-for-Size Graft Injury on Tumor Growth and Invasiveness after Liver Transplantation. *Ann. Surg.* **2008**, *247*, 1049–1057. [CrossRef]
19. Lee, H.H.; Joh, J.W.; Lee, K.W.; Kim, S.J.; Lee, D.S.; Park, J.H.; Choi, S.H.; Heo, J.S.; Hyon, W.S.; Kwak, M.S.; et al. Small-for-size graft in adult living-donor liver transplantation. *Transplant. Proc.* **2004**, *36*, 2274–2276. [CrossRef]
20. Troisi, R.; Ricciardi, S.; Smeets, P.; Petrovic, M.; Van Maele, G.; Colle, I.; Van Vlierberghe, H.; de Hemptinne, B. Effects of hemi-portocaval shunts for inflow modulation on the outcome of small-for-size grafts in living donor liver transplantation. *Am. J. Transplant.* **2005**, *5*, 1397–1404. [CrossRef]
21. Selzner, M.; Kashfi, A.; Cattral, M.S.; Selzner, N.; Greig, P.D.; Lilly, L.; McGilvray, I.D.; Therapondos, G.; Adcock, L.E.; Ghanekar, A.; et al. A graft to body weight ratio less than 0.8 does not exclude adult-to-adult right-lobe living donor liver transplantation. *Liver Transpl.* **2009**, *15*, 1776–1782. [CrossRef] [PubMed]
22. Moon, J.I.; Kwon, C.H.; Joh, J.W.; Jung, G.O.; Choi, G.S.; Park, J.B.; Kim, J.M.; Shin, M.; Kim, S.J.; Lee, S.K. Safety of small-for-size grafts in adult-to-adult living donor liver transplantation using the right lobe. *Liver Transpl.* **2010**, *16*, 864–869. [CrossRef] [PubMed]

23. Chen, P.X.; Yan, L.N.; Wang, W.T. Outcome of patients undergoing right lobe living donor liver transplantation with small-for-size grafts. *World J. Gastroenterol.* **2014**, *20*, 282–289. [CrossRef] [PubMed]
24. Vasavada, B.; Chen, C.L.; Zakaria, M. Using low graft/recipients body weight ratio graft with portal flow modulation an effective way to prevent small-for-size syndrome in living donor liver transplant: A retrospective analysis. *Exp. Clin. Transplant.* **2014**, *12*, 437–442. [PubMed]
25. Alim, A.; Erdogan, Y.; Yuzer, Y.; Tokat, Y.; Oezcelik, A. Graft-to-recipient weight ratio threshold adjusted to the model for end-stage liver disease score for living donor liver transplantation. *Liver Transpl.* **2016**, *22*, 1643–1648. [CrossRef]
26. Shoreem, H.; Gad, E.H.; Soliman, H.; Hegazy, O.; Saleh, S.; Zakaria, H.; Ayoub, E.; Kamel, Y.; Abouelella, K.; Ibrahim, T.; et al. Small for size syndrome difficult dilemma: Lessons from 10 years single centre experience in living donor liver transplantation. *World J. Hepatol.* **2017**, *28*, 930–944. [CrossRef]
27. Kim, S.H.; Lee, E.C.; Park, S.J. Impact of preserved collateral veins on small-for-size grafts in living donor liver transplantation. *Hepatol. Res.* **2018**, *48*, 295–302. [CrossRef]
28. Goja, S.; Kumar Yadav, S.; Singh Soin, A. Readdressing the Middle Hepatic Vein in Right Lobe Liver Donation: Triangle of Safety. *Liver Transpl.* **2018**, *24*, 1363–1376. [CrossRef]
29. Sethi, P.; Thillai, M.; Thankamonyamma, B.S.; Mallick, S.; Gopalakrishnan, U.; Balakrishnan, D.; Menon, R.N.; Surendran, S.; Dhar, P.; Othiyil Vayoth, S. Living Donor Liver Transplantation Using Small-for-Size Grafts: Does Size Really Matter? *J. Clin. Exp. Hepatol.* **2018**, *8*, 125–131. [CrossRef]
30. Bell, R.; Pandanaboyana, S.; Upasani, V.; Prasad, R. Impact of graft-to-recipient weight ratio on small-for-size syndrome following living donor liver transplantation. *ANZ J. Surg.* **2018**, *88*, 415–420. [CrossRef]
31. Soin, A.S.; Yadav, S.K.; Saha, S.K.; Rastogi, A.; Bhangui, P.; Srinivasan, T.; Saraf, N.; Choudhary, N.S.; Saigal, S.; Vohra, V. Is portal inflow modulation always necessary for successful utilization of small volume living donor liver grafts? *Liver Transpl.* **2019**, *25*, 1811–1821. [CrossRef] [PubMed]
32. Kusakabe, J.; Yagi, S.; Sasaki, K.; Uozumi, R.; Abe, H.; Okamura, Y.; Taura, K.; Kaido, T.; Uemoto, S. Is 0.6% Reasonable as the Minimum Requirement of the Graft-to-recipient Weight Ratio Regardless of Lobe Selection in Adult Living-donor Liver Transplantation? *Transplantation* **2021**, *105*, 2007–2017. [CrossRef] [PubMed]
33. Jo, H.S.; Yu, Y.D.; Choi, Y.J.; Kim, D.S. Left liver graft in adult-to-adult living donor liver transplantation with an optimal portal flow modulation strategy to overcome the small-for-size syndrome—A retrospective cohort study. *Int. J. Surg.* **2022**, *106*, 106953. [CrossRef] [PubMed]
34. Wong, T.C.; Fung, J.Y.Y.; Cui, T.Y.S.; Sin, S.L.; Ma, K.W.; She, B.W.H.; Chan, A.C.Y.; Chok, K.S.H.; Dai, J.W.C.; Cheung, T.T.; et al. The Risk of Going Small: Lowering GRWR and Overcoming Small-For-Size Syndrome in Adult Living Donor Liver Transplantation. *Ann. Surg.* **2021**, *274*, e1260–e1268. [CrossRef] [PubMed]
35. Hwang, S.; Lee, S.G.; Ahn, C.S.; Kim, K.H.; Moon, D.B.; Ha, T.Y.; Park, K.M.; Song, G.W.; Jung, D.H.; Kim, B.S.; et al. Small-sized liver graft does not increase the risk of hepatocellular carcinoma recurrence after living donor liver transplantation. *Transplant. Proc.* **2007**, *39*, 1526–1529. [CrossRef]
36. Hu, Z.; Zhong, X.; Zhou, J.; Xiang, J.; Li, Z.; Zhang, M.; Wu, J.; Jiang, W.; Zheng, S. Smaller grafts do not imply early recurrence in recipients transplanted for hepatocellular carcinoma: A Chinese ex- perience. *Sci. Rep.* **2016**, *26*, 26487. [CrossRef]
37. Lee, E.C.; Kim, S.H.; Shim, J.R.; Park, S.J. Small-for-size grafts increase recurrence of hepatocellular carcinoma in liver transplantation beyond milan criteria. *Liver Transpl.* **2018**, *24*, 35–43. [CrossRef]
38. Nagino, M.; DeMatteo, R.; Lang, H.; Cherqui, D.; Malago, M.; Kawakatsu, S.; DeOliveira, M.L.; Adam, R.; Aldrighetti, L.; Boudjema, K.; et al. Proposal of a New Comprehensive Notation for Hepatectomy: The "New World" Terminology. *Ann. Surg.* **2021**, *274*, 1–3. [CrossRef]
39. Rela, M.; Rammohan, A. Why Are There so Many Liver Transplants from Living Donors in Asia and so Few in Europe and the US? *J. Hepatol.* **2021**, *75*, 975–980. [CrossRef]
40. Park, M.S.; Lee, K.W.; Suh, S.W.; You, T.; Choi, Y.; Kim, H.; Hong, G.; Yi, N.J.; Kwon, C.H.D.; Joh, J.W.; et al. Living-Donor Liver Transplantation Associated with Higher Incidence of Hepatocellular Carcinoma Recurrence than Deceased-Donor Liver Transplantation. *Transplantation* **2014**, *97*, 71–77. [CrossRef]
41. Sandhu, L.; Sandroussi, C.; Guba, M.; Selzner, M.; Ghanekar, A.; Cattral, M.S.; McGilvray, I.D.; Levy, G.; Greig, P.D.; Renner, E.L.; et al. Living Donor Liver Transplantation versus Deceased Donor Liver Transplantation for Hepatocellular Carcinoma: Comparable Survival and Recurrence. *Liver Transpl.* **2012**, *18*, 315–322. [CrossRef] [PubMed]
42. Bhangui, P.; Vibert, E.; Majno, P.; Salloum, C.; Andreani, P.; Zocrato, J.; Ichai, P.; Saliba, F.; Adam, R.; Castaing, D.; et al. Intention-to-Treat Analysis of Liver Transplantation for Hepatocellular Carcinoma: Living versus Deceased Donor Transplantation. *Hepatology* **2011**, *53*, 1570–1579. [CrossRef] [PubMed]
43. Wong, T.C.L.; Ng, K.K.C.; Fung, J.Y.Y.; Chan, A.A.C.; Cheung, T.T.; Chok, K.S.H.; Dai, J.W.C.; Lo, C.M. Long-Term Survival Outcome Between Living Donor and Deceased Donor Liver Transplant for Hepatocellular Carcinoma: Intention-to-Treat and Propensity Score Matching Analyses. *Ann. Surg. Oncol.* **2019**, *26*, 1454–1462. [CrossRef] [PubMed]
44. Park, R.; Lee, S.; Sung, Y.; Yoon, J.; Suk, H.-I.; Kim, H.; Choi, S. Accuracy and Efficiency of Right-Lobe Graft Weight Estimation Using Deep-Learning-Assisted CT Volumetry for Living-Donor Liver Transplantation. *Diagnostics* **2022**, *12*, 590. [CrossRef] [PubMed]

45. Kiuchi, T.; Tanaka, K.; Ito, T.; Oike, F.; Ogura, Y.; Fujimoto, Y.; Ogawa, K. Small-for-size graft in living donor liver transplantation: How far should we go? *Liver Transpl.* **2003**, *9*, S29–S35. [CrossRef] [PubMed]
46. Yagi, S.; Iida, T.; Taniguchi, K.; Hori, T.; Hamada, T.; Fujii, K.; Mizuno, S.; Uemoto, S. Impact of portal venous pressure on regeneration and graft damage after living-donor liver transplantation. *Liver Transpl.* **2005**, *11*, 68–75. [CrossRef]
47. Man, K.; Lo, C.M.; Ng, I.O.; Wong, Y.C.; Qin, L.F.; Fan, S.T.; Wong, J. Liver transplantation in rats using small-for-size grafts: A study of hemodynamic and morphological changes. *Arch. Surg.* **2001**, *136*, 280–285. [CrossRef]
48. Kelmer Sacramento, E.; Kirkpatrick, J.M.; Mazzetto, M.; Baumgart, M.; Bartolome, A.; Di Sanzo, S.; Caterino, C.; Sanguanini, M.; Papaevgeniou, N.; Lefaki, M.; et al. Reduced proteasome activity in the aging brain results in ribosome stoichiometry loss and aggregation. *Mol. Syst. Biol.* **2020**, *16*, e9596. [CrossRef]
49. Kelly, D.M.; Demetris, A.J.; Fung, J.J.; Marcos, A.; Zhu, Y.; Subbotin, V.; Yin, L.; Totsuka, E.; Ishii, T.; Lee, M.C.; et al. Porcine partial liver transplantation: A novel model of the "small-for-size" liver graft. *Liver Transpl.* **2004**, *10*, 253–263. [CrossRef]
50. Hong, S.G.; Shin, J.; Choi, S.Y.; Powers, J.C.; Meister, B.M.; Sayoc, J.; Son, J.S.; Tierney, R.; Recchia, F.A.; Brown, M.D.; et al. Flow pattern-dependent mitochondrial dynamics regulates the metabolic profile and inflammatory state of endothelial cells. *JCI Insight.* **2022**, *7*, e159286. [CrossRef]
51. Fujiki, M.; Hashimoto, K.; Quintini, C.; Aucejo, F.; Kwon, C.H.D.; Matsushima, H.; Sasaki, K.; Campos, L.; Eghtesad, B.; Diago, T.; et al. Living Donor Liver Transplantation with Augmented Venous Outflow and Splenectomy: A Promised Land for Small Left Lobe Grafts. *Ann. Surg.* **2022**, *276*, 838–845. [CrossRef] [PubMed]
52. Martin, J.; Costa, A.; Gruszczyk, A.; Beach, T.; Allen, F.; Prag, H.; Hinchy, E.; Mahbubani, K.; Hamed, M.; Tronci, L.; et al. Succinate Accumulation Drives Ischaemia-Reperfusion Injury during Organ Transplantation. *Nat. Metab.* **2019**, *1*, 966–974. [CrossRef] [PubMed]
53. Schlegel, A.; Lesurtel, M.; Melloul, E.; Limani, P.; Tschuor, C.; Graf, R.; Humar, B.; Clavien, P.A. ALPPS: From Human to Mice Highlighting Accelerated and Novel Mechanisms of Liver Regeneration. *Ann. Surg.* **2014**, *260*, 839–846, discussion 846–847. [CrossRef] [PubMed]
54. Linecker, M.; Kambakamba, P.; Reiner, C.S.; Linh Nguyen-Kim, T.D.; Stavrou, G.A.; Jenner, R.M.; Oldhafer, K.J.; Björnsson, B.; Schlegel, A.; Györi, G.; et al. How Much Liver Needs to Be Transected in ALPPS? A Translational Study Investigating the Concept of Less Invasiveness. *Surgery* **2017**, *161*, 453–464. [CrossRef]
55. Alexandrino, H.; Rolo, A.; Teodoro, J.S.; Donato, H.; Martins, R.; Serôdio, M.; Martins, M.; Tralhão, J.G.; Caseiro Alves, F.; Palmeira, C.; et al. Bioenergetic Adaptations of the Human Liver in the ALPPS Procedure—How Liver Regeneration Correlates with Mitochondrial Energy Status. *HPB* **2017**, *19*, 1091–1103. [CrossRef] [PubMed]
56. Inomata, Y.; Kiuchi, T.; Kim, I.; Uemoto, S.; Egawa, H.; Asonuma, K.; Fujita, S.; Hayashi, M.; Tanaka, K. Auxiliary partial orthotopic living donor liver transplantation as an aid for small-for-size grafts in larger recipients. *Transplantation* **1999**, *67*, 1314–1319. [CrossRef] [PubMed]
57. Königsrainer, A.; Templin, S.; Capobianco, I.; Königsrainer, I.; Bitzer, M.; Zender, L.; Sipos, B.; Kanz, L.; Wagner, S.; Nadalin, S. Paradigm Shift in the Management of Irresectable Colorectal Liver Metastases: Living Donor Auxiliary Partial Orthotopic Liver Transplantation in Combination with Two-stage Hepatectomy (LD-RAPID). *Ann. Surg.* **2019**, *270*, 327–332. [CrossRef]
58. Balci, D.; Kirimker, E.O.; Bingol Kologlu, M.; Ustuner, E.; Goktug, U.U.; Karadag Erkoc, S.; Yilmaz, A.A.; Bayar, M.K.; Azap, A.; Er, R.E.; et al. A New Approach for Increasing Availability of Liver Grafts and Donor Safety in Living Donor Liver Transplantation: LD-RAPID Procedure in the Cirrhotic Setting with Hepatocellular Carcinoma. *Liver Transpl.* **2021**, *27*, 590–594. [CrossRef]
59. Müller, P.C.; Linecker, M.; Kirimker, E.O.; Oberkofler, C.E.; Clavien, P.A.; Balci, D.; Petrowsky, H. Induction of liver hypertrophy for extended liver surgery and partial liver transplantation: State of the art of parenchyma augmentation-assisted liver surgery. *Langenbecks Arch. Surg.* **2021**, *406*, 2201–2215. [CrossRef]
60. Liu, J.; Pang, L.; Ng, K.T.P.; Chiu, T.L.S.; Liu, H.; Liu, X.; Xu, L.; Lo, C.M.; Man, K. Compromised AMPK-PGCIα Axis Exacerbated Steatotic Graft Injury by Dysregulating Mitochondrial Homeostasis in Living Donor Liver Transplantation. *Ann. Surg.* **2022**, *276*, E483–E492. [CrossRef]
61. Teodoro, J.S.; da Silva, R.T.; Machado, I.F.; Panisello-Roselló, A.; Roselló-Catafau, J.; Rolo, A.P.; Palmeira, C.M. Shaping of Hepatic Ischemia/Reperfusion Events: The Crucial Role of Mitochondria. *Cells* **2022**, *11*, 688. [CrossRef] [PubMed]
62. Caldez, M.J.; van Hul, N.; Koh, H.W.L.; Teo, X.Q.; Fan, J.J.; Tan, P.Y.; Dewhurst, M.R.; Too, P.G.; Talib, S.Z.A.; Chiang, B.E.; et al. Metabolic Remodeling during Liver Regeneration. *Dev. Cell* **2018**, *47*, 425–438. [CrossRef] [PubMed]
63. Kong, W.; Li, W.; Bai, C.; Dong, Y.; Wu, Y.; An, W. Augmenter of Liver Regeneration-Mediated Mitophagy Protects against Hepatic Ischemia/Reperfusion Injury. *Am. J. Transplant.* **2022**, *22*, 130–143. [CrossRef]
64. ben Mosbah, I.; Duval, H.; Mbatchi, S.F.; Ribault, C.; Grandadam, S.; Pajaud, J.; Morel, F.; Boudjema, K.; Compagnon, P.; Corlu, A. Intermittent Selective Clamping Improves Rat Liver Regeneration by Attenuating Oxidative and Endoplasmic Reticulum Stress. *Cell Death Dis* **2014**, *5*, e1107. [CrossRef] [PubMed]
65. Han, L.H.; Dong, L.Y.; Yu, H.; Sun, G.Y.; Wu, Y.; Gao, J.; Thasler, W.; An, W. Deceleration of Liver Regeneration by Knockdown of Augmenter of Liver Regeneration Gene Is Associated with Impairment of Mitochondrial DNA Synthesis in Mice. *Am. J. Physiol. Gastrointest. Liver Physiol.* **2015**, *309*, G112–G122. [CrossRef]
66. Li, C.X.; Man, K.; Lo, C.M. The Impact of Liver Graft Injury on Cancer Recurrence Posttransplantation. *Transplantation* **2017**, *101*, 2665–2670. [CrossRef]

67. Mills, E.L.; Kelly, B.; Logan, A.; Costa, A.S.H.; Varma, M.; Bryant, C.E.; Tourlomousis, P.; D'britz, J.H.M.; Gottlieb, E.; Latorre, I.; et al. Succinate Dehydrogenase Supports Metabolic Repurposing of Mitochondria to Drive Inflammatory Macrophages. *Cell* **2016**, *167*, 457–470.e13. [CrossRef]
68. Schlegel, A.; Porte, R.; Dutkowski, P. Protective Mechanisms and Current Clinical Evidence of Hypothermic Oxygenated Machine Perfusion (HOPE) in Preventing Post-Transplant Cholangiopathy. *J. Hepatol.* **2022**, *76*, 1330–1347. [CrossRef]
69. van Rijn, R.; Schurink, I.; de Vries, Y.; van den Berg, A.; Cortes Cerisuelo, M.; Darwish, M.; Erdmann, J.; Gilbo, N.; de Haas, R.; Heaton, N.; et al. Hypothermic Machine Perfusion in Liver Transplantation—A Randomized Trial. *N. Engl. J. Med.* **2021**, *384*, 1391–1401. [CrossRef]
70. Panconesi, R.; Flores Carvalho, M.; Dondossola, D.; Muiesan, P.; Dutkowski, P.; Schlegel, A. Impact of Machine Perfusion on the Immune Response after Liver Transplantation—A Primary Treatment or Just a Delivery Tool. *Front. Immunol.* **2022**, *13*. [CrossRef]
71. Olumba, F.C.; Zhou, F.; Park, Y.; Chapman, W.C.; The RESTORE Investigators Group. Normothermic Machine Perfusion for Declined Livers: A Strategy to Rescue Marginal Livers for Transplantation. *J. Am. Coll. Surg.* **2023**, *236*, 614–625. [CrossRef] [PubMed]
72. Mergental, H.; Laing, R.W.; Kirkham, A.J.; Perera, M.T.P.R.; Boteon, Y.L.; Attard, J.; Barton, D.; Curbishley, S.; Wilkhu, M.; Neil, D.A.H.; et al. Transplantation of Discarded Livers Following Viability Testing with Normothermic Machine Perfusion. *Nat. Commun.* **2020**, *11*, 2939. [CrossRef] [PubMed]
73. Nasralla, D.; Coussios, C.C.; Mergental, H.; Akhtar, M.Z.; Butler, A.J.; Ceresa, C.D.L.; Chiocchia, V.; Dutton, S.J.; García-Valdecasas, J.C.; Heaton, N.; et al. A Randomized Trial of Normothermic Preservation in Liver Transplantation. *Nature* **2018**, *557*, 50–56. [CrossRef]
74. Parente, A.; Osei-Bordom, D.C.; Ronca, V.; Perera, M.T.P.R.; Mirza, D. Organ Restoration with Normothermic Machine Perfusion and Immune Reaction. *Front. Immunol.* **2020**, *11*, 565616. [CrossRef]
75. van Leeuwen, O.B.; Bodewes, S.B.; Lantinga, V.A.; Haring, M.P.D.; Thorne, A.M.; Brüggenwirth, I.M.A.; van den Berg, A.P.; de Boer, M.T.; de Jong, I.E.M.; de Kleine, R.H.J.; et al. Sequential Hypothermic and Normothermic Machine Perfusion Enables Safe Transplantation of High-Risk Donor Livers. *Am. J. Transplant.* **2022**, *22*, 1658–1670. [CrossRef] [PubMed]
76. Schlegel, A.; Mueller, M.; Muller, X.; Eden, J.; Panconesi, R.; von Felten, S.; Steigmiller, K.; Sousa Da Silva, R.X.; de Rougemont, O.; Mabrut, J.-Y.; et al. A Multicenter Randomized-Controlled Trial of Hypothermic Oxygenated Perfusion (HOPE) for Human Liver Grafts before Transplantation. *J. Hepatol.* **2023**, *78*, 783–793. [CrossRef]
77. Schlegel, A.; Muller, X.; Mueller, M.; Stepanova, A.; Kron, P.; de Rougemont, O.; Muiesan, P.; Clavien, P.A.; Galkin, A.; Meierhofer, D.; et al. Hypothermic Oxygenated Perfusion Protects from Mitochondrial Injury before Liver Transplantation. *EBioMedicine* **2020**, *60*, 103014. [CrossRef]
78. Schlegel, A.; Kron, P.; Graf, R.; Dutkowski, P.; Clavien, P.A. Warm vs. Cold Perfusion Techniques to Rescue Rodent Liver Grafts. *J. Hepatol.* **2014**, *61*, 1267–1275. [CrossRef]
79. Brüggenwirth, I.M.A.; van Leeuwen, O.B.; Müller, M.; Dutkowski, P.; Monbaliu, D.; Martins, P.N.; Porte, R.J.; de Meijer, V.E. The Importance of Adequate Oxygenation during Hypothermic Machine Perfusion. *JHEP Rep.* **2020**, *3*, 100194. [CrossRef]
80. Brüggenwirth, I.M.A.; Lantinga, V.A.; Rayar, M.; van den Berg, A.P.; Blokzijl, H.; Reyntjens, K.M.E.M.; Porte, R.J.; de Meijer, V.E. Prolonged Dual Hypothermic Oxygenated Machine Preservation (DHOPE-PRO) in Liver Transplantation: Study Protocol for a Stage 2, Prospective, Dual-Arm, Safety and Feasibility Clinical Trial. *BMJ Open Gastroenterol.* **2022**, *9*, e000842. [CrossRef]
81. Boteon, Y.L.; Laing, R.W.; Schlegel, A.; Wallace, L.; Smith, A.; Attard, J.; Bhogal, R.H.; Neil, D.A.; Hübscher, S.; Perera, M.T.P.; et al. Combined Hypothermic and Normothermic Machine Perfusion Improves Functional Recovery of Extended Criteria Donor Livers. *Liver Transplant.* **2018**, *24*, 1699–1715. [CrossRef] [PubMed]
82. Ravaioli, M.; Germinario, G.; Dajti, G.; Sessa, M.; Vasuri, F.; Siniscalchi, A.; Morelli, M.C.; Serenari, M.; del Gaudio, M.; Zanfi, C.; et al. Hypothermic Oxygenated Perfusion in Extended Criteria Donor Liver Transplantation—A Randomized Clinical Trial. *Am. J. Transplant.* **2022**, *22*, 2401–2408. [CrossRef] [PubMed]
83. Vasuri, F.; Germinario, G.; Ciavarella, C.; Carroli, M.; Motta, I.; Valente, S.; Cescon, M.; D'Errico, A.; Pasquinelli, G.; Ravaioli, M. Trophism and Homeostasis of Liver Sinusoidal Endothelial Graft Cells during Preservation, with and without Hypothermic Oxygenated Perfusion. *Biology* **2022**, *11*, 1329. [CrossRef] [PubMed]
84. Kron, P.; Schlegel, A.; Mancina, L.; Clavien, P.A.; Dutkowski, P. Hypothermic Oxygenated Perfusion (HOPE) for Fatty Liver Grafts in Rats and Humans. *J. Hepatol.* **2018**, *68*, 82–91. [CrossRef] [PubMed]
85. Jia, J.J.; Xie, H.Y.; Li, J.H.; He, Y.; Jiang, L.; He, N.; Zhou, L.; Wang, W.; Zheng, S. sen Graft Protection of the Liver by Hypothermic Machine Perfusion Involves Recovery of Graft Regeneration in Rats. *J. Int. Med. Res.* **2019**, *47*, 427–437. [CrossRef] [PubMed]
86. He, N.; Jia, J.J.; Xie, H.Y.; Li, J.H.; He, Y.; Yin, S.Y.; Liang, R.; Jiang, L.; Liu, J.; Xu, K.; et al. Partial Inhibition of HO-1 Attenuates HMP-Induced Hepatic Regeneration against Liver Injury in Rats. *Oxid. Med. Cell Longev.* **2018**, *2018*, 9108483. [CrossRef]
87. Moser, M.A.J.; Ginther, N.; Luo, Y.; Beck, G.; Ginther, R.; Ewen, M.; Matsche-Neufeld, R.; Shoker, A.; Sawicki, G. Early Experience with Hypothermic Machine Perfusion of Living Donor Kidneys—A Retrospective Study. *Transpl. Int.* **2017**, *30*, 706–712. [CrossRef]
88. He, X.; Guo, Z.; Zhao, Q.; Ju, W.; Wang, D.; Wu, L.; Yang, L.; Ji, F.; Tang, Y.; Zhang, Z.; et al. The first case of ischemia-free organ transplantation in humans: A proof of concept. *Am. J. Transplant.* **2018**, *18*, 737–744. [CrossRef]
89. Guo, Z.; Zhao, Q.; Huang, S.; Huang, C.; Wang, D.; Yang, L.; Zhang, J.; Chen, M.; Wu, L.; Zhang, Z.; et al. Ischaemia-free liver transplantation in humans: A first-in-human trial. *Lancet Reg. Health West. Pac.* **2021**, *16*, 100260. [CrossRef]

90. Tang, Y.; Wang, T.; Ju, W.; Li, F.; Zhang, Q.; Chen, Z.; Gong, J.; Zhao, Q.; Wang, D.; Chen, M.; et al. Ischemic-Free Liver Trans-plantation Reduces the Recurrence of Hepatocellular Carcinoma After Liver Transplantation. *Front. Oncol.* **2021**, *11*, 773535. [CrossRef]
91. Page, M.J.; McKenzie, J.E.; Bossuyt, P.M.; Boutron, I.; Hoffmann, T.C.; Mulrow, C.D.; Shamseer, L.; Tetzlaff, J.M.; Akl, E.A.; Brennan, S.E.; et al. The PRISMA 2020 statement: An updated guideline for reporting systematic reviews. *BMJ* **2021**, *372*, n71. [CrossRef] [PubMed]
92. Wells, G.A.; Shea, B.; O'Connell, D.; Peterson, J.; Welch, V.; Losos, M.; Tugwell, P. The Newcastle-Ottawa Scale (NOS) for Assessing the Quality of Nonrandomised Studies in Meta-Analyses. [cited 2023 Feb 06]. Available online: http://www.ohri.ca/programs/clinical_epidemiology/oxford.asp (accessed on 6 February 2023).

Disclaimer/Publisher's Note: The statements, opinions and data contained in all publications are solely those of the individual author(s) and contributor(s) and not of MDPI and/or the editor(s). MDPI and/or the editor(s) disclaim responsibility for any injury to people or property resulting from any ideas, methods, instructions or products referred to in the content.

Review

When Just One Phosphate Is One Too Many: The Multifaceted Interplay between Myc and Kinases

Dalila Boi [1,†], Elisabetta Rubini [1,†], Sara Breccia [1], Giulia Guarguaglini [2,*] and Alessandro Paiardini [1,*]

1. Department of Biochemical Sciences, Sapienza University of Rome, 00185 Rome, Italy
2. Institute of Molecular Biology and Pathology, National Research Council of Italy, Sapienza University of Rome, 00185 Rome, Italy
* Correspondence: giulia.guarguaglini@uniroma1.it (G.G.); alessandro.paiardini@uniroma1.it (A.P.)
† These authors contributed equally to this work.

Abstract: Myc transcription factors are key regulators of many cellular processes, with Myc target genes crucially implicated in the management of cell proliferation and stem pluripotency, energy metabolism, protein synthesis, angiogenesis, DNA damage response, and apoptosis. Given the wide involvement of Myc in cellular dynamics, it is not surprising that its overexpression is frequently associated with cancer. Noteworthy, in cancer cells where high Myc levels are maintained, the overexpression of Myc-associated kinases is often observed and required to foster tumour cells' proliferation. A mutual interplay exists between Myc and kinases: the latter, which are Myc transcriptional targets, phosphorylate Myc, allowing its transcriptional activity, highlighting a clear regulatory loop. At the protein level, Myc activity and turnover is also tightly regulated by kinases, with a finely tuned balance between translation and rapid protein degradation. In this perspective, we focus on the cross-regulation of Myc and its associated protein kinases underlying similar and redundant mechanisms of regulation at different levels, from transcriptional to post-translational events. Furthermore, a review of the indirect effects of known kinase inhibitors on Myc provides an opportunity to identify alternative and combined therapeutic approaches for cancer treatment.

Keywords: Myc; kinases; PLK1; Aurora-A; Aurora-B; GSK-3; PKA; PIM; BRD4; kinase inhibitors

Citation: Boi, D.; Rubini, E.; Breccia, S.; Guarguaglini, G.; Paiardini, A. When Just One Phosphate Is One Too Many: The Multifaceted Interplay between Myc and Kinases. *Int. J. Mol. Sci.* **2023**, *24*, 4746. https://doi.org/10.3390/ijms24054746

Academic Editor: Laura Paleari

Received: 24 January 2023
Revised: 19 February 2023
Accepted: 21 February 2023
Published: 1 March 2023

Copyright: © 2023 by the authors. Licensee MDPI, Basel, Switzerland. This article is an open access article distributed under the terms and conditions of the Creative Commons Attribution (CC BY) license (https://creativecommons.org/licenses/by/4.0/).

1. Introduction

c-Myc and the other protein family members (i.e., N-Myc and L-Myc), collectively known as "Myc", are ubiquitous basic helix–loop–helix–leucine zipper (bHLH-LZ) transcription factors that are critical for several cellular processes during cancer genesis and progression [1]. Indeed, Myc plays a central role among the molecular factors that drive tumour progression. c-Myc was first described as a viral oncoprotein able to induce myelocytomatosis in chickens after retroviral infection, and soon its derivation from a highly conserved vertebrate cellular gene was demonstrated [2,3]. The other members of the Myc family were subsequently discovered: N-Myc in neuroblastoma [4,5] and L-Myc in small cell lung carcinoma [6]. Since then, the Myc family members have been widely recognized as potential oncogenes and have been thoroughly studied for their structure, function, and regulation (as described in more detail by others [7,8]). Decades after its discovery, Myc continues to impress with its involvement in diverse pathways. The complexity of the Myc protein interaction network is reflected in the great heterogeneity of their transcriptional program and biomolecular activities [7–9].

Upon forming an obligate heterodimer with its protein partner Max, Myc binds to the consensus DNA sequence 5′-CACGTG-3′ (known as the E-box) and drives the expression of its target genes [10] (Figure 1A). As a mild transcriptional modulator, the effects of Myc are amplified through interactions with a wide range of cofactors, co-activators, and chromatin remodelling components [11]. Relevant functional categories of MYC-induced genes include cell cycle control, DNA replication, cell growth and adhesion, cellular bioenergetics (e.g., glycolysis and mitochondrial biogenesis), and anabolic metabolism (e.g., synthesis of amino acids, nucleotides, and lipids) [1,12,13]. In addition, Myc and HIF1 actively induce the expression of glucose transporters and glycolytic enzymes [14], and the former acts as a downstream effector of several signalling cascades: for instance, aberrant Wnt/β-catenin signalling leads to the constitutive, high transcription of the MYC gene [15,16]; similarly, potent induction of MYC transcription occurs in response to perturbations of the Sonic hedgehog [17], Notch [18,19], and Janus kinase (JAK)— signal transducer and activator of transcription 3 (STAT3) pathways [20,21].

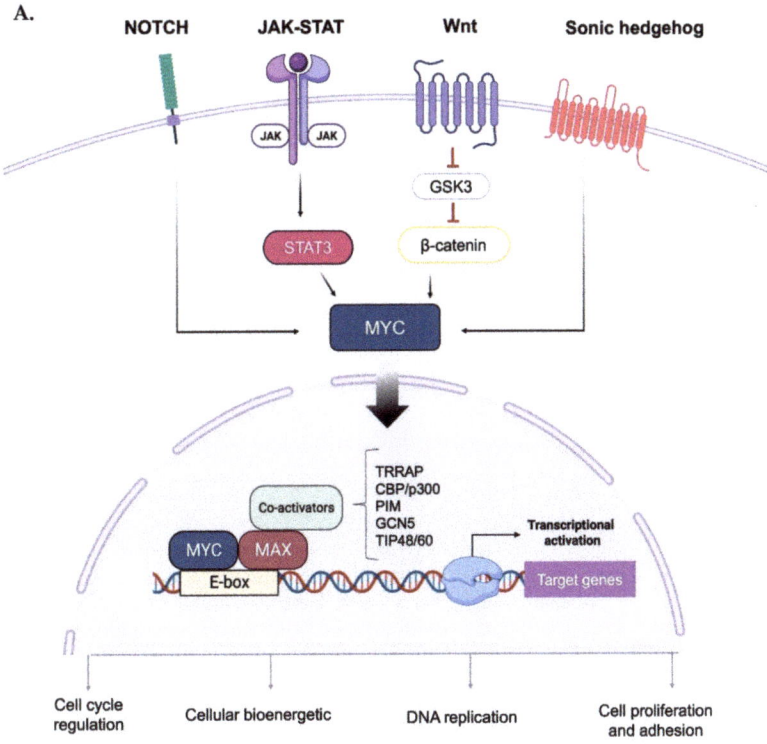

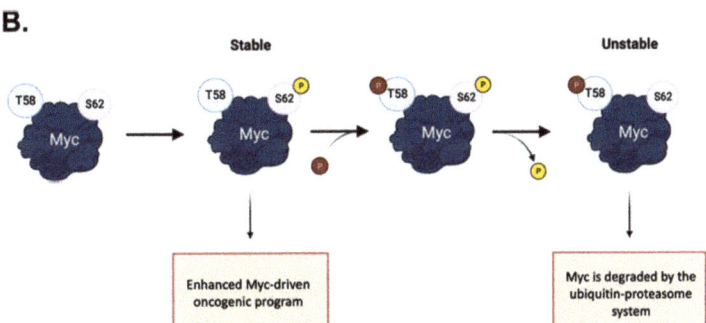

Figure 1. Activation and phospho-dependent stabilization of Myc. (**A**) Myc can be activated in response to different stimuli via several transduction pathways converging to Myc stabilization.

Stabilized Myc associates with its protein partner MAX to form a heterodimer, which, together with other co-activators, binds to the E-box elements and drives the transcription of a target gene subset involved in a wide range of cellular processes. (**B**) Phospho-dependent stabilization of Myc. Myc displays two key phosphorylation sites that undergo hierarchical phosphorylation, supervising protein stability. Phosphorylation of Myc at the S62 residue determines protein stabilization and primes the subsequent phosphorylation at T58, which induces the removal of the phosphate group at S62; the unstable, singly phosphorylated T58-Myc is then recognized by the ubiquitin ligase Fbxw7 and degraded by the ubiquitin–proteasome system. Created with BioRender.com (accessed on 20 January 2023).

The pleiotropic activities of Myc make its tight regulation mandatory in healthy cells, which need to rapidly turnover Myc proteins. However, steady Myc levels may be affected by pathways that are disrupted in cancer, leading to its enhanced expression and increased stability [22]. Myc mRNA levels are regulated by both transcription initiation (which starts from different promoters [23]) and elongation [24], yielding a transcript that is highly unstable, with a half-life of 20–30 min [25]. Additionally, Myc is rapidly degraded after its synthesis, with a turnover of 25 min [26]. Posttranslational modifications, i.e., acetylation and methylation, contribute to regulating Myc activity and stability [27]. SUMOylation, consisting of the addition of ubiquitin-like molecules to the target protein, is another intriguing observed modification in Myc. The role of the SUMOylation in regulating Myc functions was debated; Myc is SUMOylated in more than 10 lysine residues [28–30], but the mutation of all these residues did not abolish Myc SUMOylation, nor alter its activity [28–30]. In a particular case, an increased Myc transcription activity was observed: in B cell lymphoma, the SUMOylation mediated by the E3 ligase PIAS1 allows the recognition and phosphorylation of Myc in S62 by JNK1, which precludes the oncoprotein degradation [31]. Noteworthy, in this tumour, a Myc-dependent transcriptional activation of SUMO cascade enzymes was observed. The overall SUMOylation induces G2/M arrest and consequent mitotic abnormalities, which are attributed to the Aurora kinases' SUMOylation [32]. SUMOylation could also have a role in Myc protein stability, since the inhibition of the proteasome has been shown to increase the levels of SUMOylated Myc [28,33]. Moreover, the SUMO-specific protease SENP1 has been described to de-SUMOylate Myc, stabilizing the protein [33]. It is possible that the de-SUMOylation drives the subsequent de-ubiquitination on the same residues, since SUMO and ubiquitin modifications can exist on the same lysine residues [34].

The phosphorylation and dephosphorylation of key residues provide the main dynamic regulation of Myc cellular functions: its phosphorylation plays a key role in promoting Myc binding to DNA and in inducing its degradation. The conserved region of Myc called the Myc Box (MB) I is mainly involved in its ubiquitination and degradation [35] (Figure 1B). It contains the residues S62 and T58, which undergo cascade phosphorylation, regulating protein stability and functionality. S62 is a target for a variety of kinases, including ERK, CDKs, and JNK [36], and its phosphorylation correlates with the recruitment of Myc to specific promoters in response to oxidative stress. The S62A mutation has been shown to significantly alter an array of genes involved in apoptosis, proliferation, cellular bioenergetic, and signalling pathways [37]. The phosphorylation of S62 recruits the GSK-3β kinase, which, in turn, phosphorylates T58 on Myc, ensuring its recognition by the Fbxw7 ubiquitin ligase for degradation [38]. T58 and S62 are hotspots of mutation in lymphomas [39,40] and solid tumours [41,42]. The non-phosphorylatable T58A mutation leads to the accumulation of pS62 Myc, which presumably repulses Fbxw7 and increases Myc half-life to up to 120 min [40]. Additionally, Myc phosphorylated on T58 reduces the expression of BIM, which is particularly relevant as a tumour suppressor in MYC-driven B cell leukaemia [43]. The phosphorylation of other highly conserved sites, T358, S373, and T400, by the p21-activated kinase PAK2 has been reported to negatively impact on the Myc transcriptional program by interfering with the formation of the Myc–Max–DNA ternary complex [44,45].

In addition to MBI, other conserved motifs control Myc functions. MBII allows interaction with partners required for transcriptional regulation [46], while MBIII is important for cellular transformation [47]. Finally, MBIV overlaps with the nuclear localization signal [48]. Each motif is conserved among species and ensures the binding of Myc proteins with different partners. Interestingly, the MBI and the MBII motifs overlap with the transactivation domain (TAD), highlighting a well-established correlation between Myc protein stability and transcriptional activity [49].

The importance of kinases in Myc regulation goes beyond their ability to phosphorylate the protein. Some kinases can also indirectly affect Myc protein stability by inducing the degradation of the ubiquitin ligase (PLK1 and PKA, [50,51]). Additionally, some kinases physically interact with Myc, protecting it from proteasomal degradation, such as Aurora-A in neuroblastoma [52]. The cell-cycle-dependent Aurora-A and Aurora-B form a circuit with Myc, in which Aurora primes the oncogenic program of Myc and vice versa [53,54]. Aurora-A is also involved in Myc modulation in PKA-dependent tumours [55]. Other kinases are crucial for Myc transcriptional activity (i.e., PIM and BRD4 [56–58]). At the same time, Myc enhances the expression of protein kinases, creating a positive feedback loop.

In this review, we will delve into the cross-regulation between Myc and kinases, aiming at providing a comprehensive understanding of the complex signalling events between oncogenic proteins and their partners. We will identify common mechanisms of action and highlight altered oncogenic pathways. We will also describe the effects of some kinase inhibitors on Myc and explore the potential benefits of combining therapies to improve the selectivity of approaches to eradicate Myc-driven cancers.

2. Direct and Indirect Myc Regulation by Mitotic Kinases

The overexpression or deregulation of mitotic kinases is often associated with cancer progression. Defective cell cycle regulation can lead to uncontrolled cell division and chromosomal instability, making tumours hyper-proliferative and resistant to growth inhibition. As they are druggable, targeting mitotic kinases has been considered a promising strategy for anticancer intervention [59,60]. In this perspective, we describe the unusual involvement of classical mitotic kinases in promoting the survival of Myc-driven cancers.

2.1. PLK1 Kinase

PLK1 is a member of the Polo-like kinases (PLKs) family, which consists of five paralogues: PLK1, PLK2, PLK3, PLK4, and PLK5. It is a conserved serine/threonine kinase that is critical for the proper execution of mitotic events and the maintenance of genome stability during cell division [61,62]. The overexpression of PLK1 is a hallmark of many types of human cancers, including melanoma, ovarian carcinoma, breast, prostate, thyroid cancers, and glioma [63–65], and it is associated with chemoresistance and poor patient outcomes. Genetic ablation or inhibition of PLK1 has been found to affect cell cycle progression, leading to reduced proliferation, apoptosis, and increased sensitivity to chemo- and radiotherapy [66–68].

From a structural perspective, PLK1 contains an N-terminal catalytic kinase domain (KD) and a C-terminal region with two Polo-box domains (PBDI and PBDII), which are involved in the phospho-dependent recognition of substrates and are essential for determining the kinase's spatiotemporal subcellular localization [69]. In interphase, the KD and PBD establish an inhibitory interaction that suppresses PLK1 catalytic activity. During mitosis, the PBD provides a docking site for proteins with specific phosphorylated motifs [70] and undergoes conformational changes that relieve the inhibition of the KD. This allows the activation loop to be accessed by upstream kinases (such as Aurora-A and its cofactor Bora) for the phosphorylation of the key residue T210, which is necessary for the full functionality of PLK1 [71,72]. In addition to modulating the kinase activity, the phosphorylation status of PLK1 controls its nuclear translocation and its ubiquitin-mediated degradation, which helps to preserve chromosome segregation fidelity and genome integrity [73,74].

PLK1 has a complex interactome network involving various pathways beyond mitosis, such as immune response [75], epithelial-to-mesenchymal transition [76,77], and cell death signalling [78,79]. MYC-amplified tumours often have upregulated PLK1, which creates a feed-forward interaction with Myc that maintains high levels of both proteins, predicting poor prognosis [80–82]. The main regulatory role of PLK1 is to enhance Myc stability, as seen in the reduced expression of c-Myc upon PLK1 depletion [83]. PLK1 kinase activity is necessary for Myc protein accumulation, through phosphorylation of the key residue S62 [84]. In addition, following the phosphorylation of Myc at S279 by PKA [85], PLK1 can phosphorylate c-Myc at S281, allowing the ubiquitin ligase SCF$^{\beta\text{-TrCP}}$ to bind and ubiquitinylate the oncoprotein. This alternative post-translational event increases c-Myc stability and contributes to Myc-dependent cell transformation [51]. However, the main mechanism of Myc stabilization does not seem to involve the phosphorylation of Myc at S62, but rather PLK1's ability to affect the activity of Myc-degrading machinery. Evidence from the rescue of N-Myc levels in response to PLK1 pharmacological inhibition by using the proteasome inhibitor MG132 led Xiao and colleagues to propose a possible crosstalk between PLK1 and the E3 ubiquitin ligase Fbxw7; their study showed that wild-type PLK1, but not the kinase mutant K82R, suppresses Fbxw7 activity by promoting its phosphorylation and autocatalytic poly-ubiquitination in MYCN-amplified cellular models [50]. A similar PLK1/Fbxw7/Myc axis was also identified in c-MYC-driven medulloblastoma, with PLK1 antagonizing Fbxw7-mediated c-Myc degradation [86] (Figure 2).

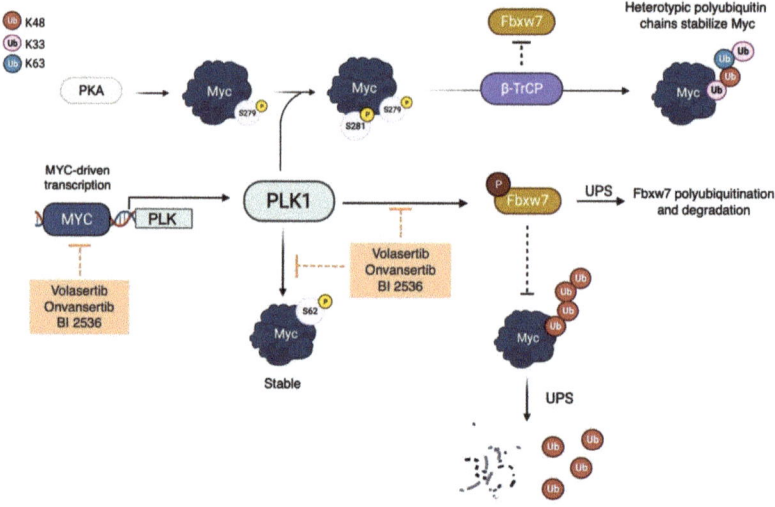

Figure 2. Various mechanisms by which PLK1 stabilizes Myc. PLK1 can directly phosphorylate Myc at S62 or indirectly promote phosphorylation at S281 in a PKA-dependent manner, resulting in the accumulation of a stable, ubiquitinylated form of Myc through β-TrCP binding (right side). Unlike Fbxw7, in fact, β-TrCP assembles K33/K63/K48 heterotypic polyubiquitin chains that do not target Myc for proteasome-mediated degradation. PLK1 also impairs Myc degradation by promoting the proteolysis of Fbxw7, thereby increasing the half-life of Myc. Stabilized Myc, in turn, enhances PLK1 transcription. These mechanisms work together to help maintain high levels of Myc in MYC-amplified tumours, which is associated with poor prognosis. The inhibitors acting on this pathway are shown (see also Section 5.1 for references). Created with BioRender.com (accessed on 20 January 2023).

Myc stabilization by PLK1 also plays a prominent role in maintaining the autophagy pathway in tumour cells: knockdown of PLK1 leads to a significant reduction in c-Myc protein levels, impacting on MYC transactivation and impairing Myc-mediated autophagy in osteosarcoma cells [87]. While PLK1 modulates Myc stability, the main contribution of Myc to the oncogenic relationship with PLK1 is through transcriptional regulation. This is supported by the presence of a Myc E-box binding site upstream of the PLK1 transcription start site, and the finding that the suppression of c-Myc leads to a corresponding reduction in both PLK1 mRNA and phospho-T210 protein levels in B lymphoma cells expressing a tetracycline-repressible MYC transgene [50,81]. ChIP experiments performed on neuroblastoma and lymphoma cell lines also demonstrated a significant recruitment of c-Myc to the PLK1 E-box, suggesting that c-Myc directly regulates the PLK1 transcriptional program [50,81]. Based on these findings, targeting PLK1 has the potential to indirectly target Myc-dependent pathways and address the current challenge of developing a therapeutic approach directed against the undruggable Myc proteins.

2.2. Aurora-A and Aurora-B Kinases

An intriguing feedback loop with Myc is described for the serine/threonine Aurora kinases. Three members of this family are present in eukaryotes: Aurora-A and Aurora-B, which are involved in the correct execution of mitosis, and Aurora-C, which is mostly implicated in meiosis [88]. Aurora-A and Aurora-B share 71% identity in their kinase domains [89] and have complementary functions in mitotic cells. The Aurora-A protein localizes at centrosomes, where it participates in the maturation and separation processes in the G2 and M phases, facilitating the recruitment of PLK1 onto CEP192 [90]. During mitosis, it interacts with its major activator, TPX2, allowing for the correct formation and orientation of the bipolar spindle [91,92]. On the other hand, Aurora-B interacts with INCENP, constituting the chromosome passenger complex (CPC) with Survivin and Borealin, which ensures chromosome cohesion, the correct attachment of microtubules to kinetochores, and cytokinesis [93,94].

Both proteins are linked to cancer [95,96]. Aurora-A is frequently overexpressed or amplified in a variety of solid and hematologic (blood-related) tumours [89,97], and according to the Cancer Genome Atlas (TCGA), it is overexpressed in almost 88% of the tumours observed [98]. Aurora-A can promote cell transformation when subjected to a suitable cellular background [99]. Additionally, the kinase promotes epithelial-to-mesenchymal transition [100,101], the expression of self-renewal genes in cancer stem cells [102], and cancer cell survival through the regulation of apoptotic modulators (reviewed by [95]). Recently, its oncogenic activities were related to its nuclear localization [54]. On the other hand, the contribution of Aurora-B in cancer development is not well understood, despite the fact that it is upregulated in most aneuploid human tumours [89] and is a poor prognosis factor in hepatocellular carcinoma, non-small cell lung carcinoma, and oral squamous cell carcinoma [103–105].

The oncogenic activities of the Aurora kinases have frequently been linked to Myc proteins. In fact, Myc-driven cancers are susceptible to Aurora kinase deprivation or inhibition. One example is c-Myc-driven B-cell lymphoma, in which the overexpression of both Aurora-A and Aurora-B sensitizes cells to pan-Aurora kinase inhibitor treatment [106]. In hepatocellular carcinoma (HCC) cells, Aurora-A interacts with and stabilizes c-Myc, promoting tumour cell survival [107]. In neuroendocrine prostate cancer, Aurora-A and N-Myc interact with each other and drive an oncogenic gene expression program [108]. In neuroblastoma, Aurora-A depletion mimics the effect of N-Myc deprivation in MYCN-amplified (MNA) but not non-MNA cells. This is due to the stabilizing effect of Aurora-A on N-Myc: the kinase binds to the oncoprotein, preventing the Fbxw7 ubiquitin ligase-mediated ubiquitination of Myc [52].

The reason why Aurora kinases' impairment is an Achilles' heel of Myc-driven tumours resides both in their importance for cell division, particularly relevant for proliferating cells, and in their ability to regulate Myc transcription and protein stability. Myc-driven expression of Aurora-A and Aurora-B favours mitotic entry and progression, supporting the active proliferation of cancer cells [106,109]. Moreover, the Aurora-A/N-Myc complex, which is observed more frequently in S phase, has been suggested to prevent the formation of replication/transcription conflicts induced by the oncogene's over-transcription activity [110].

The first physical interaction described between Myc and Aurora family members was the Aurora-A/N-Myc complex in neuroblastoma cells. Depletion of Aurora-A decreases the half-life of N-Myc from 99 min to 55 min in IMR-32 MNA neuroblastoma cells. This Aurora-A function is independent from its kinase activity, since eight different kinase-deficient mutants are able to stabilize the N-Myc oncoprotein [52]. Aurora-A stabilizes N-Myc by protecting it from recognition by the Fbxw7 ubiquitin ligase; moreover, Aurora-A leads to the accumulation of ubiquitinated N-Myc that is non-K48-linked, suggesting that Aurora-A may recruit ubiquitin ligases/deubiquitinases that create a ubiquitinated N-Myc with less degradable ubiquitin chains [52]. Although it is not clear if Aurora-A preferentially interacts with the double-phosphorylated N-Myc (as proposed by [52], whereas [111] did not observe differences between the phosphorylated and unphosphorylated forms for the binding to Aurora-A in vitro), the interplay between Aurora-A and Fbxw7 is well described. In particular, the solution of the crystal structure between the catalytic domain of Aurora-A and the 28–89 N-Myc peptide revealed that Aurora-A binds to the oncoprotein through the MB0 and MBI motifs (residues 61–89) [111]. In vitro pull-down experiments showed that Aurora-A competes with Fbxw7 for the binding of N-Myc within residues 61–89. Nevertheless, the phospho-degron of N-Myc remains able to be recognized by the ubiquitin ligase; thus, a complex involving Aurora-A, N-Myc, and the Fbxw7 is still formed [111].

The region of N-Myc found in the crystal structure is not present in c-Myc [111]. Evidence of Aurora-A/c-Myc binding is controversial, since it was reported that no binding between the two proteins occurs in liver and hepatocellular carcinoma cells [112], while on the other hand, Aurora-A and the double-phosphorylated c-Myc co-immunoprecipitated in TP53-altered (deleted or mutated) HCC cells [107]. However, Aurora-A does play a role in regulating c-Myc expression. It co-immunoprecipitates with the MYCC promoter within the NHE III1 region, which is described to be particularly important for MYCC transcription [112]. Aurora-A has also been shown to act as a co-activator of MYCC transcription in breast cancer [54,113]. Since Aurora-A lacks DNA binding abilities, Zheng and colleagues identified the ribonucleoprotein hnRNP K as the mediator that allows Aurora-A to activate transcription on the MYCC promoter. Depletion of hnRNP K impairs the recruitment of Aurora-A on the MYCC promoter, but not vice versa [54,113]. The transactivation activity of Aurora-A is independent of its kinase activity, as the administration of VX-680 and MLN8237 kinase inhibitors does not affect the expression of the oncogene, in contrast to what can be observed after Aurora-A depletion [113]. hnRNP K also transcriptionally co-activates the p53 protein [114], which negatively impacts the expression of several cell cycle genes, including MYCC [115,116]. Aurora-A may be involved in this process by phosphorylating the ribonucleoprotein at S379, leading to the disruption of its interaction with p53 [115].

On the other hand, c-Myc enhances Aurora-A expression. Mouse fibroblasts with overexpressed c-Myc have a two- to three-fold increase in Aurora-A promoter activity [106]. Using an inducible Myc-ER construct in ChIP experiments, den Hollander and colleagues found that the mouse Aurora-A gene containing two E-boxes is enriched upon induction of c-Myc, suggesting that the oncoprotein directly binds to the Aurora-A promoter, although they were not able to detect the recruitment of c-Myc on either of the Aurora kinase human genes [106]. Lu and colleagues also found a correlation between Aurora-A and c-Myc mRNA in HCC cells, and used ChIP experiments to show that c-Myc binds to the Aurora-A promoter in the highly conserved E-box regions within the CPG islands [112].

In light of their roles in modulating Myc expression and stability, it is interesting to highlight the similarity between the two Aurora kinases. For example, the specificity for the binding of Aurora-A to its activator TPX2 is determined by only one residue, i.e., G198, which is different from that in Aurora-B, N142 [117]. Moreover, the G198N substitution is sufficient to convert Aurora-A into a kinetochore-localized Aurora-B-like kinase [117–120]. Furthermore, Aurora-A/INCENP binding has also been observed [121]. This evidence suggests a sort of interchangeability between Aurora kinases A and B. However, despite the similarity between Aurora-A/Aurora-B and c-Myc/N-Myc, the interplay with each other is quite different. Aurora-A stabilizes the N-Myc protein [52], and, in turn, high levels of N-Myc directly or indirectly increase Aurora-A mRNA levels (e.g., [122]), establishing a feedback loop that feeds itself. Conversely, Aurora-B does not affect N-Myc protein levels in neuroblastoma [123,124] or retinoblastoma cells [125] upon depletion. However, Aurora-B levels do decrease after N-Myc knockdown [125]. Aurora-B is also transcriptionally activated by direct binding of N-Myc to motifs upstream of the transcription start site [124,125] and by indirect binding of c-Myc in mice, resulting in a 30-fold increase in Aurora-B promoter activity upon Myc-ER activation [106]. It is worth mentioning that Jiang and his colleagues recently identified a novel regulatory mechanism for Aurora B that can stabilize c-Myc through kinase-dependent activity in acute lymphoblastic leukaemia. Experimental results show that Aurora B directly phosphorylates c-Myc at the S67 residue, thereby promoting its stability by counteracting GSK-3-mediated T58 phosphorylation. Notably, sequence alignment did not reveal the S67 phosphorylation site in N-Myc, potentially explaining the non-regulatory role of Aurora B in MYCN-amplified cancers [53].

In conclusion, the Myc and Aurora family of proteins cross-regulate each other, establishing feedback mechanisms that are particularly relevant for cancer progression (Figure 3). In addition, Aurora-A also impacts the expression and stability of Myc proteins by participating in other processes, such as phosphorylating several members of the PI3K/AKT and Wnt/β-catenin pathways, or proteins involved in Myc stability, such as GSK3β and p53 [101,126–128]. It is clear that untangling this intricate network of interactions that affects both the cell cycle and the activity/stability of Myc will need more in-depth investigation.

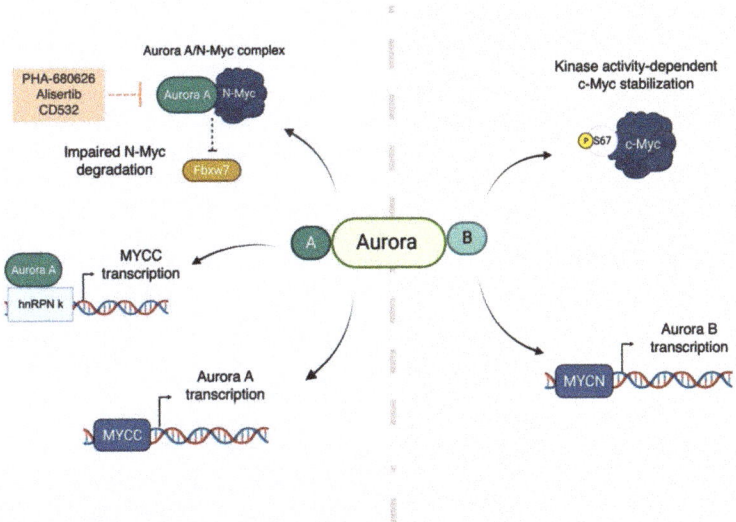

Figure 3. Schematic exemplification of the main functional interplays between Myc and Aurora kinase A (left)/B(right). Aurora-A physically interacts with N-Myc to form a complex that prevents N-Myc from the proteolytic degradation mediated by the ubiquitin ligase Fbxw7. From a transcriptional point of view, Aurora-A contributes to regulating MYCC expression in combination with hnRNP K. Furthermore, Aurora-A is a target gene of MYCC, which enhances Aurora-A transcription. A similar regulatory relationship is described for MYCN and Aurora-B, with MYCN able to bind to motifs upstream of the Aurora B transcription starting site. A kinase-dependent stabilization of Myc involves phosphorylation at serine residue 67, which is a characteristic feature of Aurora kinase B. The inhibitors acting on this pathway are shown (see also Section 5.2 for references). Created with BioRender.com (accessed on 20 January 2023).

3. Direct and Indirect Myc Regulation by Metabolic Kinases

In order to meet the bioenergetic demand of rapid proliferation, cancer cells must improve their sources of energy. To do this, they put pressure on metabolic pathways that ensure an increase in glucose, angiogenesis, protein synthesis, and DNA synthesis. Several kinases control different key metabolic pathways, such as the Wnt/β-catenin, Notch, MAPK/ERK, and PI3K/AKT pathways, and are considered interesting therapeutic targets. These signalling cascades are also important for the synthesis and stabilization of Myc proteins. Here, two examples of multitasking kinases that regulate Myc proteins are described.

3.1. GSK-3

Glycogen synthase kinase-3 (GSK-3) is a ubiquitously expressed and highly evolutionarily conserved serine/threonine kinase that exists in two isoforms encoded by distinct genes, GSK-3α and GSK-3β. GSK-3 paralogs have a bi-lobal architecture, consisting of a large C-terminal globular domain responsible for the kinase activity and a small ATP-binding N-terminal lobe. While GSK-3α and GSK-3β share some common substrates, they are differentially expressed in human tissues and have unique biological roles [129]. Unlike other kinases, GSK-3 is constitutively active under resting conditions and is inhibited by the phosphorylation at key serine residues within the N-terminal domain (S21 for GSK-3α and S9 for GSK-3), which prevents access to the active site by substrates [130]. GSK-3 is involved in several pathways (e.g., PI3K/PTEN/Akt/mTOR, Ras/Raf/MEK/ERK, Wnt/β-catenin, Hedgehog, Notch [131,132]) and plays a decisive role in a wide range of diseases [133,134].

One of the key roles of GSK-3 is in regulating the stability of Myc. The ectopic overexpression of both GSK-3α and GSK-3β was found to increase the phosphorylation of Myc at T58, while the inhibition of the endogenous kinase is associated with a significant reduction in pT58 Myc levels. In addition, there are no significant effects on Myc proteolysis or localization when Myc has a T58A mutation [38,135]. One of the best-characterized tasks of GSK-3 is its role in the destruction of the β-catenin multi-protein complex. β-catenin is a multitasking molecule that is both a core component of the cadherin complex, which coordinates cell–cell adhesion, and a crucial signal transducer in the Wnt pathway. GSK-3, along with Axin and adenomatous polyposis coli (APC), phosphorylates β-catenin at T41, S37, and S33 [136]; phosphorylated S33 and S37 interact with the β-propeller domain of β-TrCP [137], thus priming β-catenin for degradation and ultimately impacting on Myc. Interference with β-catenin proteolysis promotes its translocation into the nuclear compartment [138], where β-catenin acts as a co-activator of the TCF/LEF transcription factors, which control the expression of a wide range of genes involved in oncogenic transformation, including MYC [139] (Figure 4). Indeed, the pharmacological inhibition of GSK-3 has been reported to upregulate both Myc and β-catenin protein levels, thus enhancing Myc-mediated apoptosis in neuroblastoma and KRAS-dependent cellular models [140,141].

GSK-3 exists in different phospho-isoforms and its activity is often negatively modulated by post-translational modifications [142]. Based on this evidence, GSK-3 activation status is key for the regulation of both Myc and β-catenin stability, highlighting the importance of the crosstalk between GSK-3 and other relevant kinases. For example, Aurora kinase A was found to directly bind to and phosphorylate GSK-3β at Ser9 [127], leading to its inactivation. Higher levels of phospho-GSK-3β reduce β-catenin degradation, which accumulates in the nucleus and upregulates its downstream targets, including MYC. Knockdown of Aurora-A has been shown to reduce the amount of nuclear β-catenin in gastric cancer [127], colorectal cancer [143], and glioblastoma cells [102]. In addition to Aurora-A, other relevant Ser/Thr kinases are known to inhibit GSK-3, disrupting substrate-mediated signalling. An analogue modulatory mechanism is reported for PKA, which physically associates with GSK-3 and inhibits both the isoform α and β in a phosphorylation-dependent manner [144]. The cross-regulation between GSK-3 and other kinases also involves the PI3K/Akt signalling pathway and the effector arms of the Ras pathway [145]. When PI3K is activated in response to a wide range of stimuli, it triggers a cascade of second messengers, leading to the activation of Akt kinase, which inhibits both GSK-3α and GSK-3β through reversible phosphorylation at S21 and S9, respectively, derepressing GSK-3 substrates [146]. Blocking PI3K releases GSK from Akt-mediated inhibition, promoting T58 Myc phosphorylation and its subsequent degradation [147–149].

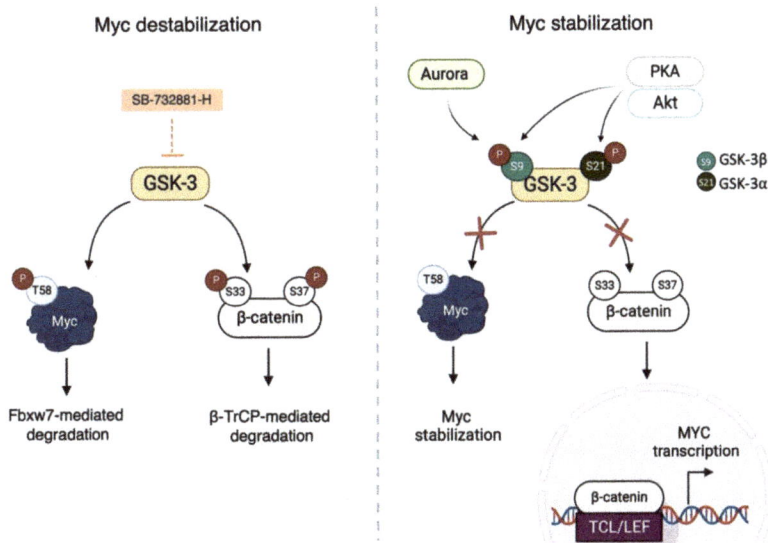

Figure 4. Main Myc regulatory mechanisms by GSK-3. GSK-3 can directly phosphorylate Myc at T58, thus destabilizing the protein and committing it to degradation. In addition, GSK-3 phosphorylates β-catenin and induces its proteolysis, preventing its nuclear translocation. The phosphorylation of GSK-3 at S9/S21 by different kinases prevents both Myc and β-catenin degradation and determines both Myc stabilization, as well as nuclear localization of β-catenin. In the nucleus, β-catenin functions as a coactivator of the TCL/LEF transcription factors and promotes MYC transcription. The main inhibitor acting on this pathway is shown (see also Section 5.3 for references). Created with BioRender.com (accessed on 20 January 2023).

The remarkable participation of GSK-3 in a wide range of cellular processes makes this kinase an interesting pharmacological target. However, its multifaceted biological functions do not allow a high degree of inhibition for translation in human diseases to be achieved [133], driving research efforts toward investigating GSK-3–substrate interactions to develop tailored therapies at the level of individual disease.

3.2. PKA

The cAMP-dependent protein kinase A (PKA) is considered the prototypical Ser/Thr kinase and regulates many cell signalling events [150,151] such as DNA synthesis, regulation of transcription, and metabolism [152]. It is not surprising that PKA dysregulation is frequently associated with cancer [153], cardiovascular disease [154,155], and neurological disorders [156–158]. The kinase is a heterotetramer consisting of two regulatory subunits (R) and two catalytic subunits (C). The cooperative binding of two molecules of cAMP to each R subunit leads to a conformational change that results in the release of the active C subunits [159,160], which can phosphorylate downstream targets. In mammalian cells, two types of R subunits (RI and RII) coexist, with two isoforms of each type (RIα, RIβ, RIIα, and RIIβ). The PKA heterotetramer containing the RI subunit is called PKA type I, while PKA type II contains the RII isoform. The regulatory subunit type confers to the holoenzyme different localization, tissue distribution, and cAMP affinity, ensuring specificity for a large number of substrates, as well as diverse sensitivity to the second messenger cascade. In addition, three isoforms of the C-subunit are described, with the two major isoforms (Cα and Cβ) having multiple splice variants that introduce diversity into the first exon [159].

PKA establishes a regulatory loop with Myc, at first stabilizing the oncoprotein, which, in turn, promotes the expression of the kinase. Inhibition of PKA with the specific inhibitor H89 or endogenous pseudosubstrate inhibitor PKI decreases c-Myc protein levels in different cell lines (PC3 and HeLa), an effect that is reversed by the proteasome inhibitor MG132 [85]. c-Myc stabilization is facilitated by PKA-mediated phosphorylation of the oncoprotein at S279, which primes a second phosphorylation mediated by PLK1 at S281 (as described above). PKA is able to phosphorylate ~50% of cellular c-Myc; the non-phosphorylatable S279A mutant is not destabilized by H89 treatment and impairs PLK1-mediated phosphorylation, suggesting a prominent role for this phosphorylation site in c-Myc protein stability [85]. c-Myc protein stability is also governed by a fine balance between ubiquitination events. The βTrcp ubiquitin ligase counteracts Fbxw7-mediated degradation of c-Myc by linking non-degradable K-Ub [51]. c-Myc recognition by βTrcp is ensured by PLK1 phosphorylation in the N-terminal of the oncoprotein sequence [51], highlighting the key influence of the PKA-PLK1-βTrcp axis on c-Myc stability (Figure 5).

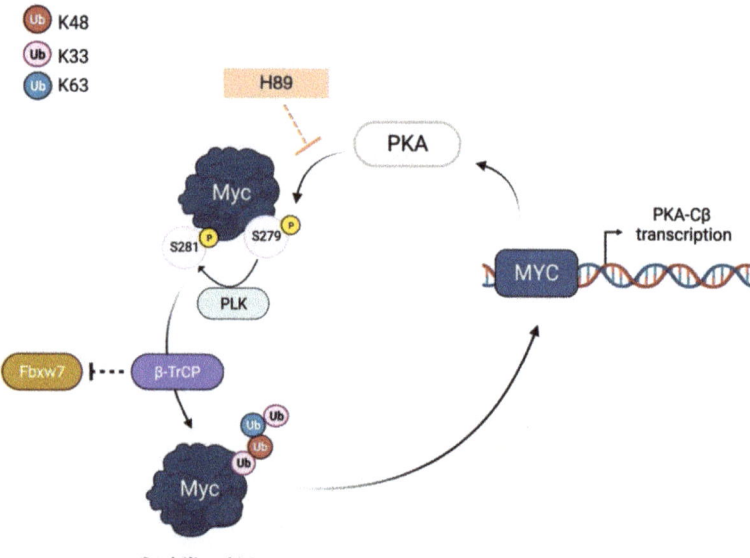

Figure 5. Myc is stabilized via a PKA/PLK1/Myc axis. Myc is stabilized via a PKA/PLK1/Myc axis, which triggers the sequential phosphorylation of the residue S279 by PKA and the residue S281 by PLK1. Stabilized Myc, in turn, enhances the transcription of PKA and, in particular, of the subunit PKA-Cβ, highlighting the relevance of PKA isoform-specific regulatory mechanisms. The main inhibitor acting on this pathway is shown (see also Section 5.4 for references). Created with BioRender.com (accessed on 20 January 2023).

Despite the clear role of PKA in aiding Myc protein stability, knockdown of PKA demonstrates isoform-specific regulation: siRNA against PKACα induces an increase in c-Myc mRNA and protein, whereas knockdown of PKACβ leads to a slight decrease in c-Myc protein and no mRNA enrichment [85]. PKACβ is overexpressed in rapidly proliferating prostate cancer cells, while PKACα overexpression was reported to induce trans-differentiation of LNCaP cells into neuroendocrine-like cells [85]. The feedback between PKACβ and Myc is well described. c-Myc transcriptionally activates PKACβ in pancreatic cancer. Wu and colleagues identified two non-canonical E-boxes within intron 1 and in the 5′ region flanking intron 1. Luciferase experiments demonstrated that c-Myc can bind the 5′/intron 1 region and activate luciferase expression four- to six-fold, comparable to other classical Myc targets [161]. PKACβ expression is abrogated upon c-Myc mutations in the TAD domain or defects in the interaction with MAX. Conversely, the role of PKACα in c-Myc stability is controversial: it is also a key regulator of the CREB transcription factor, which, together with p300, inhibits Myc expression [162,163]. Additionally, among PKACα (but not PKACβ) substrates, ENO1 and MBP1, which are splice variants described as Myc transcriptional repressors, have gained attention [85]. Some studies have observed increased levels of c- and N-Myc in cAMP stimulation conditions, while PKACα siRNA attenuates phosphorylation of its substrates and progressively decreases c- and N-Myc protein levels [55].

The overall data suggest that PKA isoforms, due to their differential substrate selectivity, have opposite effects: the Cα subunit indirectly reduces the expression of the oncogene through the activation of Myc transcriptional repressors [85,164]. At the same time, inhibition with pan-PKA inhibitors stabilizes the Myc protein, which, in turn, promotes the expression of the Cβ subunit, protecting Myc from proteasome-mediated degradation [85]. The differential cell responses to the diverse activities of PKA isoforms highlight the complexity of the cross-regulation between PKA and Myc. Indeed, Myc regulation is influenced by the cellular availability of interacting kinases, which is often dependent on tissue-specific expression, environmental conditions, or protein localization. For example, PKAI is mostly cytoplasmic and its activity is associated with cell growth and proliferation, while PKAII is anchored to subcellular structures and compartments and is mostly associated with differentiation [159,165]. This scenario emphasizes once again the concept that the activity of Myc in cells results from a very delicate balance of time- and space-regulated phosphorylation cascades.

4. Direct and Indirect Myc Regulation by Histone-Associated Kinases

Histone-associated kinases are able to phosphorylate and modify the histone proteins packaging DNA within the nucleus and can regulate various processes in the cell, including gene expression, DNA replication, and DNA repair, by modifying the structure and accessibility of the underlying DNA. PIM and BRD4 are examples of histone-associated kinases, for which we will describe the cross-regulation with Myc.

4.1. PIM

The PIM (proviral integration site for Moloney murine leukaemia virus) family of serine/threonine kinases consists of three highly similar isoforms designated as PIM-1, PIM-2, and PIM-3, which share functional redundancy and overlapping specificity for a wide range of substrates involved in cell proliferation and survival [166,167]. PIMs are overexpressed in several human haematological malignancies as well as solid tumours, frequently correlating with advanced clinical stages and poor prognosis [168–171]. In contrast to other protein kinases, PIMs lack a regulatory domain and adopt a constitutively active conformation that is maintained via the establishment of a salt bridge between acidic residues in the A-loop and a conserved arginine residue in the C-loop, keeping the ATP binding pocket open and ready for catalysis [172,173].

Although the three members of the PIM kinase family have been classified as weak oncogenes and are not sufficient alone to promote the tumour transition toward a more aggressive form, their potential is exploited through the synergistic interplay with other proteins. The cooperative partnership between PIMs and Myc is well-characterized in different types of cancers, with PIMs widely reported to play a key role in regulating Myc stability and transcriptional program [174–177]. PIM overexpression is linked to c-Myc stabilization in vivo through a phosphorylation-dependent mechanism that occurs with an isoform-specific efficiency. As described by Zhang and colleagues, PIM-1 and PIM-2 stabilize c-Myc almost to the same extent and are directed against the same phosphorylation sites, but PIM-1 preferentially phosphorylates c-Myc at S62, whereas PIM-2 is more efficient in the direct phosphorylation at S329 residue, contributing to increasing c-Myc half-life. Remarkably, phospho-S329 also plays a prominent role in enhancing c-MYC transcriptional activity [56]. Knockdown of PIMs results in a dramatic reduction in endogenous c-Myc protein levels, with a major downregulation achieved by simultaneously silencing PIM-1, -2, and -3 [178]. Interestingly, the expression of PIM-3 alone was shown to overcome some defects arising from the depletion of PIM-1 and PIM-2: PIM-3, in fact, enhances cap-dependent translation in triple-knockout mouse embryonic fibroblasts and upregulates c-Myc levels without affecting protein stability. This evidence suggests that PIM isoforms may mutually regulate each other either directly or through substrate competition [179]. Additionally, PIMs critically supervise the balance between anti- and pro-apoptotic factors in Myc-driven tumorigenesis. MYC-amplified cells, in fact, exhibit a high proliferation rate and are prone to apoptosis; notably, PIM kinases, especially PIM-2, can phosphorylate the pro-apoptotic proteins BAD and Bcl-2, thereby counteracting their function and sustaining cell proliferation [180,181].

However, one of the most important mechanisms that contribute to the oncogenic cooperation between Myc and PIM-1 is the dynamic modulation of gene transcription through epigenetic mechanisms. PIM-1 can form a ternary complex with the MYC-MAX heterodimer, which allows PIM-1 to phosphorylate S10 of histone H3 within the E-box element of Myc target genes. This contributes to the activation of a subset of genes that are under the transcriptional control of Myc [57]. The phosphorylation of H3 by PIM-1 recruits the phosphoserine-binding protein 14-3-3 to the enhancer; the interaction between 14-3-3 and the histone acetyltransferase MOF leads to the acetylation of histone H4 at residue K16 [182]. The interaction between H3S10ph and H4K16ac creates a nucleosome platform for the binding of the bromodomain protein BRD4 (see below); BRD4 stimulates the kinase activity of the positive elongation factor b (p-TEFb), which phosphorylates the C-terminal domain of RNA polymerase II promoting transcriptional elongation [182,183]. In triple-negative breast cancer models, PIM-1 knockdown results in reduced phosphorylation of c-Myc at S62 and histone H3 at S10, as well as a decrease in the total amount of Myc protein. The loss of PIM-1 mRNA after silencing is also reflected in the MYC gene signature, with 11 out of 24 MYC-target genes downregulated in cells silenced for PIM1. A large-scale gene expression profiling in HEK-293 cells that were silenced for either MYC or PIM-1 revealed that PIM-1 contributes to the regulation of 207 of the 1026 MYC-target genes, demonstrating its significant impact as a Myc-dependent chromatin modifier [184,185] (Figure 6).

Interestingly, and as previously discussed for other kinases involved in Myc regulation, the synergy between Myc and PIM kinases is reciprocal and relies on the creation of a feed-forward loop. For instance, Myc-driven lymphoma cells from transgenic mice and human patients exhibit elevated mRNA levels of the isoform PIM-3. The analysis of the nucleotide sequence of the PIM-3 locus revealed a conserved E-box that can be bound by Myc [186]. In addition, PIM-3 enhances 5′-cap-dependent translation, increasing c-Myc levels without affecting protein stability [179]. As further support for the interplay between Myc and PIM, a recent study examined their involvement in ribosomal stress, proposing a novel role for PIM-1 in prostate cancer progression. Among its diverse biological functions, PIM-1 participates in the assembly of the ribosomal subunit 40S by regulating the expression of ribosomal small subunit proteins (RPSs), especially RPS7. Luciferase-based assays indicate that the promoter of RPS7 can be directly bound by c-Myc, confirming the existence of a PIM1-c-Myc-RPS7 axis responsible for abnormal ribosomal biosynthesis and subsequent ribosomal stress [187]. Considering the pivotal role of PIM kinases in Myc regulation, their value as potential targets for the development of new drug candidates deserves further investigation.

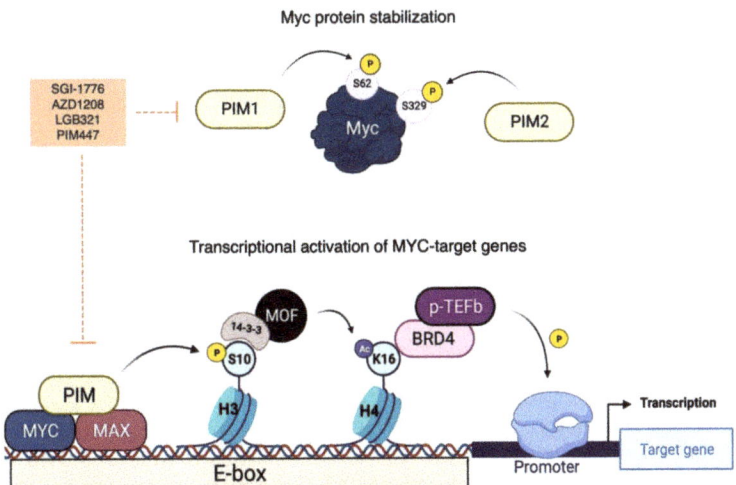

Figure 6. Regulation of Myc by PIM kinases. PIMs phosphorylate Myc at different residues with an isoform specificity and contribute to enhancing protein stability. In addition, PIMs contribute to epigenetic modulation by forming a ternary complex with MYC-MAX, thus regulating the expression of genes under MYC transcriptional control. PIM, in fact, phosphorylates the histone H3 within the E-box element of MYC-dependent genes at the serine residue 10. Phosphorylated H3 recruits the phosphoserine-binding protein 14-3-3, which interacts with the histone acetyltransferase MOF. MOF, in turn, catalyses the acetylation of the histone H4 at K16. Acetylated histone H4 provides a binding platform for BRD4 and p-TEFb, facilitating transcriptional elongation. The inhibitors acting on this pathway are shown (see also Section 5.5 for references). To simplify, dual inhibitors are not reported in the figure. Created with BioRender.com (accessed on 20 January 2023).

4.2. BRD4

BRD4 is a bromodomain and extra terminal (BET) protein, showing both acetyl-transferase activity (HAT) and kinase activity. BRD4 dysfunctions are associated with multiple diseases such as cancers, neuro-degenerative disorders, autoimmune diseases, and heart, kidney, lung, and inflammatory diseases [188,189]. BRD4 has been described to be a transcription coactivator and recruiter of chromatin remodelling factors, and it participates in mitosis, pause release of PolII, RNA splicing, and DNA damage and stress responses, most of which are functions in common with Myc [190].

BRD4 is composed of two bromodomains (BD1 and BD2) organised in tandem and an extra terminal domain (i.e., ET, important for its interaction with a wide range of proteins) at the N-terminus, as well as Motif A, Motif B, basic residue interaction domain (BID), and the Ser/Glu/Asp-rich region (SEED) [190]. In contrast, the C-terminus, which hosts the HAT catalytic domain, is an intrinsically disordered region (IDR) [190,191] and is present only in the longer of the two cellular BRD4 isoforms [192]; the disordered structure of this region ensures the flexibility required for the binding of different proteins, i.e., c-Myc [58].

BRD4 and c-Myc co-immunoprecipitate from HeLa nuclear extracts and show a strong nuclear interaction signal when examined using *is*PLAs [58]. To identify the region of interaction between the two proteins, Devaiah and colleagues performed pull-down experiments using recombinant BRD4 and c-Myc truncated proteins [58]. The TAD domain of c-Myc (residues 144-163) interacts with the C-terminus of BRD4 (aa 823-1044), but this interaction is independent of the HAT domain. The BRD4/c-Myc interaction is essential for c-Myc protein stability. In fact, the bromodomain protein phosphorylates c-Myc at T58, leading to its ubiquitination and degradation [58]. MEF cells derived from mice homozygous for floxed BRD4 alleles show decreased levels of pT58 and increased levels of c-Myc protein, whereas overexpression of BRD4 increases phosphorylation at T58 and decreases total c-Myc. Notably, in contrast to GSK-3β, BRD4 is able to phosphorylate T58 regardless of previous phosphorylation at S62, and the bromodomain protein immunoprecipitates with both T58A and S62A c-Myc constructs transfected into HeLa cells [58].

The complexity of the interaction between BRD4 and Myc is not limited to the binding of Myc to BRD4's C-terminus. BRD4 also binds the N-terminal region of the ERK1 kinase, which is responsible for stabilizing Myc through phosphorylation at S62, forming an ERK1–BRD4–Myc complex. Each of these proteins interacts with the others. Notably, ERK1 binds the region containing the kinase domain of BRD4, inhibiting its kinase activity and, therefore, its autophosphorylation and its ability to phosphorylate c-Myc [58]. This helps stabilize c-Myc and prevent its degradation, mediated by BRD4 (Figure 7). Another condition in which BRD4 is unable to phosphorylate Myc is when the bromodomain protein is bound to chromatin, due to overlap of the BD2 domain with the kinase domain [191]. However, the trimeric complex is present in both chromatin-bound and unbound fractions of the nucleus, indicating regulation at various levels. BRD4 is a Myc transcriptional activator and contributes to the expression of Myc at the end of mitosis [193–195]. Through its HAT activity, BRD4 promotes the promoter accessibility [196]. While c-Myc has no effect on BRD4 kinase activity, it does influence BRD4's ability to remodel chromatin by inhibiting its HAT activity [58,190]. Transiently transfected HeLa and U2OS cells with a c-Myc vector show reduced levels of H3K122ac by more than 40% [58]. Notably, when c-Myc is phosphorylated by both BRD4 and ERK1, it fails to inhibit BRD4 HAT activity. Indeed, this activity is dependent on phosphorylation on T58, since the T58A non-phosphorylatable mutant markedly inhibits H3K122ac [58].

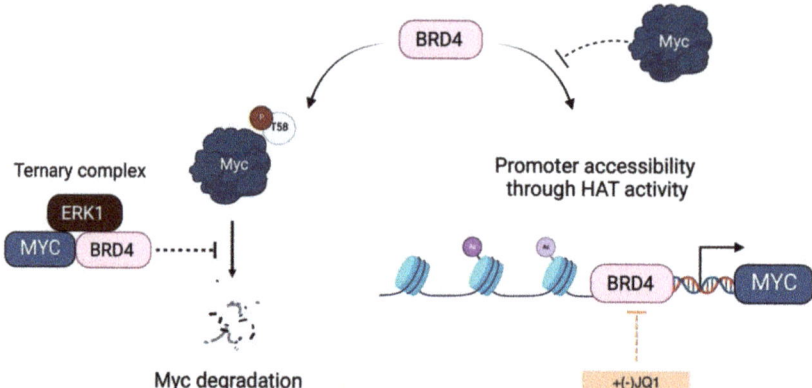

Figure 7. BRD4-dependent regulation of Myc. BRD4 directly phosphorylates Myc at T58 through its kinase activity, thus inducing its degradation and reducing its inhibitory effect on BRD4 histone acetyltransferase activity. However, the physical interaction of Myc and BRD4 with ERK1 to form a ternary complex prevents Myc from BRD4-mediated destabilization. Furthermore, BRD4 promotes MYC transcriptional activation by decondensing the chromatin around the MYC gene by means of its histone acetyltransferase activity. The main inhibitors acting on this pathway are shown (see also Section 5.6 for references). To simplify, dual inhibitors are not reported in the figure. Created with BioRender.com (accessed on 20 January 2023).

From what has been discussed so far, it appears that the regulation of the interaction between BRD4 and Myc is a complex process. BRD4 activates the transcription of Myc by remodelling the chromatin around the Myc gene locus, through its histone acetyltransferase (HAT) activity. At the same time, BRD4 phosphorylates c-Myc on T58, causing its degradation, which reduces the inhibitory effect of c-Myc on BRD4's HAT activity. In this context, ERK1, which interacts with the BRD4 kinase domain, counteracts the destabilizing effect of BRD4 on c-Myc. This intricate cross-regulation is important for the transcriptional activity of Myc. The degradation of c-Myc interferes with the acetylation of histones at Myc target genes, indicating the need to turn over c-Myc in order to ensure their transcription [196]. It has also been shown that the degradation of Myc is necessary for the pause release of RNA polymerase II [197]. Devaiah and colleagues suggest that the degradation of c-Myc induced by BRD4 influences the pause release of RNA polymerase II, and that the reduction in c-Myc protein levels is balanced by the activation of c-Myc transcription by BRD4 [58].

5. Kinase Inhibitors and Myc Proteins

Myc is deregulated and overexpressed or hyperactivated in the vast majority of human cancers. Unfortunately, because of its intrinsically disordered structure, Myc lacks a targetable binding pocket and is, thus, considered undruggable. Thus, the inhibition of Myc oncogenic functions may rely on protein–protein interaction (PPI) inhibitors; however, Myc takes on a secondary and tertiary structure only when in complex with one of its partners, making the identification of PPI inhibitors challenging [198]. Several efforts have been made in order to design disruptors of the Myc/MAX interaction, which is necessary for the Myc oncogenic activities. These PPI inhibitors are small molecules (e.g., IIA6B17 and IIA4B20 [199]; 10074-G5 and 10058-F4 [200]; Mycro1 and Mycro2 [201] and their derivatives) and cell-penetrating peptide inhibitors (Omomyc [202,203]; IDP-410 [204], even though the complete impairment of the binding of Myc on DNA is challenging, and they are found to non-specifically target other transcription factors (such as Jun). Nevertheless, an optimized Omomyc is undergoing phase I/II clinical trial (Omo-103, NCT04808362). However, the Myc pleiotropic activities might be better targeted acting at different levels.

Therefore, another potentially valid approach against MYC-driven cancers is to target the kinases that support the Myc oncogenic functions. Kinase inhibitors are widely used in cancer therapy. It is estimated that one quarter of pharmacological efforts are focused on kinase inhibitor development [205]. All the kinases described above are found to be associated with cancer, and the most promising drug candidates with reported effects on Myc are reported in Table 1.

5.1. PLK1 Inhibitors

As previously mentioned, PLK1 constitutes an appealing therapeutic target for overcoming the challenging development of pharmaceutical-grade inhibitors of the Myc protein. PLK1 inhibitors can be grouped into two classes: ATP competitors, which target the binding site of the kinase domain [229], and non-ATP competitors, which target the PBD domain [230]. Although ATP competitors have low selectivity, which can increase the risk of toxic side effects, they generally exhibit good drug-like properties. Four major PLK1 inhibitors belonging to this class have reached clinical trials: BI2536, volasertib (BI6727), onvansertib (NMS-1286937), and GSK461364.

BI2536 is a dihydropteridinone derivative that can inhibit PLK1 at nanomolar concentrations in preclinical experiments with an acceptable safety profile. However, BI2536 had no significant effects on patients with relapsed/refractory solid tumours enrolled in phase II studies [231,232], leading to the discontinuation of its development in favour of volasertib. Volasertib was developed by modifying the structure of BI2536, and has improved physicochemical and pharmacokinetic properties, including a higher distribution volume and prolonged half-life. In addition, volasertib potently inhibits PLK1 (estimated IC_{50} of 0.87 µM in cell-free in vitro assay) without affecting related kinases at concentrations up to 10 µM [233]. Volasertib underwent phase I/II clinical trials for the treatment of solid tumours and haematological malignancies; although the monotherapy seemed promising, it only showed modest antitumour effects [234], and a more complete response was achieved when volasertib was tested in combination with other therapeutic agents [235]. In MNA tumours, the inhibition of PLK1 by volasertib is correlated with a decrease in Myc protein levels. Volasertib reduces c-Myc phosphorylation at S62 and is linked to impaired cell viability as well as PARP cleavage in aggressive T-lymphoma and B-lymphoma cells [81,83]. The effects on MYC transcriptional levels may be due to the mild effect of volasertib on the BET protein BRD4 [236]. This evidence is consistent with the finding that PLK1 inhibition by volasertib leads to the stabilization of the ubiquitin ligase Fbxw7 and the degradation of its downstream target N-Myc in MYCN-overexpressing neuroblastoma cells. In the same cellular model, the combination of volasertib with the Bcl12 inhibitor ABT199 selectively induced apoptosis, suggesting that the Volasertib/ABT199 synergism may be a potential therapeutic strategy [50]. In the same study, similar effects on Myc protein levels were observed with the thiophene-amide inhibitor GSK461364. However, the impact of GSK461364 on Myc is not yet well understood, as phenotypic effects from PLK1 inhibition, which are not dependent on MYCN amplification status, have been reported in neuroblastoma cell lines [237].

Table 1. List of kinase inhibitors described to have effects on Myc.

Target Kinase	Inhibitor	Structure	Effects on Myc	Tumour Models	Phase/Status
PLK1	BI 2536		Decreased Myc protein levels, both total protein abundance and phospho-S62Myc; enhanced Myc proteolysis as a result of the stabilization of the ubiquitin ligase Fbxw7; impaired MYC transcriptional activity and downregulation of MYC-target genes	Neuroblastoma [50]	Phase II NCT00376623
	Volasertib (BI 6727)			T-lymphoma B-lymphoma Neuroblastoma Rhabdomyosarcoma [50,81,83,206]	Phase I/II NCT01121406 NCT00804856 NCT00824408 NCT00969553
	Onvansertib (PCM-075)			Medulloblastoma [207]	Phase I/II NCT03414034 NCT03829410 NCT04005690 NCT05549661 NCT04752696
Aurora-A	Alisertib (MLN8237)		Enhanced degradation of N-Myc protein as a result of the disruption of the complex with Aurora-A, which prevents N-Myc from ubiquitin-proteasome-mediated degradation	Neuroblastoma [123]	Phase I/II/III NCT01799278 NCT01898078 NCT02719691 NCT02293005
	CD532			Neuroblastoma Medulloblastoma Neuroendocrine prostate cancer [208,209]	Preclinical
	PHA-680626			Osteosarcoma Neuroblastoma [210]	Preclinical
GSK-3	SB-732881-H		Enhanced cell apoptosis in a c-Myc- and β-catenin-dependent manner	Pancreatic adenocarcinoma Lung cancer [141]	Preclinical
PKA	H89		Decreased Myc protein levels as a result of the impaired PKA-mediated stabilization of Myc	Prostate cancer Breast cancer [85]	Preclinical

Table 1. Cont.

Target Kinase	Inhibitor	Structure	Effects on Myc	Tumour Models	Phase/Status
PIM	SGI-1776		Reduction in total Myc protein abundance as well as phospho-S62Myc;	Diffuse large B-cell lymphoma Triple-negative breast cancer Chronic lymphocytic leukaemia Mantle cell lymphoma [178,211–213]	Phase I—Withdrawn due to cardiac toxicity NCT01239108 NCT00848601
	AZD1208			Acute myeloid leukaemia Prostate cancer [214,215]	Phase I NCT01489722 NCT01588548
	LGB321		Impaired MYC-driven transcriptional program	Triple-negative breast cancer [211]	Preclinical
	LGH447 (PIM447)			Diffuse large B-cell lymphoma Multiple myeloma [178,216]	Phase I NCT02370706 NCT02160951 NCT02078609 NCT02144038 NCT01456689
	SEL24/ MEN1703		Downregulation of Myc mediated at least in part by increased proteasomal degradation; Impaired MYC-driven transcriptional program	Diffuse large B-cell lymphoma [178]	Phase I/II NCT03008187
BRD4	+(-)JQ1		Myc downregulation through BRD4 displacement from MYCN promoter	MYC-amplified neuroblastoma Glioblastoma [217]	Phase I—Derivative inhibitor RO6870810 [218,219] NCT02308761 NCT03068351 NCT01987362 NCT03292172 NCT03255096
BRD4/ PLK1	9b (WNY0824)		Inhibition of N-Myc and c-Myc transcription	Acute myeloid leukaemia Castration-resistant prostate cancer Colorectal adenocarcinoma Melanoma Ovarian cancer [220,221]	Preclinical
	UMB103		Decreased N-Myc protein levels	Neuroblastoma Rhabdomyosarcoma [206]	Preclinical
	UMB160				

Table 1. Cont.

Target Kinase	Inhibitor	Structure	Effects on Myc	Tumour Models	Phase/Status
BRD4/ PI3K	SF2523		Myc downregulation through BRD4 displacement from MYCN promoter	MYCN-amplified neuroblastoma Pancreatic carcinoma [222]	Preclinical
	SF1126			Hepatocellular carcinoma [223]	Phase I/II NCT03059147 NCT02337309 NCT02644122 NCT00907205
BRD4/ HDAC	17c		Decrease in c-Myc protein levels	Colorectal cancer [224]	Preclinical
	16ae		Decrease in Myc protein levels	Acute myeloid leukaemia [225]	Preclinical
	19f		Decrease in c-Myc protein levels	Acute monocytic leukaemia [226]	Preclinical
	13a		Decrease in c-Myc mRNA and protein levels	Pancreatic cancer [227]	Preclinical
	Compound 40		c-Myc downregulation and decrease in mRNA expression	Pancreatic cancer Adenocarcinoma [227]	Preclinical
BRD4/ p38α	SB-284851-BT		Decrease in c-Myc protein levels	Immunoglobulin A Lambda myeloma [228]	Preclinical

The pyrazoloquinazoline derivative onvansertib is an oral and highly selective third-generation ATP-competitive PLK1 inhibitor with an IC50 of 36 nM [238] currently under phase I clinical study for solid tumours [239]. onvansertib was found to suppress tumour growth in both in vivo and in vitro models. In MYC-driven medulloblastoma, onvansertib caused a significant downregulation of MYC-target genes and a substantial decrease in Myc protein abundance. Additionally, PLK1 depletion in response to onvansertib stabilizes the ubiquitin ligase Fbxw7, resulting in a decrease in Myc protein in xenograft models [207]. These data are consistent with the PLK1/Fbxw7/MYC signalling axis described by Xiao and colleagues, and highlight the potential of PLK1 inhibitors as a therapeutic option for the treatment of MYC-overexpressing cancers.

5.2. Aurora-A Inhibitors

Aurora-A is the best-described kinase for targeting Myc. Aurora-A and Aurora-B kinase inhibitors are being evaluated in clinical trials for solid and hematopoietic tumours [240]. In addition, Aurora-A has been described as a critical target for MNA neuroblastoma cells. The discovery of Aurora-A's stabilizing effects on N-Myc through the physical interaction of the two proteins [52] has led to intense research aimed at destabilizing this interface. It is noteworthy that this interaction does not depend on Aurora-A kinase activity; therefore, classical ATP-competitive kinase inhibitors are not useful for disrupting the kinase/N-Myc complex. A few Aurora-A orthosteric inhibitors, however, have been shown to also act as amphosteric regulators, stabilizing Aurora-A in a conformation that is incompatible with N-Myc binding; for this reason, these compounds are hailed as conformation-disrupting (CD) inhibitors. The two best-characterized CD inhibitors are Alisertib (MLN8237) [123,241] and CD532 [208]. Recently, our group identified PHA-680626 as a novel potential CD inhibitor [210]. These compounds induce strong cell death in MNA compared to non-MNA neuroblastoma cells due to the decrease in N-Myc protein levels caused by the impairment of the Aurora-A/N-Myc complex formation.

The conformation of Aurora-A bound to CD inhibitors, which allows the disruption of the oncogenic complex, is characterized by the opening of the N- and C-lobes of the protein [242] and the flipping of the activation loop (residues 274–293) into a closed conformation. The conserved DFG motif (residues 274–276) is crucial for Aurora-A activity, and can assume two main conformations: the active DFG-In and the inactive DFG-Out. It was originally thought that the DFG-In state was strictly associated with the open conformation, but it has since been shown that CD532, a known Aurora-A CD inhibitor, stabilizes the protein in a closed conformation while maintaining the DFG-In state [208]. In fact, recent work has questioned the CD ability of a panel of kinase inhibitors and, using Förster resonance energy transfer (FRET) methodology, has revised the classification of known CD inhibitors [243].

Recently, the protein targeting chimeras (PROTAC) methodology has been explored for Aurora-A, with promising results [244–246]. Tang and colleagues developed a PROTAC from a ribociclib (CDK4/6 inhibitor) scaffold, which is able to target the Aurora-A/N-Myc complex [246]. It is worth noting that the idea of cellular pools of Aurora-A that interact in a spatially and temporally regulated manner with several partners [54] is gaining attention. In line with this idea, centrosome-localized Aurora-A is not sensitive to an MLN8237-based PROTAC (PROTAC-D)-mediated degradation [245]; moreover, deprivation of the kinase following PROTAC JB170 administration results in the accumulation of cells in the S phase, while kinase inhibition leads to G2/M enrichment in leukaemia and neuroblastoma cell lines [244], suggesting a need for further investigation into Aurora-A's non-mitotic functions.

5.3. GSK3 Inhibition

Over the last decades, GSK-3 inhibitors have been extensively investigated as potential pharmacological tools against neurodegenerative diseases and psychiatric disorders, but the well-established involvement of GSK-3 in oncogenic pathways suggests its therapeutic relevance for the development of novel anticancer therapies. Even though some compounds have reached clinical trials, most of them did not satisfy the endpoints and were discontinued due to safety issues. Currently, the most promising GSK-3 inhibitor is the compound 9-ING-41 (elraglusib), which is under clinical evaluation for the treatment of several cancers in combination with other chemotherapeutic agents (NCT03678883, NCT05239182).

One of the main concerns about the clinical application of drug candidates directed against GSK-3 still remains the controversial effects of GSK-3 inhibition, which leads to the stabilization and oncogenic activation of β-catenin. Surprisingly, the stabilization of both β-catenin and c-Myc observed in response to the dual GSK-3α/3β inhibitor SB-732881-H correlates with a paradoxical increase in c-Myc-mediated apoptosis in preclinical models of KRas-dependent human cancers, presenting a new perspective on the therapeutic implications of GSK-3 inhibitors [141]. However, a high degree of inhibition of GSK-3α/3β is mostly associated with unacceptable toxicity due to the low selectivity of available compounds. For this reason, Wagner and colleagues rationally design paralog-selective inhibitors of GSK-3 able to discriminate between the two existing isoforms. Notably, the inhibition of GSK-3α kinase function did not stabilize β-catenin in acute myeloid leukaemia cells, providing a feasible strategy to develop paralog-selective GSK-3α inhibitors enforceable to cancer therapy [247].

5.4. PKA Inhibitors

Due to its involvement in a wide range of signalling pathways, PKA dysregulation is often associated with cancer [153]. As described before, the regulative subunits of PKA are also differentially expressed in cancer, with the RIα upregulated in a series of neoplasms, whereas the RIIβ inhibits tumour growth. Another level of complexity is introduced by the AKAP, which controls the localisation of PKA near its substrates [165]. Thus, the inhibition of PKA represents a great challenge. PKA inhibitors can be divided into three main categories: ATP antagonists, cAMP antagonists, and the protein kinase inhibitor peptide (PKI).

ATP antagonists that inhibit PKA, such as H89 and KT5720, present one key issue: their IC50 depends on ATP concentration, which can vary greatly inside the cell. In addition, both H89 and KT 5720 have been shown to inhibit other kinases at similar or even lower concentrations to those used to inhibit PKA [248]. H89 has been shown to enhance the cytotoxic effects of the anticancer agent tetrandrine in several cancer cell lines. Cells with higher levels of c-Myc were more sensitive to this treatment combination, and cells resistant to treatment could be sensitized by c-Myc overexpression. These findings were confirmed in a MDA-MB-231 breast cancer murine xenograft tumour model, in which combined therapy showed a significant reduction in tumour weight compared to single treatments. These data suggest that this treatment combination could be effective for tumours expressing high levels of Myc [249]. In addition, H89-mediated inhibition of PKA has been shown to decrease the level of Myc protein in several prostate cancer cell lines due to the impairment of PKA-mediated stabilization of Myc [85].

PKI is the endogenous pseudosubstrate inhibitor of PKA. Its amino acid sequence mimics the inhibitory subunit of the kinase, impeding the binding of cAMP to the PKA catalytic subunit [250]. There are three isoforms of PKI (α, β, γ), which differ in expression pattern and Ki. PKI is more specific for PKA than ATP-competitive inhibitors, but as a peptide, it is less stable than small-molecule inhibitors and cannot pass the cell membrane. Synthetic peptide analogues have been developed, among which is PKI-(6-22)-amide, which has, however, been shown to reduce the toxicity of taxol and taxane in prostate cancer cells [251,252]. In addition, overexpression of PKI has been demonstrated to cause a shift in downstream regulation, increasing ERK phosphorylation following cAMP signalling, and the gene for PKIα has been found to be overexpressed in several cancer types. In DU145 prostate cancer cells, PKIα depletion with shRNA resulted in reduced migration and reduced tumour volume in murine xenografts [253]. PKI has been shown to decrease the stability of the Myc protein in prostate cancer, as observed for H89 [85].

Since several cAMP targets are present in the cell, cAMP analogues such as Rp-cAMP interfere with pathways that only partially overlap with those of other PKA inhibitors. As far as we know, no studies on the correlation between Rp-cAMP and Myc levels have been reported yet, but Rp-cAMP-induced PKA inhibition has been shown to enhance the stimulating effects of prostaglandin E2 on glioblastoma cell lines [254]. Considering the high degree of cAMP and PKA regulation and the ramification of their downstream effects, it should not be surprising that the exact effects of PKA inhibition have yet to be understood.

5.5. PIM Kinases

PIM inhibitors have shown in vitro activity through both Myc-dependent and Myc-independent pathways against numerous tumour cell lines ranging from haematological tumours to breast and prostate cancer [31,177,178,211,214,215], but their results in clinical studies are often disappointing.

SGI-1776 is an ATP-competitive imidazo [1,2-b]pyridazine small molecule highly selective for all three PIM isoforms, with inhibitory concentration values in the nanomolar range and without any relevant effects on cell cycle kinases. At cellular levels, SGI-1776 exhibited promising biological activity, and the SGI-1776-mediated inhibition of PIMs resulted in the reduction in phospho-S62 Myc and total Myc protein abundance in chronic lymphocytic leukaemia cells [212]. Consistent with these findings, SGI-1776 decreased phospho-S62 Myc and impaired the MYC-driven transcriptional program in mantle cell lymphoma models too [213]. Although this compound also shows evident effects in models of breast cancer and diffuse large B-cell lymphoma cell lines (DLBCL), it showed not insignificant cardiotoxic effects in a phase I study (NCT01239108), which was then terminated.

In the landscape of PIM inhibitors, an interesting candidate is NVP-LGB321. LGB321 is a potent ATP-competitive inhibitor able to bind to and prevent the activation of all the three PIM isoforms. Preclinical data demonstrated the capability of LGB321 to reduce phospho-S62 c-Myc levels, whereas no substantial effects were observed on MYC mRNA, suggesting that PIM regulates Myc via post-transcriptional mechanisms [211]. LGH447 (PIM447) is a pan-PIM inhibitor derived from the tool compound LGB231, and although it gave promising results against multiple myeloma in clinical studies, current data suggest that the true potential of this inhibitor relies upon its ability to enhance the efficacy of other anticancer drugs. Paìno and colleagues, in fact, highlighted the synergistic anticancer effects of PIM447 on a panel of different myeloma cell lines and observed a marked decrease in the phosphorylation of c-Myc at S62 and a downregulation of MYC mRNA levels [216]. Another compound, AZD1208, is an orally available and potent drug able to inhibit all three isoforms. It induced cell cycle arrest and cell death in acute myeloid lymphoma cell lines, as well as being active against prostate cancer and acute myeloid leukaemia (AML) xenograft mice models, but lacked clinical efficacy in phase I studies. Even if AZD1208 performed poorly in clinical trials, it was able to enhance the sensitivity to radiation of Myc-CaP cells and of Myc-CaP tumours in nude mice. The effectiveness of this therapy combination might lie in the fact that Pim1 is a pro-survival kinase that is upregulated in response to stressors such as radiation; PIM inhibition might then be crucial to block survival of irradiated cells. At the same time, AZD1208 downregulates p53, which is instead upregulated upon radiation.

Recently, SEL24/MEN1703 has been identified as a first-in-class dual PIM/FLT3 inhibitor and is currently under clinical evaluation for the treatment of acute myeloid leukaemia. SEL24/MEN1703 demonstrated on-target activity as a pan-PIM inhibitor, inhibiting the cell proliferation of various diffuse large B-cell lymphoma (DLBCL) cell lines and DLBCL tumour growth in murine xenograft models. Although the biological effects of SEL24/MEN1703 are not entirely dependent on its activity on Myc, there are at least three major mechanisms attributed to Myc inhibition: PLK1 inhibition, CD20 expression, and CD47 suppression. As previously discussed, PLK1 is a crucial player in Myc stabilization and regulation, and, therefore, the decrease in PLK1 following SEL24/MEN1703 administration may be crucial in thoroughly suppressing Myc-related pathways. CD20 expression was speculated to enhance the activity of anti-CD20 antibodies, and is regulated by the PIM-Myc axis; MYC, in fact, represses the gene encoding CD20. Based on these considerations, the combination of SEL24/MEN1703 and rituximab (an anti-CD20 antibody, which is the standard of care for this type of tumour) resulted in higher complement-dependent toxicity in DLBCL cells compared to rituximab alone. Indeed, increased CD20 surface expression occurred upon Myc downregulation, as well as enhanced transcription of CD20 that was not observable in cells modified to express the T58A Myc mutant. In addition to the increased expression of CD20, the action of rituximab was also enhanced by the Myc-dependent depletion of CD47, a phagocytosis inhibitory signal. These Myc-dependent effects were also shown to be true for other PIM inhibitors such as SGI-1776 and PIM447 [178,216].

In conclusion, the extensive overlapping of PIM kinase phosphorylation patterns with the targets of other kinases has, so far, made it challenging to develop effective treatments through PIM inhibition alone; combined therapies might be the turning point for the application of PIM inhibitors.

5.6. BRD4 Inhibitors

BRD4 inhibitors are also widely used to treat a variety of cancer-related and non-related disorders [188,255]. Bromodomain inhibitors (BETis) are small molecules designed to mimic acetyl-lysine, which is recognized by BRD4 through its BD domains, thereby preventing interactions between BRD4 and chromatin [256]. Given the strong correlation and cross-regulation between BRD4 and Myc, BETis are effective against Myc-driven cancers; indeed, Myc is also used as a readout to predict tumour responsiveness to BETis [257].

It is worth noting that cancer cells can acquire resistance to BETis. Cancer cells compensate for the loss of bromodomain transcriptional activity by upregulating BRD4 downstream pathways, such as Wnt/β-catenin, Hedgehog, and RAS-RAF-MAPK (reviewed by [188]), which restore Myc expression. This highlights the need for other types of interventions or combinatorial drug administration.

To improve the efficacy of BETis, they have been tested in combination with inhibitors of PI3K/AKT [258] and MAPK/ERK1-2 (MEKi) [259] signalling pathways, and with Aurora-A inhibitors in neuroblastoma [260]. Protein kinases are also frequently observed as off-targets of BETis [261]. In order to exploit the multitargeting capacity of bromodomain inhibitors, several BETis have been investigated as dual inhibitors targeting BRD4 and other proteins in preclinical and clinical trials (reviewed by [262]). Some of the dual inhibitors targeting pBRD4 and PLK1, PI3K, ALK, HDAC, p38α, and CDK9, were reported to have a strong downregulation effect on c-Myc or N-Myc in acute myeloid leukaemia, neuroblastoma, castration-resistant prostate cancer, pancreatic cancer, hepatocellular, and colorectal carcinoma (reviewed by [262]).

As observed for Aurora-A, PROTAC technologies against BRD4 result in a stronger effect compared to BET inhibition; treatment with the MZ1 degrader induced a dose-dependent decrease in neuroblastoma cell viability, more than was observed after JQ1 administration, and reduced the protein levels of both N- and c-Myc [263]. In contrast to this observation, Devaiah and colleagues observed an increase in c-Myc half-life in MZ1-treated neuroblastoma cells, due to the loss of destabilizing phosphorylation on T58, whereas treatment with JQ1 did not affect c-Myc turnover [58]. Moreover, the BET degrader dBET6 induced apoptosis in chronic myeloid leukaemia cells, whose proliferation is regulated by Myc and BRD4, and overcame osteoblast-mediated resistance of leukaemia stem cells against BCR/ABL1 tyrosine kinase inhibitors more than JQ1 [241,264].

6. Conclusions

Considering their oncogenic properties, the Myc proteins constitute an attractive therapeutic target for cancer treatment (Figure 8). However, the complex and highly dynamic nature of Myc-regulated pathways and genes presents a significant challenge for the development of small molecule inhibitors that can effectively target these proteins. A deeper understanding of the molecular mechanisms by which Myc proteins contribute to oncogenesis is necessary for the development of effective therapeutic strategies. It is clear that Myc protein levels and activity must be tightly regulated in cells, and this regulation is mediated by a variety of molecular signalling pathways. In several types of cancer, high levels of Myc lead to the uncontrolled expression of its target genes, primarily impacting metabolism and proliferation. Interestingly, Myc also upregulates proteins that help to stabilize it, creating a self-perpetuating regulatory loop. These proteins are often protein kinases involved in metabolism and the cell cycle, which are themselves targets for cancer therapy. Therefore, the availability of inhibitors of kinases that influence the stability of Myc family members may provide a promising and feasible indirect approach to target the "undruggable" Myc, with the potential to uncover new therapeutic opportunities for cancer treatment.

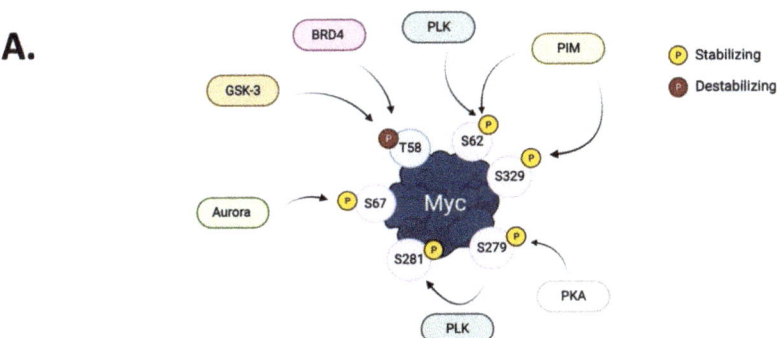

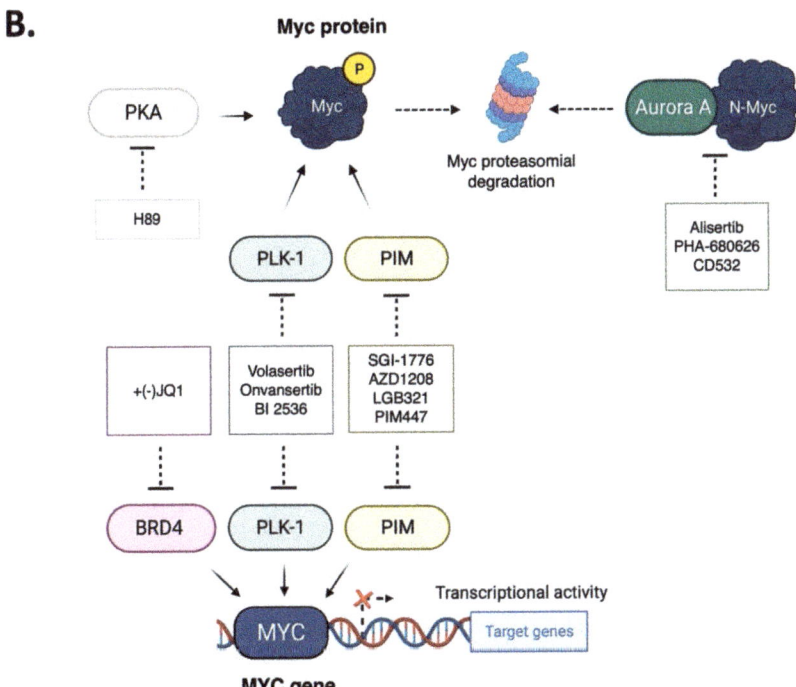

Figure 8. Graphical summary of the kinases and the key phosphorylation sites involved in Myc protein stability are displayed in panel (**A**). Panel (**B**) schematizes the principal inhibitors affecting the pathways exploited to indirectly target Myc at both the protein and gene level. Created with BioRender.com (accessed on 20 January 2023).

Author Contributions: Conceptualization, A.P., D.B. and G.G.; writing—original draft preparation, D.B., E.R. and S.B.; writing—review and editing, A.P.; visualization, G.G. All authors have made a substantial, direct, and intellectual contribution to this work and added their scientific expertise to strengthen, augment and support it. All authors have read and agreed to the published version of the manuscript.

Funding: The authors' studies and findings cited in this review were supported by grants from the Foundation AIRC for Cancer Research (project AIRC MFAG2017-ID.20447; P.I. Alessandro Paiardini and project AIRC IG2021-ID.25648; P.I. Giulia Guarguaglini).

Institutional Review Board Statement: Not applicable.

Informed Consent Statement: Not applicable.

Data Availability Statement: Not applicable.

Conflicts of Interest: The authors declare no conflict of interest.

References

1. Scafuro, M.; Capasso, L.; Carafa, V.; Altucci, L.; Nebbioso, A. Gene transactivation and transrepression in myc-driven cancers. *Int. J. Mol. Sci.* **2021**, *22*, 3458. [CrossRef] [PubMed]
2. Sheiness, D.; Bishop, J.M. DNA and RNA from Uninfected Vertebrate Cells Contain Nucleotide Sequences Related to the Putative Transforming Gene of Avian Myelocytomatosis Virus. *J. Virol.* **1979**, *31*, 514–521. [CrossRef] [PubMed]
3. Roussel, M.; Saule, S.; Lagrou, C.; Rommens, C.; Beug, H.; Graf, T.; Stehelin, D. Three new types of viral oncogene of cellular origin specific for haematopoietic cell transformation. *Nature* **1979**, *281*, 452–455. [CrossRef]
4. Schwab, M.; Alitalo, K.; Klempnauer, K.-H.; Varmus, H.E.; Bishop, J.M.; Gilbert, F.; Brodeur, G.; Goldstein, M.; Trent, J. Amplified DNA with limited homology to myc cellular oncogene is shared by human neuroblastoma cell lines and a neuroblastoma tumour. *Nature* **1983**, *305*, 245–248. [CrossRef] [PubMed]
5. Kohl, N.E.; Kanda, N.; Schreck, R.R.; Bruns, G.; Latt, S.A.; Gilbert, F.; Alt, F.W. Transposition and amplification of oncogene-related sequences in human neuroblastomas. *Cell* **1983**, *35 Pt 1*, 359–367. [CrossRef]
6. Nau, M.M.; Brooks, B.J.; Battey, J.F.; Sausville, E.; Gazdar, A.F.; Kirsch, I.R.; McBride, O.W.; Bertness, V.L.; Hollis, G.F.; Minna, J.D. L-myc, a new myc-related gene amplified and expressed in human small cell lung cancer. *Nature* **1985**, *318*, 69–73. [CrossRef]
7. Das, S.K.; Lewis, B.A.; Levens, D. MYC: A complex problem. *Trends Cell Biol.* **2023**, *33*, 235–246. Available online: http://www.ncbi.nlm.nih.gov/pubmed/35963793 (accessed on 22 February 2023). [CrossRef]
8. Lourenco, C.; Resetca, D.; Redel, C.; Lin, P.; MacDonald, A.S.; Ciaccio, R.; Kenney, T.M.G.; Wei, Y.; Andrews, D.W.; Sunnerhagen, M.; et al. MYC protein interactors in gene transcription and cancer. *Nat Rev Cancer* **2021**, *21*, 579–591. [CrossRef]
9. Conacci-Sorrell, M.; McFerrin, L.; Eisenman, R.N. An overview of MYC and its interactome. *Cold Spring Harb. Perspect. Med.* **2014**, *4*, a014357. [CrossRef]
10. Adhikary, S.; Eilers, M. Transcriptional regulation and transformation by Myc proteins. *Nat. Rev. Mol. Cell. Biol.* **2005**, *6*, 635–645. [CrossRef]
11. Hann, S.R. MYC cofactors: Molecular switches controlling diverse biological outcomes. *Cold Spring Harb. Perspect. Med.* **2014**, *4*, a014399. [CrossRef]
12. Hartl, M. The quest for targets executing MYC-dependent cell transformation. *Front. Oncol.* **2016**, *6*, 132. [CrossRef]
13. Tu, W.B.; Helander, S.; Pilstål, R.; Hickman, K.A.; Lourenco, C.; Jurisica, I.; Raught, B.; Wallner, B.; Sunnerhagen, M.; Penn, L.Z. Myc and its interactors take shape. *Biochim. Biophys. Acta-Gene Regul. Mech.* **2015**, *1849*, 469–483. [CrossRef]
14. Santinon, G.; Enzo, E.; Dupont, S. The sweet side of YAP/TAZ. *Cell Cycle* **2015**, *14*, 2543–2544. [CrossRef]
15. Rennoll, S. Regulation of MYC gene expression by aberrant Wnt/β-catenin signaling in colorectal cancer. *World J. Biol. Chem.* **2015**, *6*, 290. [CrossRef]
16. Zhang, S.; Li, Y.; Wu, Y.; Shi, K.; Bing, L.; Hao, J. Wnt/β-Catenin Signaling Pathway Upregulates c-Myc Expression to Promote Cell Proliferation of P19 Teratocarcinoma Cells. *Anat. Rec.* **2012**, *295*, 2104–2113. [CrossRef]
17. Liu, N.; Wang, S.; Cui, Y.; Shen, L.; Du, Y.; Li, G.; Zhang, G.; Wang, R. Sonic hedgehog elevates N-myc gene expression in neural stem cells. *Neural Regen. Res.* **2012**, *7*, 1703–1708.
18. Palomero, T.; Lim, W.K.; Odom, D.T.; Sulis, M.L.; Real, P.J.; Margolin, A.; Barnes, K.C.; O'Neil, J.; Neuberg, D.; Weng, A.P.; et al. NOTCH1 directly regulates c-MYC and activates a feed-forward-loop transcriptional network promoting leukemic cell growth. *Proc. Natl. Acad. Sci. USA* **2006**, *103*, 18261–18266. [CrossRef]
19. Sanchez-Martin, M.; Ferrando, A. The NOTCH1-MYC highway toward T-cell acute lymphoblastic leukemia. *Blood* **2017**, *129*, 1124–1133. [CrossRef]
20. Huang, L.; Liu, D.; Wang, N.; Ling, S.; Tang, Y.; Wu, J.; Hao, L.; Luo, H.; Hu, X.; Sheng, L.; et al. Integrated genomic analysis identifies deregulated JAK/STAT-MYC-biosynthesis axis in aggressive NK-cell leukemia. *Cell Res.* **2018**, *28*, 172–186. [CrossRef]
21. Jin, W. Role of JAK/STAT3 Signaling in the Regulation of Metastasis, the Transition of Cancer Stem Cells, and Chemoresistance of Cancer by Epithelial–Mesenchymal Transition. *Cells* **2020**, *9*, 217. [CrossRef] [PubMed]
22. Stine, Z.E.; Walton, Z.E.; Altman, B.J.; Hsieh, A.L.; Dang, C.V. MYC, metabolism, and cancer. *Cancer Discov.* **2015**, *5*, 1024–1039. [CrossRef] [PubMed]
23. Battey, J.; Moulding, C.; Taub, R.; Murphy, W.; Stewart, T.; Potter, H.; Lenoir, G.; Leder, P. The human c-myc oncogene: Structural consequences of translocation into the igh locus in Burkitt lymphoma. *Cell* **1983**, *34*, 779–787. [CrossRef] [PubMed]
24. Liu, J.; Levens, D. Making Myc. *Curr. Top. Microbiol. Immunol.* **2006**, *302*, 1–32.

25. Dani, C.; Blanchard, J.M.; Piechaczyk, M.; El Sabouty, S.; Marty, L.; Jeanteur, P. Extreme instability of myc mRNA in normal and transformed human cells. *Proc. Natl. Acad. Sci. USA* **1984**, *81*, 7046–7050. [CrossRef]
26. Hann, S.R.; Eisenman, R.N. Proteins encoded by the human c-myc oncogene: Differential expression in neoplastic cells. *Mol. Cell. Biol.* **1984**, *4*, 2486–2497.
27. Farrell, A.S.; Sears, R.C. MYC degradation. *Cold Spring Harb. Perspect. Med.* **2014**, *4*, 1–16. [CrossRef]
28. González-Prieto, R.; Cuijpers, S.A.G.; Kumar, R.; Hendriks, I.A.; Vertegaal, A.C.O. c-Myc is targeted to the proteasome for degradation in a SUMOylation-dependent manner, regulated by PIAS1, SENP7 and RNF4. *Cell Cycle* **2015**, *14*, 1859–1872. [CrossRef]
29. Kalkat, M.; Chan, P.-K.; Wasylishen, A.R.; Srikumar, T.; Kim, S.S.; Ponzielli, R.; Bazett-Jones, D.P.; Raught, B.; Penn, L.Z. Identification of c-MYC SUMOylation by mass spectrometry. *PLoS ONE* **2014**, *9*, e115337. [CrossRef]
30. Sabò, A.; Doni, M.; Amati, B. SUMOylation of Myc-family proteins. *PLoS ONE* **2014**, *9*, e91072. [CrossRef]
31. Rebello, R.J.; Kusnadi, E.; Cameron, D.P.; Pearson, H.B.; Lesmana, A.; Devlin, J.R.; Drygin, D.; Clark, A.K.; Porter, L.; Pedersen, J.; et al. The dual inhibition of RNA Pol I transcription and PIM kinase as a new therapeutic approach to treat advanced prostate cancer. *Clin. Cancer Res.* **2016**, *22*, 5539–5552. [CrossRef]
32. Höllein, A.; Fallahi, M.; Schoeffmann, S.; Steidle, S.; Schaub, F.X.; Rudelius, M.; Laitinen, I.; Nilsson, L.; Goga, A.; Peschel, C.; et al. Myc-induced SUMOylation is a therapeutic vulnerability for B-cell lymphoma. *Blood* **2014**, *124*, 2081–2090. [CrossRef]
33. Suna, X.X.; Chena, Y.; Sua, Y.; Wanga, X.; Chauhana, K.M.; Lianga, J.; Daniel, C.J.; Sears, R.C.; Dai, M.-S. SUMO protease SENP1 deSUMOylates and stabilizes c-Myc. *Proc. Natl. Acad. Sci. USA* **2018**, *115*, 10983–10988. [CrossRef]
34. Sun, X.X.; Li, Y.; Sears, R.C.; Dai, M.S. Targeting the MYC Ubiquitination-Proteasome Degradation Pathway for Cancer Therapy. *Front. Oncol.* **2021**, *11*, 679445. [CrossRef]
35. Henriksson, M.; Bakardjiev, A.; Klein, G.; Luscher, B. Phosphorylation sites mapping in the N-terminal domain of c-myc modulate its transforming potential. *Oncogene* **1993**, *8*, 3199–3209.
36. Hann, S.R. Role of post-translational modifications in regulating c-Myc proteolysis, transcriptional activity and biological function. *Semin. Cancer Biol.* **2006**, *16*, 288–302. [CrossRef]
37. Benassi, B.; Fanciulli, M.; Fiorentino, F.; Porrello, A.; Chiorino, G.; Loda, M.; Zupi, G.; Biroccio, A. c-Myc phosphorylation is required for cellular response to oxidative stress. *Mol. Cell* **2006**, *21*, 509–519. [CrossRef]
38. Gregory, M.A.; Qi, Y.; Hann, S.R. Phosphorylation by Glycogen Synthase Kinase-3 Controls c-Myc Proteolysis and Subnuclear Localization. *J. Biol. Chem.* **2003**, *278*, 51606–51612. [CrossRef]
39. Bahram, F.; Von Der Lehr, N.; Cetinkaya, C.; Larsson, L.G. c-Myc hot spot mutations in lymphomas result in inefficient ubiquitination and decreased proteasome-mediated turnover. *Blood* **2000**, *95*, 2104–2110. [CrossRef]
40. Salghetti, S.E.; Kim, S.Y.; Tansey, W.P. Destruction of Myc by ubiquitin-mediated proteolysis: Cancer-associated and transforming mutations stabilize Myc. *EMBO J.* **1999**, *18*, 717–726. [CrossRef]
41. Wang, X.; Cunningham, M.; Zhang, X.; Tokarz, S.; Laraway, B.; Troxell, M.; Sears, R.C. Phosphorylation regulates c-Myc's oncogenic activity in the mammary gland. *Cancer Res.* **2011**, *71*, 925–936. [CrossRef] [PubMed]
42. Hemann, M.T.; Bric, A.; Teruya-Feldstein, J.; Herbst, A.; Nilsson, J.A.; Cordon-Cardo, C.; Cleveland, J.L.; Tansey, W.P.; Lowe, S.W. Evasion of the p53 tumour surveillance network by tumour-derived MYC mutants. *Nature* **2005**, *436*, 807–811. [CrossRef] [PubMed]
43. Johnson, N.A. Functional and clinical impact of MYC mutations in diffuse large B cell lymphomas. *Transl Cancer Res.* **2016**, *5*, S257–S260. [CrossRef]
44. Huang, Z.; Traugh, J.A.; Bishop, J.M. Negative Control of the Myc Protein by the Stress-Responsive Kinase Pak2. *Mol. Cell. Biol.* **2004**, *24*, 1582–1594. [CrossRef]
45. Macek, P.; Cliff, M.J.; Embrey, K.J.; Holdgate, G.A.; Nissink, J.W.M.; Panova, S.; Waltho, J.P.; Davies, R.A. Myc phosphorylation in its basic helix?loop?helix region destabilizes transient-helical structures, disrupting Max and DNA binding. *J. Biol. Chem.* **2018**, *293*, 9301–9310. [CrossRef]
46. Albihn, A.; Johnsen, J.I.; Henriksson, M.A. MYC in oncogenesis and as a target for cancer therapies. *Adv. Cancer Res.* **2010**, *107*, 163–224.
47. Herbst, A.; Salghetti, S.E.; Kim, S.Y.; Tansey, W.P. Multiple cell-type-specific elements regulate Myc protein stability. *Oncogene* **2004**, *23*, 3863–3871. [CrossRef]
48. Cowling, V.H.; Chandriani, S.; Whitfield, M.L.; Cole, M.D. A Conserved Myc Protein Domain, MBIV, Regulates DNA Binding, Apoptosis, Transformation, and G 2 Arrest. *Mol. Cell. Biol.* **2006**, *26*, 4226–4239. [CrossRef]
49. Vervoorts, J.; Lüscher-Firzlaff, J.; Lüscher, B. The ins and outs of MYC regulation by posttranslational mechanisms. *J. Biol. Chem.* **2006**, *281*, 34725–34729. [CrossRef]
50. Xiao, D.; Yue, M.; Su, H.; Ren, P.; Jiang, J.; Li, F.; Hu, Y.; Du, H.; Liu, H.; Qing, G. Polo-like Kinase-1 Regulates Myc Stabilization and Activates a Feedforward Circuit Promoting Tumor Cell Survival. *Mol. Cell* **2016**, *64*, 493–506. [CrossRef]
51. Popov, N.; Schülein, C.; Jaenicke, L.A.; Eilers, M. Ubiquitylation of the amino terminus of Myc by SCFβ-TrCP antagonizes SCFFbw7-mediated turnover. *Nat. Cell Biol.* **2010**, *12*, 973–981. [CrossRef]
52. Otto, T.; Horn, S.; Brockmann, M.; Eilers, U.; Schüttrumpf, L.; Popov, N.; Kenney, A.M.; Schulte, J.H.; Beijersbergen, R.; Christiansen, H.; et al. Stabilization of N-Myc Is a Critical Function of Aurora A in Human Neuroblastoma. *Cancer Cell* **2009**, *15*, 67–78. [CrossRef]

53. Jiang, J.; Wang, J.; Yue, M.; Cai, X.; Wang, T.; Wu, C.; Su, H.; Wang, Y.; Han, M.; Zhang, Y.; et al. Direct Phosphorylation and Stabilization of MYC by Aurora B Kinase Promote T-cell Leukemogenesis. *Cancer Cell* **2020**, *37*, 200–215.e5. [CrossRef]
54. Naso, F.D.; Boi, D.; Ascanelli, C.; Pamfil, G.; Lindon, C.; Paiardini, A.; Guarguaglini, G. Nuclear localisation of Aurora-A: Its regulation and significance for Aurora-A functions in cancer. *Oncogene* **2021**, *40*, 3917–3928. [CrossRef]
55. Chan, G.K.L.; Maisel, S.; Hwang, Y.C.; Wolber, R.R.B.; Vu, P.; Patra, C.; Bouhaddou, M.; Kenerson, H.L.; Yeung, R.S.; Swaney, D.L.; et al. Oncogenic PKA signaling stabilizes MYC oncoproteins via an aurora kinase A-dependent mechanism. *bioRxiv* 2021. [CrossRef]
56. Zhang, Y.; Wang, Z.; Li, X.; Magnuson, N.S. Pim kinase-dependent inhibition of c-Myc degradation. *Oncogene* **2008**, *27*, 4809–4819. [CrossRef]
57. Zippo, A.; De Robertis, A.; Serafini, R.; Oliviero, S. PIM1-dependent phosphorylation of histone H3 at serine 10 is required for MYC-dependent transcriptional activation and oncogenic transformation. *Nat. Cell Biol.* **2007**, *9*, 932–944. [CrossRef]
58. Devaiah, B.N.; Mu, J.; Akman, B.; Uppal, S.; Weissman, J.D.; Cheng, D.; Baranello, L.; Nie, Z.; Levens, D.; Singer, D.S. MYC protein stability is negatively regulated by BRD4. *Proc. Natl. Acad. Sci. USA* **2020**, *117*, 13457–13467. [CrossRef]
59. Pérez de Castro, I.; de Cárcer, G.; Malumbres, M. A census of mitotic cancer genes: New insights into tumor cell biology and cancer therapy. *Carcinogenesis* **2007**, *28*, 899–912. [CrossRef]
60. Malumbres, M.; Barbacid, M. Cell cycle kinases in cancer. *Curr. Opin. Genet. Dev.* **2007**, *17*, 60–65. [CrossRef]
61. Colicino, E.G.; Hehnly, H. Regulating a key mitotic regulator, polo-like kinase 1 (PLK1). *Cytoskeleton* **2018**, *75*, 481–494. [CrossRef] [PubMed]
62. Combes, G.; Alharbi, I.; Braga, L.G.; Elowe, S. Playing polo during mitosis: PLK1 takes the lead. *Oncogene* **2017**, *36*, 4819–4827. [CrossRef] [PubMed]
63. Chiappa, M.; Petrella, S.; Damia, G.; Broggini, M.; Guffanti, F.; Ricci, F. Present and Future Perspective on PLK1 Inhibition in Cancer Treatment. *Front. Oncol.* **2022**, *12*, 903016. [CrossRef] [PubMed]
64. Iliaki, S.; Beyaert, R.; Afonina, I.S. Polo-like kinase 1 (PLK1) signaling in cancer and beyond. *Biochem. Pharmacol.* **2021**, *193*, 114747. [CrossRef]
65. Liu, Z.; Sun, Q.; Wang, X. PLK1, A potential target for cancer therapy. *Transl. Oncol.* **2017**, *10*, 22–32. [CrossRef]
66. Montaudon, E.; Nikitorowicz-Buniak, J.; Sourd, L.; Morisset, L.; EL Botty, R.; Huguet, L.; Dahmani, A.; Painsec, P.; Nemati, F.; Vacher, S.; et al. PLK1 inhibition exhibits strong anti-tumoral activity in CCND1-driven breast cancer metastases with acquired palbociclib resistance. *Nat. Commun.* **2020**, *11*, 4053. [CrossRef]
67. Nieto-Jimenez, C.; Galan-Moya, E.M.; Corrales-Sanchez, V.; Noblejas-Lopez, M.D.M.; Burgos, M.; Domingo, B.; Montero, J.C.; Gomez-Juarez, M.; Picazo-Martinez, M.G.; Esparis-Ogando, A.; et al. Inhibition of the mitotic kinase PLK1 overcomes therapeutic resistance to BET inhibitors in triple negative breast cancer. *Cancer Lett.* **2020**, *491*, 50–59. [CrossRef]
68. Zhang, Z.; Cheng, L.; Li, J.; Qiao, Q.; Karki, A.; Allison, D.B.; Shaker, N.; Li, K.; Utturkar, S.M.; Atallah Lanman, N.M.; et al. Targeting Plk1 sensitizes pancreatic cancer to immune checkpoint therapy. *Cancer Res.* **2022**, *82*, 3532–3548. [CrossRef]
69. Bibi, N.; Parveen, Z.; Rashid, S. Identification of Potential Plk1 Targets in a Cell-Cycle Specific Proteome through Structural Dynamics of Kinase and Polo Box-Mediated Interactions. *PLoS ONE* **2013**, *8*, e70843. [CrossRef]
70. Liu, J.; Zhang, C. The equilibrium of ubiquitination and deubiquitination at PLK1 regulates sister chromatid separation. *Cell. Mol. Life Sci.* **2017**, *74*, 2127–2134. [CrossRef]
71. Raab, M.; Matthess, Y.; Raab, C.A.; Gutfreund, N.; Dötsch, V.; Becker, S.; Sanhaji, M.; Strebhardt, K. A dimerization-dependent mechanism regulates enzymatic activation and nuclear entry of PLK1. *Oncogene* **2021**, *41*, 372–386. [CrossRef]
72. Xu, J.; Shen, C.; Wang, T.; Quan, J. Structural basis for the inhibition of Polo-like kinase 1. *Nat Struct Mol Biol* **2013**, *20*, 1047–1053. [CrossRef]
73. Beck, J.; Maerki, S.; Posch, M.; Metzger, T.; Persaud, A.; Scheel, H.; Hofmann, K.; Rotin, D.; Pedrioli, P.; Swedlow, J.R.; et al. Ubiquitylation-dependent localization of PLK1 in mitosis. *Nat. Cell Biol.* **2013**, *15*, 430–439. [CrossRef]
74. Kachaner, D.; Garrido, D.; Mehsen, H.; Normandin, K.; Lavoie, H.; Archambault, V. Coupling of Polo kinase activation to nuclear localization by a bifunctional NLS is required during mitotic entry. *Nat. Commun.* **2017**, *8*, 1701. [CrossRef]
75. Zhou, J.; Yang, Q.; Lu, L.; Tuo, Z.; Shou, Z.; Cheng, J. Plk1 inhibition induces immunogenic cell death and enhances immunity against nsclc. *Int. J. Med. Sci.* **2021**, *18*, 3516–3525. [CrossRef]
76. Fu, Z.; Wen, D. The emerging role of polo-like kinase 1 in epithelial-mesenchymal transition and tumor metastasis. *Cancers* **2017**, *9*, 131. [CrossRef]
77. Song, R.; Hou, G.; Yang, J.; Yuan, J.; Wang, C.; Chai, T.; Liu, Z. Effects of PLK1 on proliferation, invasion and metastasis of gastric cancer cells through epithelial-mesenchymal transition. *Oncol. Lett.* **2018**, *16*, 5739–5744. [CrossRef]
78. Gao, Z.; Man, X.; Li, Z.; Bi, J.; Liu, X.; Li, Z.; Li, J.; Zhang, Z.; Kong, C. PLK1 promotes proliferation and suppresses apoptosis of renal cell carcinoma cells by phosphorylating MCM3. *Cancer Gene Ther.* **2020**, *27*, 412–423. [CrossRef]
79. Luo, P.; Yan, H.; Du, J.; Chen, X.; Shao, J.; Zhang, Y.; Xu, Z.; Jin, Y.; Lin, N.; Yang, B.; et al. PLK1 (polo like kinase 1)-dependent autophagy facilitates gefitinib-induced hepatotoxicity by degrading COX6A1 (cytochrome c oxidase subunit 6A1). *Autophagy* **2021**, *17*, 3221–3237. [CrossRef]
80. Oon, M.L.; Hoppe, M.M.; Fan, S.; Phyu, T.; Phuong, H.M.; Tan, S.-Y.; Hue, S.S.-S.; Wang, S.; Poon, L.M.; Chan, H.L.E.; et al. The contribution of MYC and PLK1 expression to proliferative capacity in diffuse large B-cell lymphoma. *Leuk Lymphoma* **2019**, *60*, 3214–3224. [CrossRef]

81. Ren, Y.; Bi, C.; Zhao, X.; Lwin, T.; Wang, C.; Yuan, J.; Silva, A.S.; Shah, B.D.; Fang, B.; Li, T.; et al. PLK1 stabilizes a MYC-dependent kinase network in aggressive B cell lymphomas. *J. Clin. Investig.* **2018**, *128*, 5531–5548. [CrossRef]
82. Yu, Z.; Deng, P.; Chen, Y.; Liu, S.; Chen, J.; Yang, Z.; Chen, J.; Fan, X.; Wang, P.; Cai, Z.; et al. Inhibition of the PLK1-Coupled Cell Cycle Machinery Overcomes Resistance to Oxaliplatin in Colorectal Cancer. *Adv. Sci.* **2021**, *8*, 2100759. [CrossRef] [PubMed]
83. Murga-Zamalloa, C.; Polk, A.; Hanel, W.; Chowdhury, P.; Brown, N.; Hristov, A.C.; Bailey, N.G.; Wang, T.; Phillips, T.; Devata, S.; et al. Polo-like-kinase 1 (PLK-1) and c-myc inhibition with the dual kinase-bromodomain inhibitor volasertib in aggressive lymphomas. *Oncotarget* **2017**, *8*, 114474–114480. [CrossRef] [PubMed]
84. Tan, J.; Li, Z.; Lee, P.L.; Guan, P.; Aau, M.Y.; Lee, S.T.; Feng, M.; Lim, C.Z.; Lee, E.Y.J.; Wee, Z.N.; et al. PDK1 signaling toward PLK1-MYC activation confers oncogenic transformation, tumor-initiating cell activation, and resistance to mTOR-targeted therapy. *Cancer Discov.* **2013**, *3*, 1156–1171. [CrossRef] [PubMed]
85. Padmanabhan, A.; Li, X.; Bieberich, C.J. Protein kinase a regulates MYC protein through transcriptional and post-translational mechanisms in a catalytic subunit isoform-specific manner. *J. Biol. Chem.* **2013**, *288*, 14158–14169. [CrossRef]
86. Wang, D.; Pierce, A.; Veo, B.; Fosmire, S.; Danis, E.; Donson, A.; Venkataraman, S.; Vibhakar, R. A regulatory loop of FBXW7-MYC-PLK1 controls tumorigenesis of MYC-driven medulloblastoma. *Cancers* **2021**, *13*, 387. [CrossRef]
87. Mo, H.; He, J.; Yuan, Z.; Wu, Z.; Liu, B.; Lin, X.; Guan, J. PLK1 contributes to autophagy by regulating MYC stabilization in osteosarcoma cells. *Onco Targets Ther.* **2019**, *12*, 7527–7536. [CrossRef]
88. Quartuccio, S.M.; Schindler, K. Functions of Aurora kinase C in meiosis and cancer. *Front. Cell Dev. Biol.* **2015**, *3*, 50. [CrossRef]
89. Willems, E.; Dedobbeleer, M.; Digregorio, M.; Lombard, A.; Lumapat, P.N.; Rogister, B. The functional diversity of Aurora kinases: A comprehensive review. *Cell Div.* **2018**, *13*, 7. [CrossRef]
90. Joukov, V.; De Nicolo, A. Aurora-PLK1 cascades as key signaling modules in the regulation of mitosis. *Sci. Signal.* **2018**, *11*, eaar4195. [CrossRef]
91. Gallini, S.; Carminati, M.; De Mattia, F.; Pirovano, L.; Martini, E.; Oldani, A.; Asteriti, I.A.; Guarguaglini, G.; Mapelli, M. NuMA phosphorylation by aurora-a orchestrates spindle orientation. *Curr. Biol.* **2016**, *26*, 458–469. [CrossRef]
92. Polverino, F.; Naso, F.D.; Asteriti, I.A.; Palmerini, V.; Singh, D.; Valente, D.; Bird, A.W.; Rosa, A.; Mapelli, M.; Guarguaglini, G. The Aurora-A/TPX2 Axis Directs Spindle Orientation in Adherent Human Cells by Regulating NuMA and Microtubule Stability. *Curr. Biol.* **2021**, *31*, 658–667.e5. [CrossRef]
93. Carmena, M.; Wheelock, M.; Funabiki, H.; Earnshaw, W.C. The chromosomal passenger complex (CPC): From easy rider to the godfather of mitosis. *Nat. Rev. Mol. Cell. Biol.* **2012**, *13*, 789–803. [CrossRef]
94. Van Der Horst, A.; Lens, S.M.A. Cell division: Control of the chromosomal passenger complex in time and space. *Chromosoma* **2014**, *123*, 25–42. [CrossRef]
95. Tang, A.; Gao, K.; Chu, L.; Zhang, R.; Yang, J.; Zheng, J. Aurora kinases: Novel therapy targets in cancers. *Oncotarget* **2017**, *8*, 23937–23954. [CrossRef]
96. Gautschi, O.; Heighway, J.; Mack, P.C.; Purnell, P.R.; Lara, P.N.; Gandara, D.R. Aurora kinases as anticancer drug targets. *Clin. Cancer Res.* **2008**, *14*, 1639–1648. [CrossRef]
97. Lens, S.M.A.; Voest, E.E.; Medema, R.H. Shared and separate functions of polo-like kinases and aurora kinases in cancer. *Nat. Rev. Cancer* **2010**, *10*, 825–841. [CrossRef]
98. Mou, P.K.; Yang, E.J.; Shi, C.; Ren, G.; Tao, S.; Shim, J.S. Aurora kinase A, a synthetic lethal target for precision cancer medicine. *Exp. Mol. Med.* **2021**, *53*, 835–847. [CrossRef]
99. Asteriti, I.A.; Rensen, W.M.; Lindon, C.; Lavia, P.; Guarguaglini, G. The Aurora-A/TPX2 complex: A novel oncogenic holoenzyme? *Biochim Biophys Act-Rev. Cancer* **2010**, *1806*, 230–239. [CrossRef]
100. Wan, X.-B.; Long, Z.-J.; Yan, M.; Xu, J.; Xia, L.-P.; Liu, L.; Zhao, Y.; Huang, X.-F.; Wang, X.-R.; Zhu, X.-F.; et al. Inhibition of Aurora-A suppresses epithelial-mesenchymal transition and invasion by downregulating MAPK in nasopharyngeal carcinoma cells. *Carcinogenesis* **2008**, *29*, 1930–1937. [CrossRef]
101. Liu, X.; Li, Z.; Song, Y.; Wang, R.; Han, L.; Wang, Q.; Jiang, K.; Kang, C.; Zhang, Q. AURKA induces EMT by regulating histone modification through Wnt/β-catenin and PI3K/Akt signaling pathway in gastric cancer. *Oncotarget* **2016**, *7*, 33152–33164. [CrossRef] [PubMed]
102. Xia, Z.; Wei, P.; Zhang, H.; Ding, Z.; Yang, L.; Huang, Z.; Zhang, N. AURKA governs self-renewal capacity in glioma-initiating cells via stabilization/activation of β-catenin/Wnt Signaling. *Mol Cancer Res.* **2013**, *11*, 1101–1111. [CrossRef] [PubMed]
103. Lin, Z.-Z.; Jeng, Y.-M.; Hu, F.-C.; Pan, H.-W.; Tsao, H.-W.; Lai, P.-L.; Lee, P.-H.; Cheng, A.-L.; Hsu, H.-C. Significance of Aurora B overexpression in hepatocellular carcinoma. Aurora B Overexpression in HCC. *BMC Cancer* **2010**, *10*, 461. [CrossRef] [PubMed]
104. Vischioni, B.; Oudejans, J.; Vos, W.; Rodriguez, J.A.; Giaccone, G. Frequent overexpression of aurora B kinase, a novel drug target, in non-small cell lung carcinoma patients. *Mol. Cancer Ther.* **2006**, *5*, 2905–2913. [CrossRef]
105. Qi, G.; Ogawa, I.; Kudo, Y.; Miyauchi, M.; Siriwardena, B.S.M.S.; Shimamoto, F.; Tatsuka, M.; Takata, T. Aurora-B expression and its correlation with cell proliferation and metastasis in oral cancer. *Virchows Arch.* **2007**, *450*, 297–302. [CrossRef]
106. Den Hollander, J.; Rimpi, S.; Doherty, J.R.; Rudelius, M.; Buck, A.; Hoellein, A.; Kremer, M.; Graf, N.; Scheerer, M.; Hall, M.A.; et al. Aurora kinases A and B are up-regulated by Myc and are essential for maintenance of the malignant state. *Blood* **2010**, *116*, 1498–1505. [CrossRef]

107. Dauch, D.; Rudalska, R.; Cossa, G.; Nault, J.C.; Kang, T.-W.; Wuestefeld, T.; Hohmeyer, A.; Imbeaud, S.; Yevsa, T.; Hoenicke, L.; et al. A MYC-aurora kinase A protein complex represents an actionable drug target in p53-altered liver cancer. *Nat. Med.* **2016**, *22*, 744–753. [CrossRef]
108. Beltran, H.; Rickman, D.S.; Park, K.; Chae, S.S.; Sboner, A.; MacDonald, T.Y.; Wang, Y.; Sheikh, K.L.; Terry, S.; Tagawa, S.T.; et al. Molecular characterization of neuroendocrine prostate cancer and identification of new drug targets. *Cancer Discov.* **2011**, *1*, 487–495. [CrossRef]
109. Vader, G.; Lens, S.M.A. The Aurora kinase family in cell division and cancer. *Biochim. Biophys. Acta-Rev. Cancer* **2008**, *1786*, 60–72. [CrossRef]
110. Büchel, G.; Carstensen, A.; Mak, K.-Y.; Roeschert, I.; Leen, E.; Sumara, O.; Hofstetter, J.; Herold, S.; Kalb, J.; Baluapuri, A.; et al. Association with Aurora-A Controls N-MYC-Dependent Promoter Escape and Pause Release of RNA Polymerase II during the Cell Cycle. *Cell Rep.* **2017**, *21*, 3483–3497. [CrossRef]
111. Richards, M.W.; Burgess, S.G.; Poon, E.; Carstensen, A.; Eilers, M.; Chesler, L.; Bayliss, R. Structural basis of N-Myc binding by Aurora-A and its destabilization by kinase inhibitors. *Proc. Natl. Acad. Sci. USA* **2016**, *113*, 13726–13731. [CrossRef]
112. Lu, L.; Han, H.; Tian, Y.; Li, W.; Zhang, J.; Feng, M.; Li, Y. Aurora kinase A mediates c-Myc's oncogenic effects in hepatocellular carcinoma. *Mol. Carcinog.* **2015**, *54*, 1467–1479. [CrossRef]
113. Zheng, F.; Yue, C.; Li, G.; He, B.; Cheng, W.; Wang, X.; Yan, M.; Long, Z.; Qiu, W.; Yuan, Z.; et al. Nuclear AURKA acquires kinase-independent transactivating function to enhance breast cancer stem cell phenotype. *Nat. Commun.* **2016**, *7*, 10180. [CrossRef]
114. Hsueh, K.W.; Fu, S.L.; Huang, C.Y.F.; Lin, C.H. Aurora-A phosphorylates hnRNPK and disrupts its interaction with p53. *FEBS Lett.* **2011**, *585*, 2671–2675. [CrossRef]
115. Ho, J.S.L.; Ma, W.; Mao, D.Y.L.; Benchimol, S. p53-Dependent Transcriptional Repression of c-myc Is Required for G1 Cell Cycle Arrest. *Mol. Cell. Biol.* **2005**, *25*, 7423–7431. [CrossRef]
116. Santoro, A.; Vlachou, T.; Luzi, L.; Melloni, G.; Mazzarella, L.; D'Elia, E.; Aobuli, X.; Pasi, C.E.; Reavie, L.; Bonetti, P.; et al. p53 Loss in Breast Cancer Leads to Myc Activation, Increased Cell Plasticity, and Expression of a Mitotic Signature with Prognostic Value. *Cell Rep.* **2019**, *26*, 624–638.e8. [CrossRef]
117. Bayliss, R.; Sardon, T.; Ebert, J.; Lindner, D.; Vernos, I.; Conti, E. Determinants for Aurora-A activation and Aurora-B discrimination by TPX2. *Cell Cycle* **2004**, *3*, 402–405. [CrossRef]
118. Fu, J.; Bian, M.; Liu, J.; Jiang, Q.; Zhang, C. A single amino acid change converts Aurora-A into Aurora-B-like kinase in terms of partner specificity and cellular function. *Proc. Natl. Acad. Sci. USA* **2009**, *106*, 6939–6944. [CrossRef]
119. Eyers, P.A.; Churchill, M.E.A.; Maller, J.L. The Aurora A and Aurora B protein kinases: A single amino acid difference controls intrinsic activity and activation by TPX2. *Cell Cycle* **2005**, *4*, 784–789. [CrossRef]
120. Hans, F.; Skoufias, D.A.; Dimitrov, S.; Margolis, R.L. Molecular distinctions between Aurora A and B: A single residue change transforms Aurora A into correctly localized and functional Aurora B. *Mol. Biol. Cell* **2009**, *20*, 3491–3502. [CrossRef]
121. DeLuca, K.F.; Meppelink, A.; Broad, A.J.; Mick, J.E.; Peersen, O.B.; Pektas, S.; Lens, S.M.; DeLuca, J.G. Aurora A kinase phosphorylates Hec1 to regulate metaphase kinetochore-microtubule dynamics. *J. Cell Biol.* **2018**, *217*, 163–177. [CrossRef] [PubMed]
122. Berwanger, B.; Hartmann, O.; Bergmann, E.; Bernard, S.; Nielsen, D.; Krause, M.; Kartal, A.; Flynn, D.; Wiedemeyer, R.; Schwab, M.; et al. Loss of a FYN-regulated differentiation and growth arrest pathway in advanced stage neuroblastoma. *Cancer Cell* **2002**, *2*, 377–386. [CrossRef] [PubMed]
123. Brockmann, M.; Poon, E.; Berry, T.; Carstensen, A.; Deubzer, H.E.; Rycak, L.; Jamin, Y.; Thway, K.; Robinson, S.P.; Roels, F.; et al. Small Molecule Inhibitors of Aurora-A Induce Proteasomal Degradation of N-Myc in Childhood Neuroblastoma. *Cancer Cell* **2013**, *24*, 75–89. [CrossRef] [PubMed]
124. Bogen, D.; Wei, J.S.; Azorsa, D.O.; Ormanoglu, P.; Buehler, E.; Guha, R.; Keller, J.M.; Griner, L.A.M.; Ferrer, M.; Song, Y.K.; et al. Aurora B kinase is a potent and selective target in MYCN-driven neuroblastoma. *Oncotarget* **2015**, *6*, 35247–35262. [CrossRef] [PubMed]
125. Borah, N.A.; Sradhanjali, S.; Barik, M.R.; Jha, A.; Tripathy, D.; Kaliki, S.; Rath, S.; Raghav, S.K.; Patnaik, S.; Mittal, R.; et al. Aurora kinase B expression, its regulation and therapeutic targeting in human retinoblastoma. *Investig. Ophthalmol. Vis. Sci.* **2021**, *62*, 16. [CrossRef]
126. Li, M.; Sun, C.; Bu, X.; Que, Y.; Zhang, L.; Zhang, Y.; Zhang, L.; Lu, S.; Huang, J.; Zhu, J.; et al. ISL1 promoted tumorigenesis and EMT via Aurora kinase A-induced activation of PI3K/AKT signaling pathway in neuroblastoma. *Cell Death Dis.* **2021**, *12*, 620. [CrossRef]
127. Dar, A.A.; Belkhiri, A.; El-Rifai, W. The aurora kinase A regulates GSK-3β in gastric cancer cells. *Oncogene* **2009**, *28*, 866–875. [CrossRef]
128. Katayama, H.; Sasai, K.; Kawai, H.; Yuan, Z.-M.; Bondaruk, J.; Suzuki, F.; Fujii, S.; Arlinghaus, R.B.; Czerniak, B.A.; Sen, S. Phosphorylation by aurora kinase A induces Mdm2-mediated destabilization and inhibition of p53. *Nat. Genet.* **2004**, *36*, 55–62. [CrossRef]
129. Kaidanovich-Beilin, O.; Woodgett, J.R. GSK-3: Functional Insights from Cell Biology and Animal Models. *Front. Mol. Neurosci.* **2011**, *4*, 40. [CrossRef]
130. Doble, B.W.; Woodgett, J.R. GSK-3: Tricks of the trade for a multi-tasking kinase. *J. Cell Sci.* **2003**, *116*, 1175–1186. [CrossRef]

131. Duda, P.; Akula, S.M.; Abrams, S.L.; Steelman, L.S.; Martelli, A.M.; Cocco, L.; Ratti, S.; Candido, S.; Libra, M.; Montalto, G.; et al. Targeting GSK3 and Associated Signaling Pathways Involved in Cancer. *Cells* **2020**, *9*, 1110. [CrossRef]
132. McCubrey, J.A.; Rakus, D.; Gizak, A.; Steelman, L.S.; Abrams, S.L.; Lertpiriyapong, K.; Fitzgerald, T.L.; Yang, L.V.; Montalto, G.; Cervello, M.; et al. Effects of mutations in Wnt/β-catenin, hedgehog, Notch and PI3K pathways on GSK-3 activity—Diverse effects on cell growth, metabolism and cancer. *Biochim. Biophys. Acta-Mol. Cell Res.* **2016**, *1863*, 2942–2976. [CrossRef]
133. Beurel, E.; Grieco, S.F.; Jope, R.S. Glycogen synthase kinase-3 (GSK3): Regulation, actions, and diseases. *Pharmacol. Ther.* **2015**, *148*, 114–131. [CrossRef]
134. Papadopoli, D.; Pollak, M.; Topisirovic, I. The role of GSK3 in metabolic pathway perturbations in cancer. *Biochim. Biophys. Acta-Mol. Cell Res.* **2021**, *1868*, 119059. [CrossRef]
135. Sears, R.; Nuckolls, F.; Haura, E.; Taya, Y.; Tamai, K.; Nevins, J.R. Multiple Ras-dependent phosphorylation pathways regulate Myc protein stability. *Genes Dev.* **2000**, *14*, 2501–2514. [CrossRef]
136. Shah, K.; Kazi, J.U. Phosphorylation-Dependent Regulation of WNT/Beta-Catenin Signaling. *Front. Oncol.* **2022**, *12*, 858782. [CrossRef]
137. Stamos, J.L.; Weis, W.I. The β-catenin destruction complex. *Cold Spring Harb. Perspect. Biol.* **2013**, *5*, a007898. [CrossRef]
138. Jamieson, C.; Sharma, M.; Henderson, B.R. Regulation of β-catenin nuclear dynamics by GSK-3β involves a LEF-1 positive feedback loop. *Traffic* **2011**, *12*, 983–999. [CrossRef]
139. Jaworski, T.; Banach-Kasper, E.; Gralec, K. GSK-3β at the intersection of neuronal plasticity and neurodegeneration. *Neural Plast.* **2019**, *2019*, 4209475. [CrossRef]
140. Duffy, D.J.; Krstic, A.; Schwarzl, T.; Higgins, D.G.; Kolch, W. GSK3 inhibitors regulate MYCN mRNA levels and reduce neuroblastoma cell viability through multiple mechanisms, including p53 and Wnt signaling. *Mol. Cancer Ther.* **2014**, *13*, 454–467. [CrossRef]
141. Kazi, A.; Xiang, S.; Yang, H.; Delitto, D.; Trevino, J.; Jiang, R.H.Y.; Ayaz, M.; Lawrence, H.R.; Kennedy, P.; Sebti, S.M. GSK3 suppression upregulates β-catenin and c-Myc to abrogate KRas-dependent tumors. *Nat. Commun.* **2018**, *9*, 5154. [CrossRef] [PubMed]
142. Mccubrey, J.A.; Steelman, L.S.; Bertrand, F.E.; Davis, N.M.; Sokolosky, M.; Abrams, S.L.; Montalto, G.; Antonino, B.; Libra, M.; Nicoletti, F.; et al. GSK-3 as potential target for therapeutic irvention in cancer. *Oncotarget* **2014**, *5*, 2881–2911. [CrossRef] [PubMed]
143. Jacobsen, A.; Bosch, L.J.W.; Martens-De Kemp, S.R.; Carvalho, B.; Sillars-Hardebol, A.H.; Dobson, R.J.; De Rinaldis, E.; Meijer, G.A.; Abeln, S.; Heringa, J.; et al. Aurora kinase A (AURKA) interaction with Wnt and Ras-MAPK signalling pathways in colorectal cancer. *Sci. Rep.* **2018**, *8*, 7522. [CrossRef] [PubMed]
144. Fang, X.; Yu, S.X.; Lu, Y.; Bast, R.C.; Woodgett, J.R.; Mills, G.B. Phosphorylation and inactivation of glycogen synthase kinase 3 by protein kinase A. *Proc. Natl. Acad. Sci. USA* **2000**, *97*, 11960–11965. [CrossRef] [PubMed]
145. Hoxhaj, G.; Manning, B.D. The PI3K–AKT network at the interface of oncogenic signalling and cancer metabolism. *Nat. Rev. Cancer* **2020**, *20*, 74–88. [CrossRef]
146. Hermida, M.A.; Dinesh Kumar, J.; Leslie, N.R. GSK3 and its interactions with the PI3K/AKT/mTOR signalling network. *Adv. Biol. Regul.* **2017**, *65*, 5–15. [CrossRef]
147. Gaggianesi, M.; Mangiapane, L.R.; Modica, C.; Pantina, V.D.; Porcelli, G.; Di Franco, S.; Iacono, M.L.; D'Accardo, C.; Verona, F.; Pillitteri, I.; et al. Dual Inhibition of Myc Transcription and PI3K Activity Effectively Targets Colorectal Cancer Stem Cells. *Cancers* **2022**, *14*, 673. [CrossRef]
148. Sun, K.; Atoyan, R.; Borek, M.A.; Dellarocca, S.; Samson, M.E.S.; Ma, A.W.; Xu, G.X.; Patterson, T.; Tuck, D.P.; Viner, J.L.; et al. Dual HDAC and PI3K inhibitor CUDC-907 down regulates MYC and suppresses growth of MYC-dependent cancers. *Mol. Cancer Ther.* **2017**, *16*, 285–299. [CrossRef]
149. Swords, M.F.R.T.; Schenk, T.; Stengel, S.; Gil, V.S.; Petrie, K.R.; Perez, A.; Ana, R.; Watts, J.M.; Vargas, F.; Elias, R.; et al. Inhibition of the PI3K/AKT/mTOR Pathway Leads to Down-Regulation of c-Myc and Overcomes Resistance to ATRA in Acute Myeloid Leukemia. *Blood* **2015**, *126*, 1363. [CrossRef]
150. Taylor, S.S.; Zhang, P.; Steichen, J.M.; Keshwani, M.M.; Kornev, A.P. PKA: Lessons learned after twenty years. *Biochim. Biophys. Acta-Proteins Proteom.* **2013**, *1834*, 1271–1278. [CrossRef]
151. Reikhardt, B.A.; Shabanov, P.D. Catalytic Subunit of PKA as a Prototype of the Eukaryotic Protein Kinase Family. *Biochem* **2020**, *85*, 409–424. [CrossRef]
152. Taskén, K.; Aandahl, E.M. Localized Effects of cAMP Mediated by Distinct Routes of Protein Kinase A. *Physiol. Rev.* **2004**, *84*, 137–167. [CrossRef]
153. Turnham, R.E.; Scott, J.D. Protein kinase A catalytic subunit isoform PRKACA. History, function and physiology. *Gene* **2016**, *577*, 101–108. [CrossRef]
154. Zhu, Y.R.; Jiang, X.X.; Zheng, Y.; Xiong, J.; Wei, D.; Zhang, D.M. Cardiac function modulation depends on the A-kinase anchoring protein complex. *J. Cell. Mol. Med.* **2019**, *23*, 7170–7179. [CrossRef]
155. Colombe, A.S.; Pidoux, G. Cardiac camp-pka signaling compartmentalization in myocardial infarction. *Cells* **2021**, *10*, 922. [CrossRef] [PubMed]
156. Dell'Acqua, M.L.; Smith, K.E.; Gorski, J.A.; Horne, E.A.; Gibson, E.S.; Gomez, L.L. Regulation of neuronal PKA signaling through AKAP targeting dynamics. *Eur. J. Cell Biol.* **2006**, *85*, 627–633. [CrossRef]

157. Greggio, E.; Bubacco, L.; Russo, I. Cross-talk between LRRK2 and PKA: Implication for Parkinson's disease? *Biochem. Soc. Trans.* **2017**, *45*, 261–267. [CrossRef]
158. Gao, F.; Yang, S.; Wang, J.; Zhu, G. cAMP-PKA cascade: An outdated topic for depression? *Biomed Pharm.* **2022**, *150*, 113030. [CrossRef]
159. Skålhegg, B.S.; Tasken, K. Institute for Nutrition Reserarch, and 2 Institute of Medical Biochemistry, University of Oslo, Norway. *Nutrition* **2000**, *5*, 678–693.
160. Daniel, P.B.; Walker, W.H.; Habener, J.F. Cyclic amp signaling and gene regulation. *Annu. Rev. Nutr.* **1998**, *18*, 353–383. [CrossRef] [PubMed]
161. Wu, K.J.; Mattioli, M.; Morse, H.C.; Dalla-Favera, R. c-MYC activates protein kinase A (PKA) by direct transcriptional activation of the PKA catalytic subunit beta (PKA-Cβ) gene. *Oncogene* **2002**, *21*, 7872–7882. [CrossRef]
162. Rajabi, H.N.; Baluchamy, S.; Kolli, S.; Nag, A.; Srinivas, R.; Raychaudhuri, P.; Thimmapaya, B. Effects of depletion of CREB-binding protein on c-Myc regulation and cell cycle G1-S transition. *J. Biol. Chem.* **2005**, *280*, 361–374. [CrossRef]
163. Kolli, S.; Buchmann, A.M.; Williams, J.; Weitzman, S.; Thimmapaya, B. Antisense-mediated depletion of p300 in human cells leads to premature exit and up-regulation of c-MYC. *Proc. Natl. Acad. Sci. USA* **2001**, *98*, 4646–4651. [CrossRef]
164. Johannessen, M.; Moens, U. Multisite phosphorylation of the cAMP response element-binding protein (CREB) by a diversity of protein kinases. *Front. Biosci.* **2007**, *12*, 1814–1832.
165. Naviglio, S.; Caraglia, M.; Abbruzzese, A.; Chiosi, E.; Di Gesto, D.; Marra, M.; Romano, M.; Sorrentino, A.; Sorvillo, L.; Spina, A.; et al. Protein kinase A as a biological target in cancer therapy. *Expert Opin. Ther. Targets* **2009**, *13*, 83–92. [CrossRef]
166. Santio, N.M.; Koskinen, P.J. PIM kinases: From survival factors to regulators of cell motility. *Int. J. Biochem. Cell Biol.* **2017**, *93*, 74–85. [CrossRef]
167. Warfel, N.A.; Kraft, A.S. PIM kinase (and Akt) biology and signaling in tumors. *Pharmacol. Ther.* **2015**, *151*, 41–49. [CrossRef]
168. Chen, J.; Tang, G. PIM-1 kinase: A potential biomarker of triple-negative breast cancer. *OncoTargets Ther.* **2019**, *12*, 6267–6273. [CrossRef]
169. Eerola, S.K.; Kohvakka, A.; Tammela, T.L.J.; Koskinen, P.J.; Latonen, L.; Visakorpi, T. Expression and ERG regulation of PIM kinases in prostate cancer. *Cancer Med.* **2021**, *10*, 3427–3436. [CrossRef]
170. Mondello, P.; Cuzzocrea, S.; Mian, M. PIM kinases in hematological malignancies: Where are we now and where are we going? *J. Hematol. Oncol.* **2014**, *7*, 95. [CrossRef]
171. Zhang, X.; Song, M.; Kundu, J.K.; Lee, M.-H.; Liu, Z.-Z. PIM Kinase as an Executional Target in Cancer. *J. Cancer Prev.* **2018**, *23*, 109–116. [CrossRef] [PubMed]
172. Bullock, A.N.; Debreczeni, J.; Amos, A.L.; Knapp, S.; Turk, B.E. Structure and substrate specificity of the Pim-1 kinase. *J. Biol. Chem.* **2005**, *280*, 41675–41682. [CrossRef] [PubMed]
173. Qian, K.C.; Wang, L.; Hickey, E.R.; Studts, J.; Barringer, K.; Peng, C.; Kronkaitis, A.; Li, J.; White, A.; Mische, S.; et al. Structural basis of constitutive activity and a unique nucleotide binding mode of human Pim-1 kinase. *J. Biol. Chem.* **2005**, *280*, 6130–6137. [CrossRef] [PubMed]
174. Arrouchi, H.; Lakhlili, W.; Ibrahimi, A. A review on PIM kinases in tumors. *Bioinformation* **2019**, *15*, 40–45. [CrossRef]
175. Julson, J.R.; Marayati, R.; Beierle, E.A.; Stafman, L.L. The Role of PIM Kinases in Pediatric Solid Tumors. *Cancers* **2022**, *14*, 3565. [CrossRef]
176. Wang, J.; Kim, J.; Roh, M.; E Franco, O.; Hayward, S.W.; Wills, M.L.; A Abdulkadir, S. Pim1 kinase synergizes with c-MYC to induce advanced prostate carcinoma. *Oncogene* **2010**, *29*, 2477–2487. [CrossRef]
177. Wu, J.; Chu, E.; Kang, Y. Pim kinases in multiple myeloma. *Cancers* **2021**, *13*, 4304. [CrossRef]
178. Szydłowski, M.; Garbicz, F.; Jabłońska, E.; Górniak, P.; Komar, D.; Pyrzyńska, B.; Bojarczuk, K.; Prochorec-Sobieszek, M.; Szumera-Ciećkiewicz, A.; Rymkiewicz, G.; et al. Inhibition of PIM Kinases in DLBCL Targets MYC Transcriptional Program and Augments the Efficacy of Anti-CD20 Antibodies. *Cancer Res.* **2021**, *81*, 6029–6043. [CrossRef]
179. Beharry, Z.; Mahajan, S.; Zemskova, M.; Lin, Y.-W.; Tholanikunnel, B.G.; Xia, Z.; Smith, C.D.; Kraft, A.S. The Pim protein kinases regulate energy metabolism and cell growth. *Proc. Natl. Acad. Sci. USA* **2011**, *108*, 528–533. [CrossRef]
180. Fox, C.J.; Hammerman, P.S.; Cinalli, R.M.; Master, S.R.; Chodosh, L.A.; Thompson, C.B. The serine/threonine kinase Pim-2 is a transcriptionally regulated apoptotic inhibitor. *Genes Dev.* **2003**, *17*, 1841–1854. [CrossRef]
181. Yan, B.; Zemskova, M.; Holder, S.; Chin, V.; Kraft, A.; Koskinen, P.J.; Lilly, M. The PIM-2 Kinase Phosphorylates BAD on Serine 112 and Reverses BAD-induced Cell Death. *J. Biol. Chem.* **2003**, *278*, 45358–45367. [CrossRef]
182. Lee, J.S.; Smith, E.; Shilatifard, A. The Language of Histone Crosstalk. *Cell* **2010**, *142*, 682–685. [CrossRef]
183. Zippo, A.; Serafini, R.; Rocchigiani, M.; Pennacchini, A.; Krepelova, A.; Oliviero, S. Histone Crosstalk between H3S10ph and H4K16ac Generates a Histone Code that Mediates Transcription Elongation. *Cell* **2009**, *138*, 1122–1136. [CrossRef]
184. Brasó-Maristany, F.; Filosto, S.; Catchpole, S.; Marlow, R.; Quist, J.; Francesch-Domenech, E.; A Plumb, D.; Zakka, L.; Gazinska, P.; Liccardi, G.; et al. PIM1 kinase regulates cell death, tumor growth and chemotherapy response in triple-negative breast cancer. *Nat. Med.* **2016**, *22*, 1303–1313. [CrossRef]
185. Chandriani, S.; Frengen, E.; Cowling, V.H.; Pendergrass, S.A.; Perou, C.M.; Whitfield, M.L.; Cole, M.D. A core MYC gene expression signature is prominent in basal-like breast cancer but only partially overlaps the core serum response. *PLoS ONE* **2009**, *4*, e6693. [CrossRef]

186. Forshell, L.P.; Li, Y.; Forshell, T.Z.P.; Rudelius, M.; Nilsson, L.; Keller, U.; Nilsson, J. The direct Myc target Pim3 cooperates with other Pim kinases in supporting viability of Myc-induced B-cell lymphomas. *Oncotarget* **2011**, *2*, 448–460. [CrossRef]
187. Zhang, C.; Qie, Y.; Yang, T.; Wang, L.; Du, E.; Liu, Y.; Xu, Y.; Qiao, B.; Zhang, Z. Kinase PIM1 promotes prostate cancer cell growth via c-Myc-RPS7-driven ribosomal stress. *Carcinogenesis* **2019**, *40*, 52–60. [CrossRef]
188. Liang, Y.; Tian, J.; Wu, T. BRD4 in physiology and pathology: "BET" on its partners. *BioEssays* **2021**, *43*, 2100180. [CrossRef]
189. Wang, N.; Wu, R.; Tang, D.; Kang, R. The BET family in immunity and disease. *Signal Transduct. Target. Ther.* **2021**, *6*, 23. [CrossRef]
190. Kotekar, A.; Singh, A.K.; Devaiah, B.N. BRD4 and MYC: Power couple in transcription and disease. *FEBS J.* **2022**, 1–23. [CrossRef]
191. Weissman, J.D.; Singh, A.K.; Devaiah, B.N.; Schuck, P.; LaRue, R.C.; Singer, D.S. The intrinsic kinase activity of BRD4 spans its BD2-B-BID domains. *J. Biol. Chem.* **2021**, *297*. [CrossRef] [PubMed]
192. Zhang, S.; Roeder, R.G. The Long and the Short of BRD4: Two Tales in Breast Cancer. *Mol. Cell* **2020**, *78*, 993–995. [CrossRef] [PubMed]
193. Ba, M.; Long, H.; Yan, Z.; Wang, S.; Wu, Y.; Tu, Y.; Gong, Y.; Cui, S. BRD4 promotes gastric cancer progression through the transcriptional and epigenetic regulation of c-MYC. *J. Cell Biochem.* **2018**, *119*, 973–982. [CrossRef] [PubMed]
194. Zhao, R.; Nakamura, T.; Fu, Y.; Lazar, Z.; Spector, D.L. Gene bookmarking accelerates the kinetics of post-mitotic transcriptional re-activation. *Nat. Cell Biol.* **2011**, *13*, 1295–1304. [CrossRef]
195. Dey, A.; Nishiyama, A.; Karpova, T.; McNally, J.; Ozato, K. Brd4 marks select genes on mitotic chromatin and directs postmitotic transcription. *Mol. Biol. Cell* **2009**, *20*, 4899–4909. [CrossRef]
196. Devaiah, B.N.; Case-Borden, C.; Gegonne, A.; Hsu, C.H.; Chen, Q.; Meerzaman, D.; Dey, A.; Ozato, K.; Singer, D.S. BRD4 is a histone acetyltransferase that evicts nucleosomes from chromatin. *Nat. Struct. Mol. Biol.* **2016**, *23*, 540–548. [CrossRef]
197. Jaenicke, L.A.; von Eyss, B.; Carstensen, A.; Wolf, E.; Xu, W.; Greifenberg, A.K.; Geyer, M.; Eilers, M.; Popov, N. Ubiquitin-Dependent Turnover of MYC Antagonizes MYC/PAF1C Complex Accumulation to Drive Transcriptional Elongation. *Mol. Cell* **2016**, *61*, 54–67. [CrossRef]
198. Beaulieu, M.; Castillo, F.; Soucek, L. Structural and Biophysical Insights into the Function of the Intrinsically Disordered Myc Oncoprotein. *Cells* **2020**, *9*, 1038. [CrossRef]
199. Berg, T.; Cohen, S.B.; Desharnais, J.; Sonderegger, C.; Maslyar, D.J.; Goldberg, J.; Boger, D.L.; Vogt, P.K. Small-molecule antagonists of Myc/Max dimerization inhibit Myc-induced transformation of chicken embryo fibroblasts. *Proc. Natl. Acad. Sci. USA* **2002**, *99*, 3830–3835. [CrossRef]
200. Yin, X.; Giap, C.; Lazo, J.S.; Prochownik, E.V. Low molecular weight inhibitors of Myc-Max interaction and function. *Oncogene* **2003**, *22*, 6151–6159. [CrossRef]
201. Kiessling, A.; Sperl, B.; Hollis, A.; Eick, D.; Berg, T. Selective Inhibition of c-Myc/Max Dimerization and DNA Binding by Small Molecules. *Chem. Biol.* **2006**, *13*, 745–751. [CrossRef]
202. Beaulieu, M.-E.; Jauset, T.; Massó-Vallés, D.; Martínez-Martín, S.; Rahl, P.; Maltais, L.; Zacarias-Fluck, M.F.; Casacuberta-Serra, S.; del Pozo, E.S.; Fiore, C.; et al. Intrinsic cell-penetrating activity propels omomyc from proof of concept to viable anti-myc therapy. *Sci. Transl. Med.* **2019**, *11*, aar5012. [CrossRef]
203. Soucek, L.; Helmer-Citterich, M.; Sacco, A.; Jucker, R.; Cesareni, G.; Nasi, S. Design and properties of a Myc derivative that efficiently homodimerizes. *Oncogene* **1998**, *17*, 2463–2472. [CrossRef]
204. Gargini, R.; Segura-Collar, B.; Garranzo-Asensio, M.; Hortigüela, R.; Iglesias-Hernández, P.; Lobato-Alonso, D.; Moreno-Raja, M.; Esteban-Martin, S.; Sepúlveda-Sánchez, J.M.; Nevola, L.; et al. IDP-410: A Novel Therapeutic Peptide that Alters N-MYC Stability and Reduces Angiogenesis and Tumor Progression in Glioblastomas. *Neurotherapeutics* **2022**, *19*, 408–420. [CrossRef]
205. Roskoski, R. Properties of FDA-approved small molecule protein kinase inhibitors: A 2021 update. *Pharmacol. Res.* **2021**, *165*, 105463. [CrossRef]
206. Timme, N.; Han, Y.; Liu, S.; Yosief, H.O.; García, H.D.; Bei, Y.; Klironomos, F.; MacArthur, I.; Szymansky, A.; von Stebut, J.; et al. Small-Molecule Dual PLK1 and BRD4 Inhibitors are Active Against Preclinical Models of Pediatric Solid Tumors. *Transl. Oncol.* **2020**, *13*, 221–232. [CrossRef]
207. Wang, D.; Veo, B.; Pierce, A.; Fosmire, S.; Madhavan, K.; Balakrishnan, I.; Donson, A.; Alimova, I.; Sullivan, K.D.; Joshi, M.; et al. A novel PLK1 inhibitor onvansertib effectively sensitizes MYC-driven medulloblastoma to radiotherapy. *Neuro Oncol.* **2022**, *24*, 414–426. [CrossRef]
208. Gustafson, W.C.; Meyerowitz, J.G.; Nekritz, E.A.; Chen, J.; Benes, C.; Charron, E.; Simonds, E.F.; Seeger, R.; Matthay, K.K.; Hertz, N.T.; et al. Drugging MYCN through an Allosteric Transition in Aurora Kinase A. *Cancer Cell* **2014**, *26*, 414–427. [CrossRef]
209. Lee, J.K.; Phillips, J.W.; Smith, B.A.; Park, J.W.; Stoyanova, T.; McCaffrey, E.F.; Baertsch, R.; Sokolov, A.; Meyerowitz, J.G.; Mathis, C.; et al. N-Myc Drives Neuroendocrine Prostate Cancer Initiated from Human Prostate Epithelial Cells. *Cancer Cell* **2016**, *29*, 536–547. [CrossRef]
210. Boi, D.; Souvalidou, F.; Capelli, D.; Polverino, F.; Marini, G.; Montanari, R.; Pochetti, G.; Tramonti, A.; Contestabile, R.; Trisciuoglio, D.; et al. PHA-680626 Is an Effective Inhibitor of the Interaction between Aurora-A and N-Myc. *Int. J. Mol. Sci.* **2021**, *22*, 13122. [CrossRef]
211. Horiuchi, D.; Camarda, R.; Zhou, A.; Yau, C.; Momcilovic, O.; Balakrishnan, S.; Corella, A.N.; Eyob, H.; Kessenbrock, K.; Lawson, D.A.; et al. PIM1 kinase inhibition as a targeted therapy against triple-negative breast tumors with elevated MYC expression. *Nat. Med.* **2016**, *22*, 1321–1329. [CrossRef] [PubMed]

212. Chen, L.S.; Redkar, S.; Bearss, D.; Wierda, W.G.; Gandhi, V. Pim kinase inhibitor, SGI-1776, induces apoptosis in chronic lymphocytic leukemia cells. *Blood* **2009**, *114*, 4150–4157. [CrossRef] [PubMed]
213. Yang, Q.; Chen, L.S.; Neelapu, S.S.; Miranda, R.N.; Medeiros, L.J.; Gandhi, V. Transcription and translation are primary targets of Pim kinase inhibitor SGI-1776 in mantle cell lymphoma. *Blood* **2012**, *120*, 3491–3500. [CrossRef] [PubMed]
214. Keeton, E.K.; McEachern, K.; Dillman, K.S.; Palakurthi, S.; Cao, Y.; Grondine, M.R.; Kaur, S.; Wang, S.; Chen, Y.; Wu, A.; et al. AZD1208, a potent and selective pan-Pim kinase inhibitor, demonstrates efficacy in preclinical models of acute myeloid leukemia. *Blood* **2014**, *123*, 905–913. [CrossRef]
215. Kirschner, A.N.; Wang, J.; Van Der Meer, R.; Anderson, P.D.; Franco-Coronel, O.E.; Kushner, M.H.; Everett, J.H.; Hameed, O.; Keeton, E.K.; Ahdesmaki, M.; et al. PIM kinase inhibitor AZD1208 for treatment of MYC-driven prostate cancer. *J. Natl. Cancer Inst.* **2015**, *107*, dju407. [CrossRef]
216. Paíno, T.; González-Méndez, L.; San-Segundo, L.; Corchete, L.; Hernández-García, S.; Díaz-Tejedor, A.; Algarín, E.; Mogollón, P.; Martín-Sánchez, M.; Gutiérrez, N.; et al. Protein translation inhibition is involved in the activity of the pan-pim kinase inhibitor pim447 in combination with pomalidomide-dexamethasone in multiple myeloma. *Cancers* **2020**, *12*, 2743. [CrossRef]
217. Puissant, A.; Frumm, S.M.; Alexe, G.; Bassil, C.F.; Qi, J.; Chanthery, Y.H.; Nekritz, E.A.; Zeid, R.; Gustafson, W.C.; Greninger, P.; et al. Targeting MYCN in neuroblastoma by BET bromodomain inhibition. *Cancer Discov.* **2013**, *3*, 308–323. [CrossRef]
218. Dickinson, M.; Briones, J.; Herrera, A.F.; González-Barca, E.; Ghosh, N.; Cordoba, R.; Rutherford, S.C.; Bournazou, E.; Labriola-Tompkins, E.; Franjkovic, I.; et al. Phase 1b study of the BET protein inhibitor RO6870810 with venetoclax and rituximab in patients with diffuse large B-cell lymphoma. *Blood Adv.* **2021**, *5*, 4762–4770. [CrossRef]
219. Shapiro, G.I.; LoRusso, P.; Dowlati, A.; TDo, K.; Jacobson, C.A.; Vaishampayan, U.; Weise, A.; Caimi, P.F.; Eder, J.P.; French, C.A.; et al. A Phase 1 study of RO6870810, a novel bromodomain and extra-terminal protein inhibitor, in patients with NUT carcinoma, other solid tumours, or diffuse large B-cell lymphoma. *Br. J. Cancer* **2021**, *124*, 744–753. [CrossRef]
220. Wang, N.Y.; Xu, Y.; Xiao, K.J.; Zuo, W.Q.; Zhu, Y.X.; Hu, R.; Wang, W.L.; Shi, Y.J.; Yu, L.T.; Liu, Z.H. Design, synthesis, and biological evaluation of 4,5-dihydro-[1,2,4]triazolo [4,3-f]pteridine derivatives as novel dual-PLK1/BRD4 inhibitors. *Eur. J. Med. Chem.* **2020**, *191*, 112152. [CrossRef]
221. Xu, Y.; Wang, Q.; Xiao, K.; Liu, Z.; Zhao, L.; Song, X.; Hu, X.; Feng, Z.; Gao, T.; Zuo, W.; et al. Novel dual BET and PLK1 inhibitor WNY0824 exerts potent antitumor effects in CRPC by inhibiting transcription factor function and inducing mitotic abnormality. *Mol. Cancer Ther.* **2020**, *19*, 1221–1231. [CrossRef]
222. Andrews, F.H.; Singh, A.R.; Joshi, S.; Smith, C.A.; Morales, G.A.; Garlich, J.R.; Durden, D.L.; Kutateladze, T.G. Dual-activity PI3K-BRD4 inhibitor for the orthogonal inhibition of MYC to block tumor growth and metastasis. *Proc. Natl. Acad. Sci. USA* **2017**, *114*, E1072–E1080. [CrossRef]
223. Joshi, S.; Singh, A.R.; Liu, K.X.; Pham, T.V.; Zulcic, M.; Skola, D.; Chun, H.B.; Glass, C.K.; Morales, G.A.; Garlich, J.R.; et al. SF2523: Dual PI3K/BRD4 inhibitor blocks tumor immunosuppression and promotes adaptive immune responses in cancer. *Mol. Cancer Ther.* **2019**, *18*, 1036–1044. [CrossRef]
224. Pan, Z.; Li, X.; Wang, Y.; Jiang, Q.; Jiang, L.; Zhang, M.; Zhang, N.; Wu, F.; Liu, B.; He, G. Discovery of Thieno [2,3- d]pyrimidine-Based Hydroxamic Acid Derivatives as Bromodomain-Containing Protein 4/Histone Deacetylase Dual Inhibitors Induce Autophagic Cell Death in Colorectal Carcinoma Cells. *J. Med. Chem.* **2020**, *63*, 3678–3700. [CrossRef]
225. Shao, M.; He, L.; Zheng, L.; Huang, L.; Zhou, Y.; Wang, T.; Chen, Y.; Shen, M.; Wang, F.; Yang, Z.; et al. Structure-based design, synthesis and in vitro antiproliferative effects studies of novel dual BRD4/HDAC inhibitors. *Bioorganic Med. Chem. Lett.* **2017**, *27*, 4051–4055. [CrossRef]
226. Cheng, G.; Wang, Z.; Yang, J.; Bao, Y.; Xu, Q.; Zhao, L.; Liu, D. Design, synthesis and biological evaluation of novel indole derivatives as potential HDAC/BRD4 dual inhibitors and anti-leukemia agents. *Bioorganic Chem.* **2019**, *84*, 410–417. [CrossRef]
227. He, S.; Dong, G.; Li, Y.; Wu, S.; Wang, W.; Sheng, C. Potent Dual BET/HDAC Inhibitors for Efficient Treatment of Pancreatic Cancer. *Angew. Chem.-Int. Ed.* **2020**, *59*, 3028–3032. [CrossRef]
228. Divakaran, A.; Talluri, S.K.; Ayoub, A.M.; Mishra, N.K.; Cui, H.; Widen, J.C.; Berndt, N.; Zhu, J.-Y.; Carlson, A.S.; Topczewski, J.J.; et al. Molecular Basis for the N-Terminal Bromodomain-and-Extra-Terminal-Family Selectivity of a Dual Kinase-Bromodomain Inhibitor. *J. Med. Chem.* **2018**, *61*, 9316–9334. [CrossRef]
229. Eckerdt, F. Polo-like kinase 1 inhibitors SBE13 and BI 2536 induce different responses in primary cells. *Cell Cycle* **2011**, *10*, 1030–1031. [CrossRef]
230. Reindl, W.; Yuan, J.; Krämer, A.; Strebhardt, K.; Berg, T. Inhibition of Polo-like Kinase 1 by Blocking Polo-Box Domain-Dependent Protein-Protein Interactions. *Chem. Biol.* **2008**, *15*, 459–466. [CrossRef]
231. Awad, M.M.; Chu, Q.S.-C.; Gandhi, L.; Stephenson, J.J.; Govindan, R.; Bradford, D.S.; Bonomi, P.D.; Ellison, D.M.; Eaton, K.D.; Fritsch, H.; et al. An open-label, phase II study of the polo-like kinase-1 (Plk-1) inhibitor, BI 2536, in patients with relapsed small cell lung cancer (SCLC). *Lung Cancer* **2017**, *104*, 126–130. [CrossRef] [PubMed]
232. Mross, K.; Dittrich, C.; Aulitzky, W.E.; Strumberg, D.; Schutte, J.; Schmid, R.M.; Hollerbach, S.; Merger, M.; Munzert, G.; Fleischer, F.; et al. A randomised phase II trial of the Polo-like kinase inhibitor BI 2536 in chemo-nave patients with unresectable exocrine adenocarcinoma of the pancreas-a study within the Central European Society Anticancer Drug Research (CESAR) collaborative network. *Br. J. Cancer* **2012**, *107*, 280–286. [CrossRef] [PubMed]
233. Gjertsen, B.T.; Schöffski, P. Discovery and development of the Polo-like kinase inhibitor volasertib in cancer therapy. *Leukemia* **2015**, *29*, 11–19. [CrossRef]

234. Stadler, W.M.; Vaughn, D.J.; Sonpavde, G.; Vogelzang, N.J.; Tagawa, S.T.; Petrylak, D.P.; Rosen, P.; Lin, C.-C.; Mahoney, J.; Modi, S.; et al. An open-label, single-arm, phase 2 trial of the polo-like kinase inhibitor volasertib (BI 6727) in patients with locally advanced or metastatic urothelial cancer. *Cancer* **2014**, *120*, 976–982. [CrossRef] [PubMed]
235. Ellis, P.M.; Leighl, N.B.; Hirsh, V.; Reaume, M.N.; Blais, N.; Wierzbicki, R.; Sadrolhefazi, B.; Gu, Y.; Liu, D.; Pilz, K.; et al. A randomized, open-label phase II trial of volasertib as monotherapy and in combination with standard-dose pemetrexed compared with pemetrexed monotherapy in second-line treatment for non-small-cell lung cancer. *Clin. Lung Cancer* **2015**, *16*, 457–465. [CrossRef]
236. Ciceri, P.; Müller, S.; O'mahony, A.; Fedorov, O.; Filippakopoulos, P.; Hunt, J.P.; Lasater, E.A.; Pallares, G.; Picaud, S.; Wells, C.; et al. Dual kinase-bromodomain inhibitors for rationally designed polypharmacology. *Nat. Chem. Biol.* **2014**, *10*, 305–312. [CrossRef]
237. Pajtler, K.W.; Sadowski, N.; Ackermann, S.; Althoff, K.; Schönbeck, K.; Batzke, K.; Schäfers, S.; Odersky, A.; Heukamp, L.; Astrahantseff, K.; et al. The GSK461364 PLK1 inhibitor exhibits strong antitumoral activity in preclinical neuroblastoma models. *Oncotarget* **2017**, *8*, 6730–6741. [CrossRef]
238. Beria, I.; Bossi, R.T.; Brasca, M.G.; Caruso, M.; Ceccarelli, W.; Fachin, G.; Fasolini, M.; Forte, B.; Fiorentini, F.; Pesenti, E.; et al. NMS-P937, a 4,5-dihydro-1H-pyrazolo [4,3-h]quinazoline derivative as potent and selective Polo-like kinase 1 inhibitor. *Bioorganic Med. Chem. Lett.* **2011**, *21*, 2969–2974. [CrossRef]
239. Weiss, G.J.; Jameson, G.; Von Hoff, D.D.; Valsasina, B.; Davite, C.; Di Giulio, C.; Fiorentini, F.; Alzani, R.; Carpinelli, P.; Di Sanzo, A.; et al. Phase I dose escalation study of NMS-1286937, an orally available Polo-Like Kinase 1 inhibitor, in patients with advanced or metastatic solid tumors. *Investig. New Drugs* **2018**, *36*, 85–95. [CrossRef]
240. Jing, X.L.; Chen, S.W. Aurora kinase inhibitors: A patent review (2014–2020). *Expert. Opin. Ther. Pat.* **2021**, *31*, 625–644. [CrossRef]
241. Pérez Fidalgo, J.A.; Roda, D.; Roselló, S.; Rodríguez-Braun, E.; Cervantes, A. Aurora kinase inhibitors: A new class of drugs targeting the regulatory mitotic system. *Clin. Transl. Oncol.* **2009**, *11*, 787–798. [CrossRef]
242. Meyerowitz, J.G.; Weiss, W.A.; Gustafson, W.C. A new "angle" on kinase inhibitor design: Prioritizing amphosteric activity above kinase inhibition. *Mol. Cell Oncol.* **2015**, *2*, 10–12. [CrossRef]
243. Lake, E.W.; Muretta, J.M.; Thompson, A.R.; Rasmussen, D.M.; Majumdar, A.; Faber, E.B.; Ruff, E.F.; Thomas, D.D.; Levinson, N.M. Quantitative conformational profiling of kinase inhibitors reveals origins of selectivity for Aurora kinase activation states. *Proc. Natl. Acad. Sci. USA* **2018**, *115*, E11894–E11903. [CrossRef]
244. Adhikari, B.; Bozilovic, J.; Diebold, M.; Schwarz, J.D.; Hofstetter, J.; Schröder, M.; Wanior, M.; Narain, A.; Vogt, M.; Stankovic, N.D.; et al. PROTAC-mediated degradation reveals a non-catalytic function of AURORA-A kinase. *Nat. Chem. Biol.* **2020**, *16*, 1179–1188. [CrossRef]
245. Wang, R.; Ascanelli, C.; Abdelbaki, A.; Fung, A.; Rasmusson, T.; Michaelides, I.; Roberts, K.; Lindon, C. Selective targeting of non-centrosomal AURKA functions through use of a targeted protein degradation tool. *Commun. Biol.* **2021**, *4*, 640. [CrossRef]
246. Tang, J.; Moorthy, R.; Demir, Ö.; Baker, Z.D.; Naumann, J.A.; Jones, K.F.; Grillo, M.J.; Haefner, E.S.; Shi, K.; Levy, M.J.; et al. Targeting N-Myc in Neuroblastoma with Selective Aurora Kinase A Degraders. *bioRxiv* **2022**. [CrossRef]
247. Wagner, F.F.; Benajiba, L.; Campbell, A.J.; Weïwer, M.; Sacher, J.R.; Gale, J.P.; Ross, L.; Puissant, A.; Alexe, G.; Conway, A.; et al. Exploiting an asp-glu "switch" in glycogen synthase kinase 3 to design paralog-selective inhibitors for use in acute myeloid leukemia. *Sci. Transl. Med.* **2018**, *10*, eaam8460. [CrossRef]
248. Andrew, J. Murray. Pharmacological PKA Inhibition:All May Not Be What It Seems. *Sci. Signal.* **2008**, *1*, re4.
249. Yu, M.; Liu, T.; Chen, Y.; Li, Y.; Li, W. Combination therapy with protein kinase inhibitor H89 and Tetrandrine elicits enhanced synergistic antitumor efficacy. *J Exp Clin. Cancer Res.* **2018**, *37*, 114. [CrossRef]
250. Dalton, G.D.; Dewey, W.L. Protein kinase inhibitor peptide (PKI): A family of endogenous neuropeptides that modulate neuronal cAMP-dependent protein kinase function. *Neuropeptides* **2006**, *40*, 23–34. [CrossRef]
251. Zynda, E.R.; Matveev, V.; Makhanov, M.; Chenchik, A.; Kandel, E.S. Protein kinase a type II-α regulatory subunit regulates the response of prostate cancer cells to taxane treatment. *Cell Cycle* **2014**, *13*, 3292–3301. [CrossRef] [PubMed]
252. Liu, C.; Ke, P.; Zhang, J.; Zhang, X.; Chen, X. Protein Kinase Inhibitor Peptide as a Tool to Specifically Inhibit Protein Kinase A. *Front. Physiol.* **2020**, *11*, 574030. [CrossRef] [PubMed]
253. Hoy, J.J.; Salinas Parra, N.; Park, J.; Kuhn, S.; Iglesias-Bartolome, R. Protein kinase A inhibitor proteins (PKIs) divert GPCR-Gαs-cAMP signaling toward EPAC and ERK activation and are involved in tumor growth. *FASEB J.* **2020**, *34*, 13900. [CrossRef]
254. Tian, Y.; Yang, T.; Yu, S.; Liu, C.; He, M.; Hu, C. Prostaglandin E2 increases migration and proliferation of human glioblastoma cells by activating transient receptor potential melastatin 7 channels. *J. Cell Mol. Med.* **2018**, *22*, 6327–6337. [CrossRef] [PubMed]
255. Xu, Y.; Vakoc, C.R. Targeting cancer cells with BET bromodomain inhibitors. *Cold Spring Harb. Perspect. Med.* **2017**, *7*, a026674. [CrossRef]
256. Liu, Z.; Wang, P.; Chen, H.; Wold, E.A.; Tian, B.; Brasier, A.R.; Zhou, J. Drug Discovery Targeting Bromodomain-Containing Protein 4. *J. Med. Chem.* **2017**, *60*, 4533–4558. [CrossRef]
257. Mertz, J.A.; Conery, A.R.; Bryant, B.M.; Sandy, P.; Balasubramanian, S.; Mele, D.A.; Bergeron, L.; Sims, R.J., 3rd. Targeting MYC dependence in cancer by inhibiting BET bromodomains. *Proc. Natl. Acad. Sci. USA* **2011**, *108*, 16669–16674. [CrossRef]
258. Zuo, W.; Zhu, Y.; Liu, Z.; Xia, Y.; Xu, Y.; Peng, C.; Yu, L.; Wang, N. BRD4 inhibition sensitizes aggressive non-Hodgkin lymphomas to PI3Kδ inhibitors by suppressing PI3K reactivation and c-MYC expression. *Am J Cancer Res.* **2021**, *11*, 215–235.

259. Schafer, J.M.; Lehmann, B.D.; Gonzalez-Ericsson, P.I.; Marshall, C.B.; Beeler, J.S.; Redman, L.N.; Jin, H.; Sanchez, V.; Stubbs, M.C.; Scherle, P.; et al. Targeting MYCN-expressing triple-negative breast cancer with BET and MEK inhibitors. *Sci. Transl. Med.* **2020**, *12*, aaw8275. [CrossRef]
260. Yi, J.S.; Sias-Garcia, O.; Nasholm, N.; Hu, X.; Iniguez, A.B.; Hall, M.D.; Davis, M.; Guha, R.; Moreno-Smith, M.; Barbieri, E.; et al. The synergy of BET inhibitors with aurora A kinase inhibitors in MYCN-amplified neuroblastoma is heightened with functional TP53. *Neoplasia* **2021**, *23*, 624–633. [CrossRef]
261. Schwalm, M.P.; Knapp, S. BET bromodomain inhibitors. *Curr. Opin. Chem. Biol.* **2022**, *68*, 102148. [CrossRef]
262. Jin, W.; Tan, H.; Wu, J.; He, G.; Liu, B. Dual-target inhibitors of bromodomain-containing protein 4 (BRD4) in cancer therapy: Current situation and future directions. *Drug Discov. Today* **2022**, *27*, 246–256. [CrossRef]
263. Zhang, X.; Guo, X.; Zhuo, R.; Tao, Y.; Liang, W.; Yang, R.; Chen, Y.; Cao, H.; Jia, S.; Yu, J.; et al. BRD4 inhibitor MZ1 exerts anti-cancer effects by targeting MYCN and MAPK signaling in neuroblastoma. *Biochem. Biophys. Res. Commun.* **2022**, *604*, 63–69. [CrossRef]
264. Peter, B.; Eisenwort, G.; Sadovnik, I.; Bauer, K.; Willmann, M.; Rülicke, T.; Berger, D.; Stefanzl, G.; Greiner, G.; Hoermann, G.; et al. BRD4 degradation blocks expression of MYC and multiple forms of stem cell resistance in Ph+ chronic myeloid leukemia. *Am. J. Hematol.* **2022**, *97*, 1215–1225. [CrossRef]

Disclaimer/Publisher's Note: The statements, opinions and data contained in all publications are solely those of the individual author(s) and contributor(s) and not of MDPI and/or the editor(s). MDPI and/or the editor(s) disclaim responsibility for any injury to people or property resulting from any ideas, methods, instructions or products referred to in the content.

Article

Expression of RBMS3 in Breast Cancer Progression

Tomasz Górnicki [1,*], Jakub Lambrinow [1], Monika Mrozowska [1], Hanna Romanowicz [2], Beata Smolarz [2], Aleksandra Piotrowska [1], Agnieszka Gomułkiewicz [1], Marzena Podhorska-Okołów [3], Piotr Dzięgiel [1] and Jędrzej Grzegrzółka [1]

[1] Division of Histology and Embryology, Department of Human Morphology and Embryology, Wroclaw Medical University, 50-368 Wroclaw, Poland
[2] Laboratory of Cancer Genetics, Department of Pathology, Polish Mother's Memorial Hospital Research Institute, Rzgowska 281/289, 93-338 Lodz, Poland
[3] Division of Ultrastructure Research, Department of Human Morphology and Embryology, Wroclaw Medical University, 50-368 Wroclaw, Poland
* Correspondence: tomasz.gornicki@student.umw.edu.pl

Abstract: The aim of the study was to evaluate the localization and intensity of RNA-binding motif single-stranded-interacting protein 3 (RBMS3) expression in clinical material using immunohistochemical (IHC) reactions in cases of ductal breast cancer (in vivo), and to determine the level of RBMS3 expression at both the protein and mRNA levels in breast cancer cell lines (in vitro). Moreover, the data obtained in the in vivo and in vitro studies were correlated with the clinicopathological profiles of the patients. Material for the IHC studies comprised 490 invasive ductal carcinoma (IDC) cases and 26 mastopathy tissues. Western blot and RT-qPCR were performed on four breast cancer cell lines (MCF-7, BT-474, SK-BR-3 and MDA-MB-231) and the HME1-hTERT (Me16C) normal immortalized breast epithelial cell line (control). The Kaplan–Meier plotter tool was employed to analyze the predictive value of overall survival of *RBMS3* expression at the mRNA level. Cytoplasmatic RBMS3 IHC expression was observed in breast cancer cells and stromal cells. The statistical analysis revealed a significantly decreased RBMS3 expression in the cancer specimens when compared with the mastopathy tissues ($p < 0.001$). An increased expression of RBMS3 was corelated with HER2(+) cancer specimens ($p < 0.05$) and ER($-$) cancer specimens ($p < 0.05$). In addition, a statistically significant higher expression of RBMS3 was observed in cancer stromal cells in comparison to the control and cancer cells ($p < 0.0001$). The statistical analysis demonstrated a significantly higher expression of *RBMS3* mRNA in the SK-BR-3 cell line compared with all other cell lines ($p < 0.05$). A positive correlation was revealed between the expression of RBMS3, at both the mRNA and protein levels, and longer overall survival. The differences in the expression of RBMS3 in cancer cells (both in vivo and in vitro) and the stroma of breast cancer with regard to the molecular status of the tumor may indicate that RBMS3 could be a potential novel target for the development of personalized methods of treatment. RBMS3 can be an indicator of longer overall survival for potential use in breast cancer diagnostic process.

Keywords: RNA-binding protein 3 (RBMS3); carcinogenesis; cancer prevention; target discovery; target therapy; epithelial–mesenchymal transition (EMT)

1. Introduction

According to the WHO GLOBOCAN 2020 report data, breast cancer (BC) is the most commonly diagnosed cancer, with nearly 2.3 million new cases worldwide in 2020 [1]. It is also the fifth leading cause of cancer mortality, responsible for 6.9% of all cancer-related deaths in 2020 [1]. Based on the Global Cancer Observatory forecast, by the year 2030, the number of new cases will increase to 2.7 million per year [2]. Genetic and molecular analyses have allowed researchers to identify four main intrinsic subtypes of BC: luminal A, luminal B, HER2-enriched and triple-negative breast cancer (TNBC, also called basal-like) [3]. As indicated in Table 1, each subtype differs in its expression of biomarkers, especially estrogen receptor (ER), progesterone receptor (PR), and human epidermal growth factor receptor 2 (HER2) [4,5]. The expression of these biomarkers plays a significant role, among other anatomical features, in estimating the prognosis of BC [3].

Table 1. Different molecular subtypes of breast cancer. Each molecular subtype is defined by the expression of three main receptors: estrogen receptor, progesterone receptor, and human epidermal growth factor receptor 2 [4].

Molecular Subtype of Breast Cancer	Receptor Status		
	Estrogen Receptor (ER)	Progesterone Receptor (PR)	Human Epidermal Growth Factor Receptor 2 (HER2)
Luminal A	+	≥20%	-
Luminal B	+	<20%	+/-
HER2-enriched	-	-	+
Triple-negative breast cancer (TNBC)	-	-	-

RBMS3 is a glycine-rich protein that belongs to the family of c-Myc gene single-strand binding proteins (MSSPs). RBMS3, similarly to other MSSPs, is involved in processes that are crucial for cell life, such as cell-cycle progression and apoptosis [6,7]. RBMS3 participates in various processes, both physiological and pathological, e.g., embryogenesis or liver fibrosis [8,9]. Published papers indicate that RBMS3 can be viewed as a regulating factor of carcinogenesis in various cancers, including ovarian and nasopharyngeal cancers [10–12]. RBMS3 is postulated to regulate the progression of nasopharyngeal cancer by influencing the expression of the p53 protein, becoming a potential regulator of the cell cycle in this type of cancer [12]. In ovarian cancer, it can be involved in drug-resistance mechanisms [10]. It has been reported that RBMS3 participates in the carcinogenesis process of breast cancer, since it is often described as a suppressor protein. In recent studies, authors have provided data that point to the fact that a certain level of RBMS3 is necessary for cancer progression [13–18]. The currently postulated mechanisms explaining the role of RBMS3 in the progression of breast cancer include involvement in the epithelial–mesenchymal transition (EMT) by inhibiting the Wnt/β-catenin signaling pathway and other EMT-related transcription factors, such as TWIST1 or PRRX1 [12,17,19,20]. Another mechanism of influence of RBMS3 in breast cancer is its presence in the miR-141-3p/RBMS3 axis that inhibits proliferation and promotes apoptosis in breast cancer cells [15]. Another study reported data related to RBMS3's suppression leading to downregulation of cell programmed death ligand-1 (PD-L1) in TNBC, resulting in increased anti-tumor immune activities [18]. There is also evidence that the expression of RBMS3 in the stroma cells of breast cancer could have an impact on the progression of BC [16]. Although RBMS3 seems to play a major role in carcinogenesis, there remains a need for extensive research because of its complex influence on breast cancer.

The aim of this study is to discuss the role of RBMS3 in breast cancer. Using immunohistochemical staining performed on paraffin-embedded blocks of breast cancer samples and molecular analysis performed with breast cancer cells from cell-line cultures, we showed the correlation between RBMS3 levels and particular intrinsic subtypes of BC. A further aim of this study is to discuss RBMS3 as a novel potential therapeutic target and biomarker of overall survival in breast cancer.

2. Results

2.1. The Immunohistochemical Intensity of RBMS3's Expression Varies in Cancer Cells, the Stroma of the Tumor, and the Control Mastopathy Cases, Exhibiting a Dependence on the Expression of Crucial Breast Cancer Receptors

The analysis of the immunohistochemical expression of RBMS3 in 490 cases of IDC and 26 cases of mastopathy showed a statistically significant decrease in RBMS3 expression in the cancer specimens compared to the mastopathy samples (Mann–Whitney test $p < 0.001$, Figure 1a, Figure 2a,b). Furthermore, the statistical analysis of the clinical data together with the immunohistochemical expression of RBMS3 showed a significantly increased expression of RBMS3 in the cancer cells of the HER2 positive cases (Mann–Whitney test $p < 0.05$, Figure 1b). Meanwhile, increased expression of RBMS3 correlated with the negative status of the estrogen receptor (Mann–Whitney test $p < 0.05$, Figure 1c). However, there was no statistically significant difference in the expression of RBMS3 in cancer cells between progesterone-positive and -negative cases.

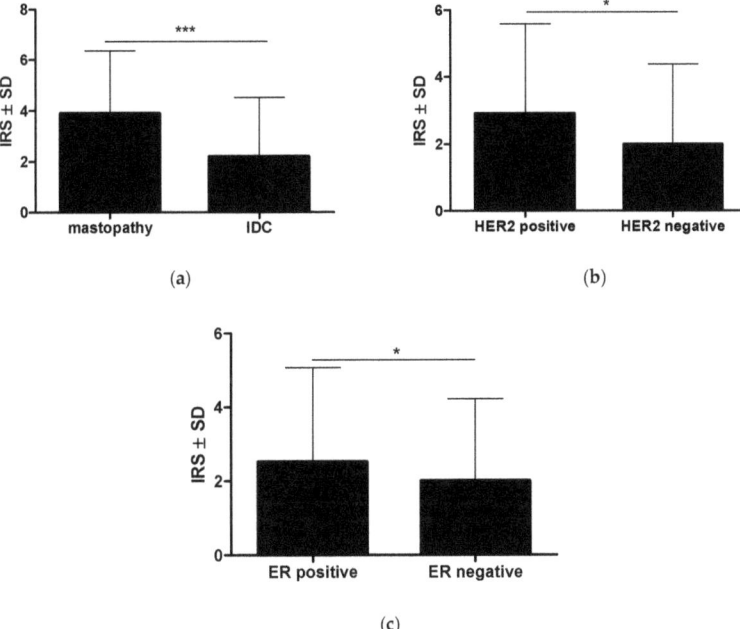

Figure 1. (a) The statistical analysis revealed a significantly higher RBMS3 expression as assessed by the immunoreactive score in the control mastopathy cases compared with the cancer cells in breast cancer. (b) The breast cancer cells with positive expression of HER2 and (c) estrogen receptors presented a higher expression of RBMS3 compared with tumors lacking expression of these receptors (Mann–Whitney test * $p < 0.05$; *** $p < 0.001$) (IDC—invasive ductal carcinoma, IRS—immunoreactive score).

Expression of RBMS3 in the stroma of the cancer cases was significantly higher than in the control specimens (Mann–Whitney test $p < 0.0001$, Figure 3a, Figure 2c,d). Moreover, RBMS3 expression in TNBC samples was significantly lower than in the other molecular types (Mann–Whitney test $p < 0.001$, Figure 3b). Further investigation in the stroma of breast cancer showed significant increases in RBMS3 expression in the specimens with positive expression of the progesterone receptor and samples with positive expression of the estrogen receptor (respectively, Mann–Whitney test $p < 0.01$ and $p < 0.001$, Figure 3c,d). On the other hand, we observed no correlation of RBMS3 expression with expression of the

HER2 receptor. RBMS3 expression in the stroma of IDC was significantly higher than in the cancer cells (Mann–Whitney test $p < 0.0001$, Figure 3e).

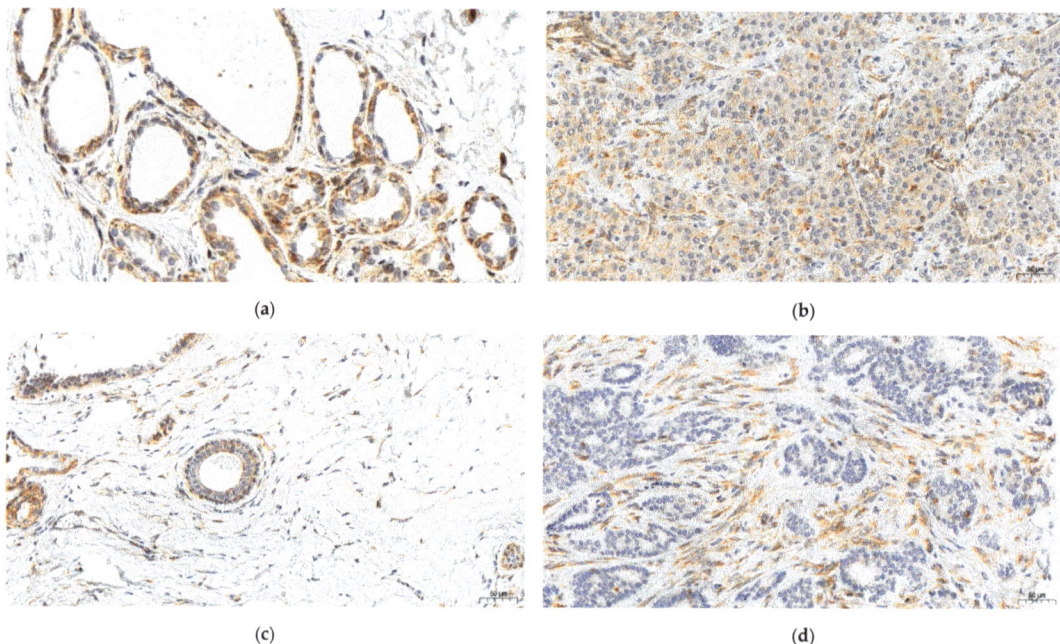

Figure 2. Immunohistochemical visualization of RBMS3 expression. (a) High cytoplasmic expression of RBMS3 in mastopathy cases. (b) Low expression of RBMS3 in breast cancer cells. (c) Low cytoplasmic expression of RBMS3 in the stroma of mastopathy cases and (d) high expression in the stroma cells of the breast cancer cases. Magnification ×200.

There were no statistically significant differences in the expression of RBMS3 with regard to the grade, TNM, and stage of the cancer. This absence of difference was observed in the cancer cells and the stroma.

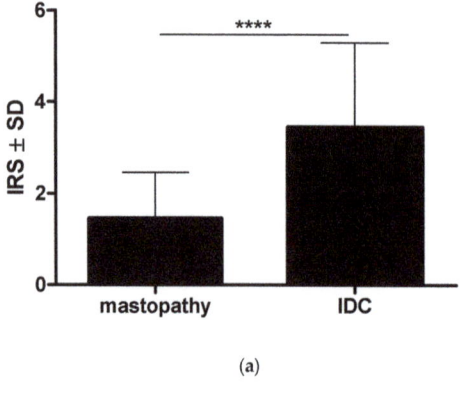
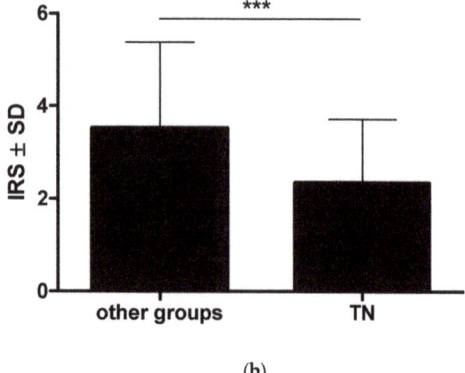

Figure 3. Cont.

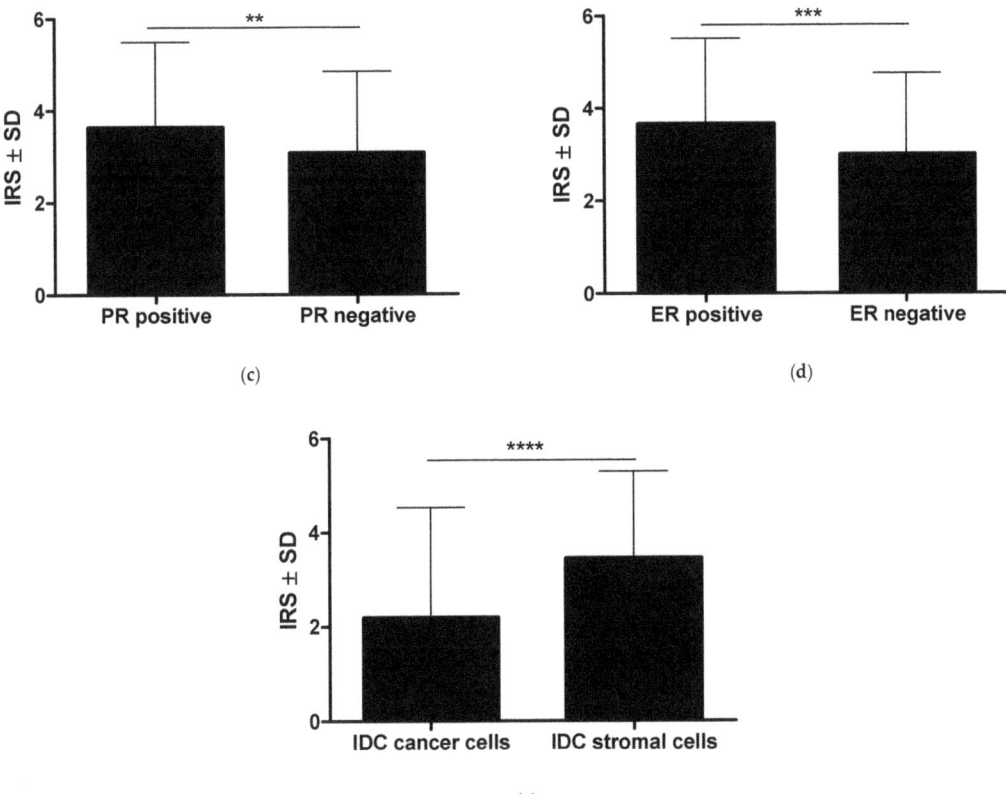

Figure 3. Analysis of RBMS3 expression in the stroma of breast cancer: (**a**) RBMS3 expression in the stroma of breast cancer was statistically higher than in the mastopathy cases; (**b**) Triple-negative (TN) cases of breast cancer displayed lower expression of RBMS3 in the stroma than the other molecular types of breast cancer combined. (**c**) Significant increase in RBMS3 expression in the specimens with positive expression of the progesterone receptor and (**d**) with positive expression of the estrogen receptor. The expression of RBMS3 was statistically lower in the cancer cells than in the stromal cells of the breast cancer specimens (**e**). (Mann–Whitney test ** $p < 0.01$, *** $p < 0.001$, **** $p < 0.0001$) (IDC—invasive ductal carcinoma).

2.2. In Vitro Analysis of RBMS3 Expression Differs from RBMS3 Expression in Clinical Material

For further investigation of the difference in RBMS3 expression in the different molecular types of breast cancer, we performed an RT-qPCR analysis of *RBMS3* expression at the mRNA level in the chosen cell lines representing the various molecular types of breast cancer. When compared to the control Me16C cell line (ANOVA and Bonferroni's multiple comparison test $p < 0.05$, Figure 4a), the expression of *RBMS3* was significantly different (mostly lower) in all the examined cell lines, with the only exception being the SK-BR-3 cell line. *RBMS3* expression was highest in the SK-BR-3 cell line among all the investigated cell lines. The Western blot analysis of the protein expression showed a significantly higher expression of RBMS3 in the control cell line than in the MCF-7 and BT-474 cancer cell lines (ANOVA and Bonferroni's multiple comparison test $p < 0.05$, Figure 4b) There is a visible and statistically significant trend that the more aggressive types of breast cancer, including TNBC and HER-2-positive cancers, presented higher expression of RBMS3 than their benign counterparts.

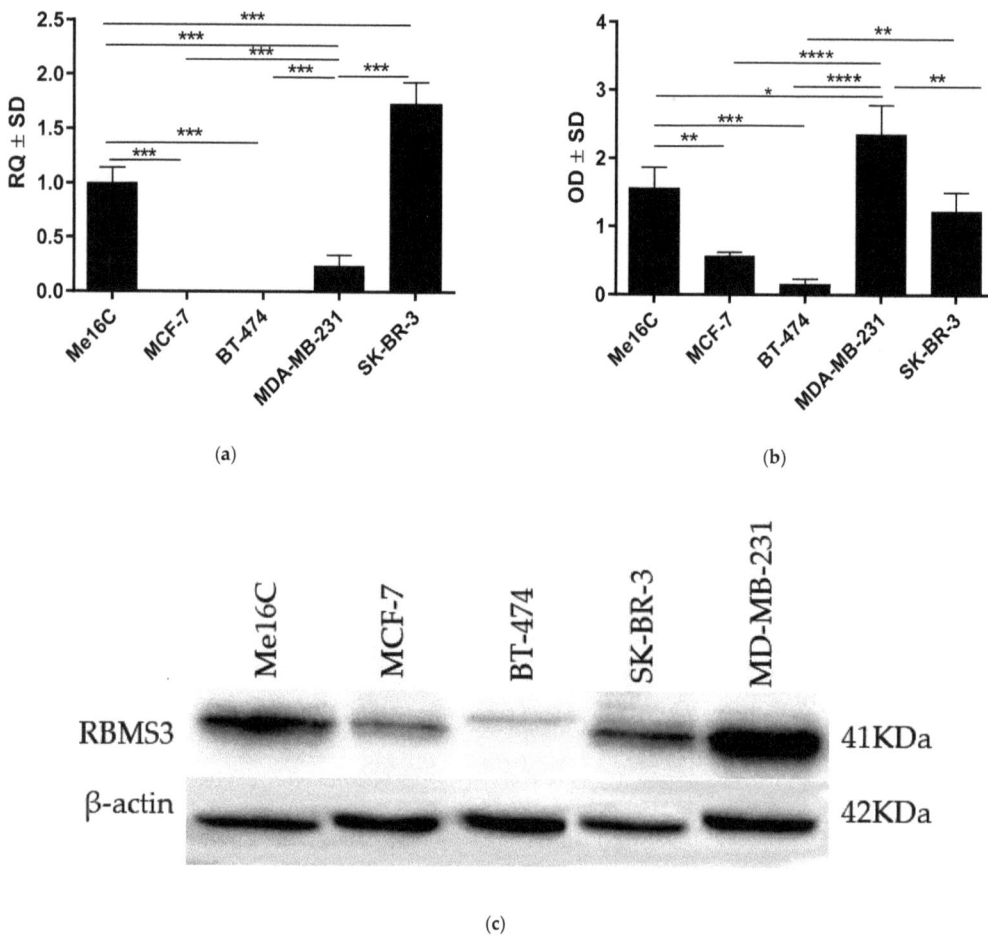

Figure 4. In vitro analysis of RBMS3 expression in breast cancer cell lines representing the four main molecular types of breast cancer and a control cell line (Me16C). (**a**) The statistical analysis of *RBMS3*'s expression at the mRNA level showed a significantly different expression of *RBMS3* in all the examined cell lines in comparison to the control Me16C cell line. (**b**,**c**) Analysis at the protein level showed a significantly higher expression of RBMS3 in the MDA-MB-231 and SK-BR-3 cell lines than in the MCF-7 and BT-474 cell lines; ((**a**) Bonferroni's multiple comparison test, (**b**) Bonferroni's multiple comparison test, * $p < 0.05$, ** $p < 0.01$, *** $p < 0.001$, **** $p < 0.0001$).

2.3. RBMS3 Expression May Be an Indicator of Longer Overall Survival

The analysis of the clinical data regarding the survival of patients showed shorter overall survival in the group of patients without an IHC expression of RBMS3 (Gehan–Breslow–Wilcoxon test $p = 0.051$, Figure 5). The univariate and multivariate Cox analyses of the overall survival indicated that only G, pT, and pN were independent prognostic factors (Table 2).

Additionally, using the Kaplan–Meier estimator we performed an analysis of the *RBMS3* mRNA expression of 2976 cases of breast cancer. This revealed that the group of patients with lower *RBMS3* expression (cut-off point: median) had statistically significant shorter overall survival ($p < 0.0001$, Figure 6) [21].

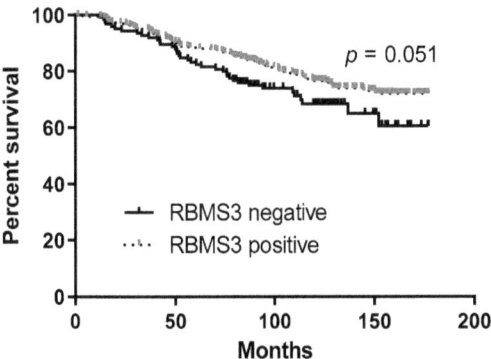

Figure 5. The analysis of the clinical data regarding the survival of patients showed shorter overall survival in the group of patients without IHC expression of RBMS3 (Gehan–Breslow–Wilcoxon test $p = 0.051$).

Table 2. Univariate and multivariate Cox analyses of overall survival in cases of invasive ductal carcinoma.

Characteristics	Univariate Cox Analysis of Survival				Multivariate Cox Analysis of Survival			
	p-Value	Hazard Ratio	HR 95% CI Lower	HR 95% CI Upper	p-Value	Hazard Ratio	HR 95% CI Lower	HR 95% CI Upper
G1 vs. G2-G3	<0.0100	3.0873	1.5179	6.2792	<0.0100	2.5309	1.2509	5.1208
pT1 vs. pT2-pT4	<0.0001	2.4469	1.7123	3.4966	<0.0010	2.0371	1.4201	2.9221
pN0 vs. pN1-pN3	<0.0001	2.6544	1.8541	3.8001	<0.0001	2.1583	1.4997	3.1062
ER negative vs. ER positive	0.2260	0.7987	0.5550	1.1493				
PR negative vs. PR positive	0.1416	0.7626	0.5313	1.0946				
HER2 0-HER2 2 vs. HER2 3	0.4485	1.3206	0.64338	2.7105				
Triple-negative vs. other groups	0.3742	1.3843	0.67566	2.8361				
RBMS3 IRS stromal: 0 vs. 1–12	0.3196	0.6548	0.2844	1.5075				
RBMS3 IRS cancer: 0 vs. 1–12	<0.0500	0.6470	0.4470	0.9365	0.1429	0.7576	0.5226	1.0983

ER—estrogen receptor, PR—progesterone receptor, RBMS3—RNA-binding motif single-stranded-interacting protein 3, IRS—immunoreactive scale, HR—hazard ratio, CI—confidential interval.

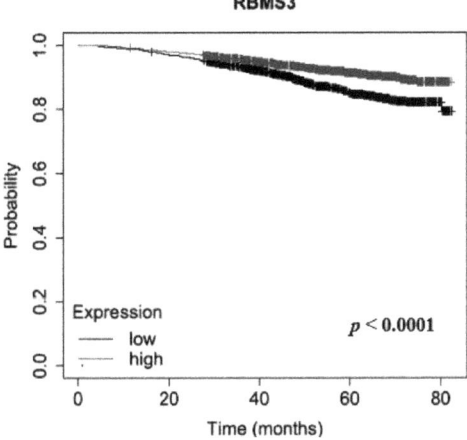

Figure 6. Analysis of *RBMS3* mRNA expression in 2976 cases of breast cancer, using the Kaplan–Meier estimator. The analysis revealed a significant positive correlation between the expression of *RBMS3* and overall survival ($p < 0.0001$) [21].

3. Discussion

RBMS3 is reported to be deregulated in many different types of neoplastic processes, for example, gastric cancer, esophageal squamous cell carcinoma, breast cancer, or gall bladder carcinoma [13,22–24]. In this study, we have discussed the role of RBMS3 in the progression of breast cancer with particular emphasis on receptor expression and the molecular type. We provided an analysis of RBMS3 expression in clinical material and cell lines, and presented experimental data supporting the statement of the potential role of RBMS3 expression in tumor stromal cells.

The results of our experiments apparently support previous studies' results that indicate the downregulation of RBMS3 expression in breast cancer cells [13,19] and its correlation with negative estrogen-receptor status [14]. In addition, we discovered another potential interaction of RBMS3 with a positive HER2-receptor status, supported by an immunostaining analysis and the high expression of *RBMS3* at the mRNA level in the SK-BR-3 line (representing the HER2-enriched subtype). At the protein level, the expression levels of RBMS3 in the SK-BR-3 and MDA-MB-231 cell lines were the highest among all the examined breast cancer cell lines and were significantly higher than in MCF-7 and BT-474 cell lines. A significantly higher expression of RBMS3 in more aggressive types of tumors characterized by the lack of estrogen-receptor expression, and in the case of the SK-BR-3 cell line the presence of the HER2 receptor, may indicate that a certain level of RBMS3 expression is necessary for specific types of cancer progression and their ability to create metastasis [20,25]. RBMS3′s anticancer function could be related to the mechanisms that regulate adhesiveness and invasiveness, which are also associated with the EMT process in cancer. These findings are in partial agreement with recent reports that provide evidence of RBMS3 knockdown resulting in the impairment of in vivo tumor growth and a decreased level of angiogenesis [17,18]. It is important to mention that the research carried out by Block et.al and Zhu et.al was conducted only on the triple-negative type of breast cancer cells. The results provided in this study support the claim that RBMS3 expression in the TNBC and HER-2-enriched types is similarly high as in the control cell lines, meaning that RBMS3 could possibly act as a suppressor in Lum-A and Lum-B types of breast cancer. They may also suggest that a normal level of RBMS3 expression is necessary for the growth of TNBC and HER-2-enriched types of breast cancer; this observation requires more detailed investigation.

The role of the tumor microenvironment (TME) is a topic of rapidly increasing interest among scientists [26]. The TME is the unique environment in which the tumor develops. It consists of an extracellular matrix, blood vessels, signaling molecules, and multiple types of cells that play a pivotal role in tumor cancerogenesis by stimulating and facilitating uncontrolled cell proliferation [27,28]. Stromal cells are an integral part of the TME. Alongside other elements, they play a part in the maintenance of cancer stemness by promoting angiogenesis, invasion, metastasis, and chronic inflammation [29]. The transcriptomic analysis of the *RBMS3* gene's expression in the stromal cells of breast cancer provides evidence of its being gradually downregulated through all three grades of breast cancer [16]. The results that we present suggest a higher expression of the RBMS3 protein in the stroma cells of breast cancer compared with the mastopathy control cases or the cancer cells, with no significant differences between grades. Together, these data suggest that RBMS3′s deregulation in the stroma of the tumor may influence the role of stromal cells in breast cancer through currently unknown mechanisms. Furthermore, there may exist a currently unknown post-transcriptional mechanism regulating the expression of RBMS3 in the stroma of the tumor, which could explain the grade-dependent expression of *RBMS3* and the lack of grade dependency at the protein level. A negative correlation of RBMS3 expression in the stromal cells with TNBC, and a positive one with ER- and PR-receptor status of the tumor, may indicate that there is a possibility for RBMS3 to display an antitumor effect depending on the molecular characteristics of the tumor. Specifically, a negative correlation with TNBC may indicate the tumor-suppressor role of RBMS3 in breast cancer stroma. A higher expression of RBMS3 in the stroma of breast cancer may indicate the potentially important role of the TME in the progression of IDC.

In addition to its potential antitumor properties, the expression of RBMS3 may be an indicator of overall survival. These capabilities were reported by scientists researching lung squamous cell carcinoma and gastric cancer [23,24,30]. In this current study, we provide evidence of RBMS3's potential use as a positive prognostic marker of overall survival in breast cancer. The results of our clinical data analysis are consistent with the findings of Wang et al. [14]. The analysis of *RBMS3* mRNA expression in samples from the GEO and EGA data repositories also supports the suggestion that RBMS3 may be a useful tool for breast cancer diagnosis. The analysis of RBMS3 expression can be included as a supplementary category in defining prognosis of patient survival based on the molecular characteristics of the tumor, increasing the accuracy of predictions. The correlation of RBMS3 expression with TNBC and the expression of progesterone receptor may also lead to the distinction of new molecular subtypes of breast cancer based on the analysis of combined biomarkers.

Taking into consideration all the results presented in this study, we provide evidence of a potential novel explanation of RBMS3's role in breast cancer. Currently available reports have tried to explain RBMS3's anticancer activity in all types of breast cancer through the inhibition of the Wnt/β-catenin pathway and the inhibition of the epithelial–mesenchymal transition process (EMT), mainly by impacting on TWIST, PRRX1, or MMP2 [14,17,19]. Our results suggest that further studies should be conducted to consider the differences in RBMS3 expression correlated with receptor expression in cancer cells and stromal cells. We distinguished a positive correlation with overall survival, supporting the idea of a potential tumor-suppressing role for the expression of RBMS3 in breast cancer stroma. These findings open the way for further studies to unveil the exact role and mechanisms of these correlations.

Although further studies on the exact molecular mechanisms underlying the role of RBMS3 in breast cancer are required, RBMS3 may be potentially used in the development of novel therapeutic and diagnostic approaches in breast cancer. These may target not only cancer cells but also tumor stroma cells, making these therapies more complex and potentially adaptive to the patient's type of tumor, which would translate into a more personalized approach to patient treatment.

4. Materials and Methods

4.1. Patients' Characteristics

The clinical material consisted of 524 paraffin blocks with clinical data from patients operated on for IDC. The clinical and pathological characteristics of the patients are presented in Table 3. Additionally, 26 paraffin blocks and clinical data from cases of mastopathy were analyzed as a control for the breast cancer cases. Patients' clinical material was obtained from the Division of Pathomorphology of the Polish Mother's Memorial Hospital Research Institute. The experiment was performed in accordance with the ethical standards and following the approval of the Ethics Committee of Wroclaw Medical University (decision no. KB 625/2022 25.08.2022).

Table 3. Clinical and pathological characteristics of studied patients.

Parameters	Patients	
	IHC n = 524	%
Age		
≤60	165	31.49
>60	359	68.51
Tumor grade		
G1	87	16.60
G2	342	65.27
G3	92	17.56
No data	3	0.57
Tumor size		
pT1	325	62.02
pT2	168	32.06
pT3	3	0.57
pT4	9	1.72
No data	19	3.63
Lymph nodes		
pN0	314	59.92
pN1-pN3	180	34.35
pNx	30	5.73
Stage		
I	224	42.75
II	257	49.05
III	18	3.44
IV	0	0.00
ER		
Neg.	177	33.78
Pos.	344	65.65
No data	3	0.57
PR		
Neg.	183	34.92
Pos.	338	64.50
No data	3	0.57
HER2		
Neg.	272	51.91
Pos.	36	6.87
No data	216	41.22
Molecular tumor types		
Triple-negative	34	6.49
Other types	487	92.94
No data	3	0.57

4.2. Tissue Microarrays (TMAs)

A total of 21 TMAs were prepared from 524 cases of IDC and 26 cases of mastopathy. Prior to performing TMA blocks, the histological slides stained with hematoxylin and eosin were obtained from whole samples of breast cancer and mastopathy cases stored in the form of paraffin blocks (donor blocks). The slides were scanned using the Pannoramic Midi II histological scanner (3DHISTECH Ltd, Budapest, Hungary). After that, using the Pannoramic Viewer program 1.15.4 (3DHISTECH Ltd.), representative areas from the entire sections where selected. In addition, to increase the representativeness of each case, 3 representative cores each with a size of 1.5 mm were selected from the donor blocks and transferred to the TMA 'recipient' block using the TMA Grand Master system 2.6.6.69657 (3DHISTECH Ltd.).

4.3. Immunohistochemistry

The paraffin blocks with the breast cancer and mastopathy cases were cut into 4-μm sections. The immunohistochemical reactions were performed using anti-RBMS3 rabbit polyclonal antibody (Catalog # PA5-57028, Invitrogen, Thermo Fisher Scientific, Waltham, MA, USA) in a 1:200 dilution. The immunohistochemical reactions were performed using a Dako Autostainer Link 48 (Dako, Glostrup, Denmark). The visualization of the reactions was carried out using EnVision™ FLEX High pH (Link, Glostrup, Denmark) reagents (Dako), according to the manufacturer's instructions. The IHC reactions for 490 cases of IDC were suitable for the further analysis. The IHC reaction for RBMS3 antigen was assessed using the immunoreactive scale (IRS) by Remmele and Stegner [31], that evaluates the percentage of positive cancer cells (A) and the intensity of color reaction (B). The final score is the product of the values A and B (see Table 4).

Table 4. Graphic presentation of Remmele and Stegner scale showing the available values. The final score is the multiplication of A and B values (A × B) [31].

Points	Percentage of Positive Cancer Cells (A)	Intensity of Color Reaction (B)
0	0%	No color reaction
1	<10%	Mild reaction
2	10–50%	Moderate reaction
3	51–80%	Strong reaction
4	81–100%	

4.4. Kaplan–Meier Plotter

The Kaplan–Meier plotter tool was used for correlation of *RBMS3* mRNA expression with overall survival [21]. This is a tool for Kaplan–Meier plot generation based on data from GEO, EGA, and TCGA. *RBMS3* mRNA expression data was split into two groups for analysis: "high expression" and "low expression" using the median as the cut-off value.

4.5. Cell Lines

Four breast cancer cell lines were used in the experiments, representing types of tumors of increasing aggressiveness (MCF-7: luminal A, BT-474: luminal B, SK-BR-3: HER2-enriched, and MDA-MB-231: triple-negative), along with a normal cell line: immortalized breast epithelial cell line (HME1-hTERT) (Me16C). All cell lines were provided by ATCC (American Type Culture Collection ATCC®, Old Town Manassas, VA, USA). Respective culture media were used to provide optimal conditions for cell growth: MEBM (Lonza, Basel, Switzerland) for the Me16C cell line, αMEM (Lonza) for the MCF-7 and BT-474 cell lines, McCoy's (ATCC) for the SK-BR-3 cell line, L-15 (Lonza) for the MDA-MB-231 cell line. All media contained 1% l-glutamine and penicillin-streptomycin solution, as well as 10% fetal bovine serum (Sigma-Aldrich®, St. Louis, MO, USA). The cells were passaged with the use of TrypLE™ (Gibco, Thermo Fisher Scientific, Waltham, MA, USA) when they were at approximately 70% confluence.

4.6. Real-Time PCR

Real-time PCR was applied to determine the relative level of *RBMS3* mRNA expression in the analyzed cell lines (MDA-MB-231, SK-BR-3, BT-474, MCF-7, Me16C). Total RNA was isolated with the use of a RNeasy mini kit (Qiagen, Hilden, Germany), according to the manufacturer's instructions. Reverse transcription reactions were performed with the use of iScript™ cDNA synthesis kit (Bio-Rad, Hercules, CA, USA). The conditions of the reactions were as follows: priming for 5 min at 25 °C, reverse transcription for 20 min at 46 °C, and final inactivation of reverse transcriptase for 1 min at 95 °C. RT-qPCR was carried out in 20-μL volumes using the TaqMan Universal PCR MasterMix (Applied Biosystems, Foster City, CA, USA). The reactions were performed using a 7500 Real-time PCR system and iTaq™ Universal Probes Supermix (Bio-Rad), according to the manufacturer's instructions.

The TaqMan probes employed were Hs01104892_m1 for RBMS3 (Applied Biosystems) and endogenous control gene Hs99999903_m1 for β-actin (Applied Biosystems), further used for normalization purposes. The experiments were run in triplicate. The reactions were carried out under the following conditions: initial denaturation for 2 min at 94 °C, followed by 40 cycles of denaturation (15 s, 94 °C) and annealing with elongation (1 min, 60 °C). The relative *RBMS3* mRNA expression levels were calculated using the ΔΔCt method.

4.7. Western Blotting

Whole cell lysates were obtained from the BC cell lines (MDA-MB-231, SK-BR-3, BT-474, and MCF-7) and the control cell line (Me16C) using CelLytic™ MT Cell Lysis Reagent (Sigma-Aldrich) with the addition of Halt™ Protease Inhibitor Cocktail 100x (Thermo Fisher Scientific) and 2 mM PMSF (phenylmethylsulphonyl fluoride) (Sigma-Aldrich). The protein level was determined through colorimetric analysis with the use of bicinchoninic acid (Pierce BCA Protein Assay Kit) and NanoDrop 1000 (Thermo Scientific). The lysates were mixed with 4X SDS-PAGE gel-loading buffer (200 mM Tris-HCl—pH 6.8, 400 mM DTT, 8% SDS, 0.4% bromophenol blue, 40% glycerol) for 10 min at 95 °C, loaded onto 10% acrylamide gel and separated by SDS-PAGE under reducing conditions, then transferred onto a PVDF membrane in the XCell SureLock™ Mini gel electrophoresis system (Thermo Fisher Scientific). After the protein transfer, the membrane was incubated in a blocker solution (4% BSA in TBST buffer) for 1 h at RT, followed by overnight incubation at 4 °C with anti-RBMS3 monoclonal rabbit antibody, (Catalog # PA5-57028, Invitrogen, Thermo Fisher Scientific). Subsequently, the membrane was washed with TBST with 0.1% Tween-20 and incubated for 1 h at RT with secondary antibody (Jackson ImmunoResearch, Mill Valley, CA, USA) diluted at 1:3000, then rinsed and treated with Luminata Classico (Merck KGaA, Darmstadt, Germany) chemiluminescent substrate. Rabbit anti-human β-actin monoclonal antibody (#4970; Cell Signaling Technology, Danvers, MA, USA) diluted 1:1000 was used as an internal control. The Western blotting results were analyzed using the ChemiDoc MP system (Bio-Rad). The experiments were run in triplicate.

4.8. Statistical Analysis

The Kolmogorov–Smirnov test was applied to evaluate the normality assumption of the groups examined. The Mann–Whitney and ANOVA with Bonferroni's multiple comparison post hoc tests were conducted to compare the differences in the expression of the examined markers in all groups of patients in vitro and in the clinicopathological data. Additionally, the Spearman's correlation test was applied to analyze the existing correlations. The Kaplan–Meier method was used to construct survival curves. The Gehan–Breslow–Wilcoxon method was applied and univariate and multivariate Cox analyses of survival were performed to evaluate the survival analysis. All statistical analyses were conducted using Prism 9.0 (GraphPad Software) and Statistica 13.3 (Tibco Software, Inc.). The results were considered statistically significant when $p < 0.05$.

Supplementary Materials: The following supporting information can be downloaded at: https://www.mdpi.com/article/10.3390/ijms24032866/s1.

Author Contributions: Conceptualization, T.G., J.L. and J.G.; methodology, M.M., A.P., A.G., H.R. and B.S.; software, T.G.; validation, J.G. and M.M.; writing—original draft preparation, T.G., J.L. and M.M.; writing—review and editing, J.G. and P.D.; visualization, M.M. and A.P.; supervision, J.G., P.D. and M.P.-O.; project administration, T.G. and J.G.; funding acquisition, J.G., P.D. and M.P.-O. All authors have read and agreed to the published version of the manuscript.

Funding: This research received no external funding.

Institutional Review Board Statement: The study was conducted in accordance with the Declaration of Helsinki and approved by the Ethics Committee of the Wroclaw Medical University (decision no. KB 625/2022 25.08.2022) for studies involving humans.

Informed Consent Statement: Informed consent was obtained from all subjects involved in the study.

Data Availability Statement: The data presented in this study are available in this article and supplementary materials attached to it.

Conflicts of Interest: The authors declare no conflict of interest.

References

1. Sung, H.; Ferlay, J.; Siegel, R.L.; Laversanne, M.; Soerjomataram, I.; Jemal, A.; Bray, F. Global Cancer Statistics 2020: GLOBOCAN Estimates of Incidence and Mortality Worldwide for 36 Cancers in 185 Countries. *CA Cancer J. Clin.* **2021**, *71*, 209–249. [CrossRef] [PubMed]
2. Ferlay, J.; Laversanne, M.; Ervik, M.; Lam, F.; Colombet, M.; Mery, L.; Piñeros, M.; Znaor, A.; Soerjomataram, I.; Bray, F. *Global Cancer Observatory: Cancer Tomorrow*; International Agency for Research on Cancer: Lyon, France, 2018; Available online: https://gco.iarc.fr/tomorrow (accessed on 8 October 2018).
3. Łukasiewicz, S.; Czeczelewski, M.; Forma, A.; Baj, J.; Sitarz, R.; Stanisławek, A. Breast Cancer—Epidemiology, Risk Factors, Classification, Prognostic Markers, and Current Treatment Strategies—An Updated Review. *Cancers* **2021**, *13*, 4287. [CrossRef] [PubMed]
4. Tsang, J.Y.S.; Tse, G.M. Molecular Classification of Breast Cancer. *Adv. Anat. Pathol.* **2020**, *27*, 27–35. [CrossRef] [PubMed]
5. Vuong, D.; Simpson, P.; Green, B.; Cummings, M.; Lakhani, S.R. Molecular classification of breast cancer. *Virchows Arch.* **2014**, *465*, 1–14. [CrossRef]
6. Penkov, D.; Ni, R.; Else, C.; Piñol-Roma, S.; Ramirez, F.; Tanaka, S. Cloning of a human gene closely related to the genes coding for the c-myc single-strand binding proteins. *Gene* **2000**, *243*, 27–36. [CrossRef]
7. Niki, T.; Izumi, S.; Saëgusa, Y.; Taira, T.; Takai, T.; Iguchi-Ariga, S.M.M.; Ariga, H. MSSP promotes ras/myc cooperative cell transforming activity by binding to c-Myc. *Genes Cells* **2000**, *5*, 127–141. [CrossRef]
8. Lu, C.-K.; Lai, Y.-C.; Chen, H.-R.; Chiang, M.-K. Rbms3, an RNA-Binding Protein, Mediates the Expression of *Ptf1a* by Binding to Its 3′UTR During Mouse Pancreas Development. *DNA Cell Biol.* **2012**, *31*, 1245–1251. [CrossRef]
9. Fritz, D.; Stefanovic, B. RNA-binding Protein RBMS3 Is Expressed in Activated Hepatic Stellate Cells and Liver Fibrosis and Increases Expression of Transcription Factor Prx1. *J. Mol. Biol.* **2007**, *371*, 585–595. [CrossRef]
10. Wu, G.; Cao, L.; Zhu, J.; Tan, Z.; Tang, M.; Li, Z.; Hu, Y.; Yu, R.; Zhang, S.; Song, L.; et al. Loss of RBMS3 Confers Platinum Resistance in Epithelial Ovarian Cancer via Activation of miR-126-5p/β-catenin/CBP signaling. *Clin. Cancer Res.* **2019**, *25*, 1022–1035. [CrossRef]
11. Zhang, Y.; Xu, Y.; Feng, L.; Li, F.; Sun, Z.; Wu, T.; Shi, X.; Li, J.; Li, X. Comprehensive characterization of lncRNA-mRNA related ceRNA network across 12 major cancers. *Oncotarget* **2016**, *7*, 64148–64167. [CrossRef]
12. Górnicki, T.; Lambrinow, J.; Mrozowska, M.; Podhorska-Okołów, M.; Dzięgiel, P.; Grzegrzółka, J. Role of RBMS3 Novel Potential Regulator of the EMT Phenomenon in Physiological and Pathological Processes. *Int. J. Mol. Sci.* **2022**, *23*, 10875. [CrossRef]
13. Yang, Y.; Quan, L.; Ling, Y. RBMS3 Inhibits the Proliferation and Metastasis of Breast Cancer Cells. *Oncol. Res. Featur. Preclin. Clin. Cancer Ther.* **2018**, *26*, 9–15. [CrossRef] [PubMed]
14. Wang, C.; Wu, Y.; Liu, Y.; Pan, F.; Zeng, H.; Li, X.; Yu, L. Tumor Suppressor Effect of RBMS3 in Breast Cancer. *Technol. Cancer Res. Treat.* **2021**, *20*. [CrossRef] [PubMed]
15. Dong, S.; Ma, M.; Li, M.; Guo, Y.; Zuo, X.; Gu, X.; Zhang, M.; Shi, Y. LncRNA MEG3 regulates breast cancer proliferation and apoptosis through miR-141-3p/RBMS3 axis. *Genomics* **2021**, *113*, 1689–1704. [CrossRef] [PubMed]
16. Uddin, N.; Wang, X. Identification of key tumor stroma-associated transcriptional signatures correlated with survival prognosis and tumor progression in breast cancer. *Breast Cancer* **2022**, *29*, 541–561. [CrossRef] [PubMed]
17. Block, C.J.; Mitchell, A.V.; Wu, L.; Glassbrook, J.; Craig, D.; Chen, W.; Dyson, G.; DeGracia, D.; Polin, L.; Ratnam, M.; et al. RNA binding protein RBMS3 is a common EMT effector that modulates triple-negative breast cancer progression via stabilizing PRRX1 mRNA. *Oncogene* **2021**, *40*, 6430–6442. [CrossRef]
18. Zhou, Y.; Liang, Z.; Xia, Y.; Li, S.; Liang, J.; Hu, Z.; Tang, C.; Zhao, Q.; Gong, Q.; Ouyang, Y. Disruption of RBMS3 suppresses PD-L1 and enhances antitumor immune activities and therapeutic effects of auranofin against triple-negative breast cancer. *Chem. Biol. Interact.* **2023**, *369*, 110260. [CrossRef]
19. Zhu, L.; Xi, P.-W.; Li, X.-X.; Sun, X.; Zhou, W.-B.; Xia, T.-S.; Shi, L.; Hu, Y.; Ding, Q.; Wei, J.-F. The RNA binding protein RBMS3 inhibits the metastasis of breast cancer by regulating Twist1 expression. *J. Exp. Clin. Cancer Res.* **2019**, *38*, 105, Erratum in *J. Exp. Clin. Cancer Res.* **2020**, *39*, 21. [CrossRef]
20. Grzegrzolka, J.; Biala, M.; Wojtyra, P.; Kobierzycki, C.; Olbromski, M.; Gomulkiewicz, A.; Piotrowska, A.; Rys, J.; Podhorska-Okolow, M.; Dziegiel, P. Expression of EMT Markers SLUG and TWIST in Breast Cancer. *Anticancer. Res.* **2015**, *35*, 3961–3968.
21. Győrffy, B. Survival analysis across the entire transcriptome identifies biomarkers with the highest prognostic power in breast cancer. *Comput. Struct. Biotechnol. J.* **2021**, *19*, 4101–4109. [CrossRef]
22. Wu, Y.; Meng, D.; You, Y.; Sun, R.; Yan, Q.; Bao, J.; Sun, Y.; Yun, D.; Li, Y.; Sun, D. Increased expression of RBMS3 predicts a favorable prognosis in human gallbladder carcinoma. *Oncol. Rep.* **2020**, *44*, 55–68. [CrossRef] [PubMed]
23. Zhang, T.; Wu, Y.; Fang, Z.; Yan, Q.; Zhang, S.; Sun, R.; Khaliq, J.; Li, Y. Low expression of RBMS3 and SFRP1 are associated with poor prognosis in patients with gastric cancer. *Am. J. Cancer Res.* **2016**, *6*, 2679–2689. [PubMed]

24. Li, Y.; Chen, L.; Nie, C.-J.; Zeng, T.-T.; Liu, H.; Mao, X.; Qin, Y.; Zhu, Y.-H.; Fu, L.; Guan, X.-Y. Downregulation of RBMS3 Is Associated with Poor Prognosis in Esophageal Squamous Cell Carcinoma. *Cancer Res* **2011**, *71*, 6106–6115. [CrossRef] [PubMed]
25. Weigelt, B.; Geyer, F.C.; Reis-Filho, J.S. Histological types of breast cancer: How special are they? *Mol. Oncol.* **2010**, *4*, 192–208. [CrossRef]
26. Anderson, N.M.; Simon, M.C. The tumor microenvironment. *Curr. Biol.* **2020**, *30*, R921–R925. [CrossRef]
27. Arneth, B. Tumor Microenvironment. *Medicina* **2019**, *56*, 15. [CrossRef]
28. Ratajczak-Wielgomas, K.; Grzegrzolka, J.; Piotrowska, A.; Gomulkiewicz, A.; Witkiewicz, W.; Dziegiel, P. Periostin expression in cancer-associated fibroblasts of invasive ductal breast carcinoma. *Oncol. Rep.* **2016**, *36*, 2745–2754. [CrossRef]
29. Denton, A.E.; Roberts, E.W.; Fearon, D.T. Stromal Cells in the Tumor Microenvironment. *Adv. Exp. Med. Biol.* **2018**, *1060*, 99–114. [CrossRef]
30. Liang, Y.-N.; Liu, Y.; Meng, Q.; Li, X.; Wang, F.; Yao, G.; Wang, L.; Fu, S.; Tong, D. RBMS3 is a tumor suppressor gene that acts as a favorable prognostic marker in lung squamous cell carcinoma. *Med. Oncol.* **2015**, *32*, 30. [CrossRef]
31. Remmele, W.; Stegner, H.E. Vorschlag zur einheitlichen Definition eines Immunreaktiven Score (IRS) für den immunhistochemischen Ostrogenrezeptor-Nachweis (ER-ICA) im Mammakarzinomgewebe [Recommendation for uniform definition of an immunoreactive score (IRS) for immunohistochemical estrogen receptor detection (ER-ICA) in breast cancer tissue]. *Pathologe* **1987**, *8*, 138–140. (In German)

Disclaimer/Publisher's Note: The statements, opinions and data contained in all publications are solely those of the individual author(s) and contributor(s) and not of MDPI and/or the editor(s). MDPI and/or the editor(s) disclaim responsibility for any injury to people or property resulting from any ideas, methods, instructions or products referred to in the content.

MDPI AG
Grosspeteranlage 5
4052 Basel
Switzerland
Tel.: +41 61 683 77 34

International Journal of Molecular Sciences Editorial Office
E-mail: ijms@mdpi.com
www.mdpi.com/journal/ijms

Disclaimer/Publisher's Note: The statements, opinions and data contained in all publications are solely those of the individual author(s) and contributor(s) and not of MDPI and/or the editor(s). MDPI and/or the editor(s) disclaim responsibility for any injury to people or property resulting from any ideas, methods, instructions or products referred to in the content.

www.ingramcontent.com/pod-product-compliance
Lightning Source LLC
LaVergne TN
LVHW070413100526
838202LV00014B/1452